Geometric Quantization

SECOND EDITION

N.M.J. Woodhouse
Mathematical Institute
University of Oxford

CLARENDON PRESS · OXFORD

Oxford University Press, Great Clarendon Street, Oxford OX2 6DP

Oxford New York
Athens Auckland Bangkok Bogota Bombay
Buenos Aires Calcutta Cape Town Dar es Salaam
Delhi Florence Hong Kong Istanbul Karachi
Kuala Lumpur Madras Madrid Melbourne
Mexico City Nairobi Paris Singapore
Taipei Tokyo Toronto Warsaw

and associated companies in
Berlin Ibadan

Oxford is a trade mark of Oxford University Press

Published in the United States
by Oxford University Press Inc., New York

First published 1991
First published in paperback 1997

A catalogue record for this book is available from the British Library

Library of Congress Cataloging-in-Publication Data
(Data available)

ISBN 0 19 853673 9 (Hbk)
ISBN 0 19 850270 2 (Pbk)

Printed in Great Britain by
Bookcraft Ltd.,
Midsomer Norton, Avon

PREFACE

GEOMETRIC quantization continues to be a popular research topic. A large number of papers on the subject have been published during the past twelve years, as well as several books, including, notably, those by Śniatycki (1980), Hurt (1982), and Guillemin and Sternberg (1984). The original applications have been refined, and new ones have been found, most recently in conformal field theory and the geometry of knots (Axelrod, Della Pietra, and Witten 1991, Atiyah 1990, Verlinde and Verlinde 1990).

In this completely revised edition, I have simplified the presentation, added a number of new examples, and tried to remove the many errors in the original. I have also rearranged the material in a more logical order. I hope that the new edition will prove useful as a complement to other books on geometric quantization, as well as providing background to the more recent developments.

I am very grateful to many colleagues for comments on the first edition. I particularly thank R. Baston, J. Ellergaard-Anderson, and L. J. Mason for their help in preparing the new one.

Wadham College, Oxford N.M.J.W.
June 1991

PREFACE TO THE FIRST EDITION

ACCORDING to the Copenhagen philosophy, the physical predictions of a quantum theory must be formulated in terms of classical concepts. Thus, in addition to the traditional structure of Hilbert space, unitary transformations, and self-adjoint operators, a sensible quantum theory must contain a prescription for going over to the classical limit and for relating the quantum mechanical observables to those of the corresponding classical system.

However, as Dirac (1925) stressed in the early days of quantum theory, the correspondence between a classical theory and a quantum theory should be based not so much on a coincidence between their predictions in the limit $\hbar \to 0$, as on an analogy between their mathematical structures: the primary role of the classical theory is not in approximating the quantum theory, but in providing a framework for its interpretation.

In the simple systems that one first meets in elementary quantum mechanics, the correspondence is based on the canonical quantization: a classical observable—represented by some function $f(p_a, q^b)$ of the canonical coordinates—is associated with the quantum mechanical observable represented by the operator

$$f\left(-i\hbar\frac{\partial}{\partial q^a}, q^b\right).$$

This formal substitution raises many problems: for example, except in the simplest cases, the quantum observable depends on the ordering of the ps and qs in the classical expression for f; the quantization depends critically on the initial choice of the coordinates and it is not invariant under general canonical transformations; and the domain of the operator is left undetermined by such a formal expression. Nonetheless, judiciously supplemented with physical intuition, canonical quantization and its various generalizations have been remarkably successful.

The mathematical questions remain, however, and although they are relatively easy to answer for simple systems in Euclidean space, they are very much harder when the classical system involves constraints or contains particles with internal degrees of freedom.

The particular question that geometric quantization attempts to answer is: to what extent is canonical quantization a well-defined mathematical procedure and to what extent does it depend on the choice of canonical coordinates?

At one level, one can regard geometric quantization as a straightforward analysis of the various structures needed for the quantization of a classical system; for example, these might be preferred symmetries or special classes of coordinate systems. The aim is not to introduce new physical ideas, but to unify and clarify the various forms of canonical quantization and to make precise the analogies between the structures of classical and quantum theories. Starting with a classical phase space, represented by a symplectic manifold, one looks for a geometric, coordinate-free construction for the Hilbert space and observables of the underlying quantum theory: with no explicit dependence on a particular coordinate system, such a construction can be expected to give a very clear insight into the ambiguities involved in passing from the classical to the quantum domain.

At a more ambitious level, on the other hand, one can apply geometric quantization to systems with no special symmetries, in which the traditional forms of canonical quantization cannot be used in the obvious way; for example, to systems in curved space-time. Here a slightly different interpretation is needed: there is no one preferred quantization, but a whole family; and these agree with each other, and, presumably, with the underlying true theory, only in the semiclassical limit. In other words, here one must think of quantization as an approximation, giving only an incomplete picture of the real physics.

This book is a survey of the constructions and applications of geometric quantization, beginning with an account of symplectic geometry and the geometric formulation of Hamiltonian mechanics. (I assume that my reader has some familiarity with coordinate-free differential geometry and at least a passing acquaintance with quantum theory and Hamiltonian mechanics; the notation and some of the less familiar mathematical ideas are explained in the Appendix.)

I should like to stress three points: first that this is a book about quantization and not about quantum theory. The geometric method is developed not as a substitute for the normal analytic method of quantum theory, but as a means of understanding more clearly the relationship between classical and quantum mechanics. Secondly, this is a work of applied, not pure mathematics. Although the mathematical ideas are fairly sophisticated, I have made no attempt to develop them beyond their immediate applications and, without, I hope, glossing over any essential difficulties, I have made no attempt to be completely rigorous. Also, I have tried not to use abstract geometric arguments in places where coordinate-based calculations are simpler and quicker. Thirdly, I should stress that the ideas presented here are not original. However, they seem to me to be sufficiently important to justify this account of them, which, I hope, will complement the work of others. The principal published sources are: Blattner (1973), Guillemin and Sternberg (1977), Kirillov (1976), Kostant (1970a), Mackey (1963), Onofri and Pauri (1972), Segal (1960), Simms (1968), Souriau (1966, 1970), and Weinstein (1977). Also I have been heavily influenced by unpublished lectures and notes by F. A. E. Pirani (in Chapter 2), D. J Simms (in Chapter 3), B. Kostant (in Chapters 4 and 5), J. H. Rawnsley (in Chapter 10), and R. Geroch (in Chapter 6).

Finally, I should like to thank M. G. Eastwood, K. C. Hannabuss, L. P. Hughston, R. Penrose, S. P. Pratt, J. H. Rawnsley, and D. J. Simms for helpful comments and suggestions, and Ina Godwin for her help in typing.

Wadham College, Oxford N.M.J.W.
October 1979

CONTENTS

1

SYMPLECTIC GEOMETRY

1.1 Symplectic manifolds

THE basic object in the geometric formulation of classical mechanics is the *symplectic manifold.* This is a pair (M, ω) in which M is a smooth manifold and ω is a closed nondegenerate 2-form defined everywhere on M. In other words,

$$d\omega = 0$$

and the map[1]

$$T_m M \to T_m^* M : X \mapsto X \lrcorner \omega \qquad (1.1.1)$$

is a linear isomorphism at each $m \in M$. It is sometimes useful to think of ω as an antisymmetric metric on M; then (1.1.1) corresponds to 'lowering the index' on X.

Finite-dimensional symplectic manifolds arise as the phase spaces of classical systems with a finite number of degrees of freedom, where the symplectic form ω is the geometric structure that underlies the definitions of Hamiltonian vector fields and Poisson brackets. Infinite-dimensional symplectic manifolds arise in field theory. They are less straightforward objects and their study is complicated by some subtle analytical questions. There are, for example, a number of different forms of the nondegeneracy condition on ω, and for infinite-dimensional manifolds they are not equivalent. For the moment we shall concentrate on the finite-dimensional case.

1.2 Symplectic vector spaces

We shall begin the formal development of finite-dimensional symplectic geometry by looking at some basic facts about symplectic vector spaces and their symmetry groups.[2]

Definition (1.2.1). A symplectic vector space (V, ω) is a vector space V together with an antisymmetric, nondegenerate bilinear form ω on V. That is $\omega(X, Y) = -\omega(Y, X)$ for every $X, Y \in V$; and $X \lrcorner \omega = 0$ only if $X = 0$.

For the moment, V will be real and finite-dimensional, but the definitions and results that follow also apply to finite-dimensional complex symplectic vector spaces. The main example to keep in mind is the tangent space at some point of a symplectic manifold, although symplectic vector spaces are also of direct interest as phase spaces of linear systems, and we shall often want to think of (V, ω) as a special example of a symplectic manifold (V is made into a manifold by using components in some basis as coordinates; ω then becomes a closed 2-form).

Example. Let Q be an n-dimensional real vector space and put $V = Q^* \oplus Q$, where Q^* is the dual space. Define ω on V by

$$\omega\big((p,q),(p',q')\big) = \tfrac{1}{2}\big(p(q') - p'(q)\big),$$

where $p, p' \in Q^*$ and $q, q' \in Q$. ∎

This is a fundamental example since every symplectic vector space can be represented in this way (but not, of course, uniquely)—as follows from Proposition (1.2.5) below.

Let (V, ω) be a symplectic vector space and let $F \subset V$ be a subspace. The *symplectic complement*[3] of F is the subspace

$$F^\perp = \{X \in V \mid \omega(X,Y) = 0 \;\; \forall\, Y \in F\}.$$

The properties of the symplectic complement are summarized in the following lemma, the proof of which is straightforward.

Lemma (1.2.2). Let F and G be subspaces of a symplectic vector space (V, ω). Then

(a) $F^\perp \supset G^\perp$ whenever $F \subset G$

(b) $(F^\perp)^\perp = F$

(c) $(F + G)^\perp = F^\perp \cap G^\perp$

(d) $(F \cap G)^\perp = F^\perp + G^\perp$

(e) $\dim F^\perp = \dim V - \dim F$.

Definition (1.2.3). A subspace F of a symplectic vector space is said to be

isotropic whenever $F \subset F^\perp$

coisotropic whenever $F^\perp \subset F$

symplectic whenever $F \cap F^\perp = \{0\}$

Lagrangian whenever $F = F^\perp$.

Note that if F is isotropic then $F^\perp$ is coisotropic, and conversely. It follows from (1.2.2) that a Lagrangian subspace of V is the same as an isotropic subspace of dimension $\frac{1}{2}\dim V$. The categories are not exhaustive: a general subspace is neither isotropic, nor coisotropic, nor symplectic.

Lemma (1.2.4). Every finite-dimensional symplectic vector space has even dimension and contains a Lagrangian subspace.

Proof. Let F be a k-dimensional isotropic subspace of (V, ω). Then $F \subset F^\perp$ and so $\dim F \leq \dim F^\perp = \dim V - \dim F$. It follows that $k \leq \frac{1}{2}\dim V$. If $k \neq \frac{1}{2}\dim V$, then $F \neq F^\perp$, and so we can construct an isotropic subspace F' of dimension $k+1$ by taking the linear span of F and some vector $X \in F^\perp - F$. By continuing in this way, we construct a sequence of isotropic subspaces of increasing dimension. The sequence can certainly be started (since every one-dimensional subspace is isotropic); and it must terminate, at which point the subspace has dimension $k = \frac{1}{2}\dim V$ and is Lagrangian. ■

Proposition (1.2.5). Let (V, ω) be a $2n$-dimensional symplectic vector space. Then V has a basis $\{X^1, X^2, \ldots, X^n, Y_1, Y_2, \ldots, Y_n\}$ such that

$$\omega(X^a, X^b) = 0, \qquad 2\omega(X^a, Y_b) = \delta^a_b, \qquad \omega(Y_a, Y_b) = 0,$$

where $a, b = 1, 2, \ldots, n$.

Such a basis is called a *symplectic frame.* The positioning of the indices needs some explanation. Suppose that $\{X^a, Y_b\}$ is a symplectic frame. Let P be the subspace of V spanned by the Xs and let Q be the subspace spanned by the Ys. Since V is the direct sum $P \oplus Q$, the symplectic form gives a natural identification of P with Q^* by mapping $Z \in P$ to the linear form $2\omega(Z, \cdot)$ on Q. This is an isomorphism since $2\omega(Z, \cdot)$ vanishes on Q only if $Z \in Q^\perp = Q$, which implies that $Z = 0$ since $P \cap Q = 0$. If $Z = p_a X^a$, then the value of the corresponding linear form on $Y = q^a Y_a \in Q$ is $2\omega(Z, q^a Y_a) = p_a q^a$. The use of superscripts and subscripts is designed to allow the consistent application of the Einstein conventions. Vectors in Q have components with upper (contravariant) indices, while those in the dual space P have components with lower (covariant) indices. The repetition of an index (once as a superscript and once as a subscript) in an expression indicates a sum over $a = 1, 2, \ldots, n$.

Proof. Let Q be a Lagrangian subspace of V and let W be some other n-dimensional subspace such that $V = W \oplus Q$. Then ω identifies W with Q^* by mapping $Z \in W$ to $2\omega(Z, \cdot) \in Q^*$.

Let $\{Y_1, \ldots, Y_n\}$ be a basis for Q and let $\{Z^1, \ldots, Z^n\}$ be the dual basis for $W = Q^*$. Then $\omega(Y_a, Y_b) = 0$ and $2\omega(Z^a, Y_b) = \delta^a_b$. Put $\lambda^{ab} = \omega(Z^a, Z^b)$ and $X^a = Z^a + \lambda^{ab}Y_b$ (with the summation convention). Since $\lambda^{ab} = -\lambda^{ba}$,

$$\omega(X^a, X^b) = \omega(Z^a + \lambda^{ac}Y_c, Z^b + \lambda^{bd}Y_d) = \lambda^{ab} + \tfrac{1}{2}\lambda^{ba} - \tfrac{1}{2}\lambda^{ab} = 0$$

$$2\omega(X^a, Y_b) = \omega(Z^a + \lambda^{ac}Y_c, Y_b) = \delta^a_b.$$

Therefore $\{X^a, Y_b\}$ is a symplectic frame. ∎

We shall make frequent use of the following 'reduction' process.

Proposition (1.2.6). Let F be a subspace of a symplectic vector space (V, ω) and let $V' = F/(F \cap F^\perp)$. Then ω projects onto a bilinear form ω' on V', and (V', ω') is a symplectic vector space.

Proof. Let $\pi : F \to V'$ denote the projection along $F \cap F^\perp$. Define ω' by

$$\omega'(X', Y') = \omega(X, Y)$$

where $X, Y \in F$, and $X' = \pi(X)$ and $Y' = \pi(Y)$. This is clearly well defined, skew-symmetric, and nondegenerate. ∎

If $P \subset W$ is a Lagrangian subspace, then $P' = \pi(F \cap P)$ is an isotropic subspace of V'. The following lemma from Weinstein (1977) shows that Lagrangian subspaces behave nicely under the reduction of coisotropic subspaces.

Lemma (1.2.7). Let F be a coisotropic subspace of a symplectic vector space (V, ω) and let $\pi : F \to V' = F/F^\perp$ be the projection. Let P be a Lagrangian subspace of V. Then $P' = \pi(P \cap F)$ is a Lagrangian subspace of V'.

Proof. Note that in this case $F \cap F^\perp = F^\perp$. We have to show that $P'^\perp \subset P'$. Let $X' \in P'^\perp$. Then $X' = \pi(X)$ for some $X \in F$ such that

$$\omega(X, Y) = 0 \quad \forall\, Y \in F \cap P.$$

That is, $X \in (F \cap P)^\perp = F^\perp + P^\perp = F^\perp + P$. Therefore, $\pi(X) \in P'$. ∎

1.3 The symplectic group

Definition (1.3.1). A linear *canonical transformation* of a symplectic vector space (V, ω) is a linear map $\rho : V \to V$ such that

$$\omega(\rho X, \rho Y) = \omega(X, Y)$$

for every $X, Y \in V$. The group[4] of all such transformations is denoted by $SP(V, \omega)$.

When V is real and of dimension $2n$, we can introduce a symplectic frame $\{X^a, Y_b\}$ and identify V with $\mathbb{R}^{2n}$. Then $SP(V, \omega)$ is identified with the symplectic group $SP(n, \mathbb{R})$. This is the subgroup of $GL(2n, \mathbb{R})$ of matrices ρ such that

$$\rho^t \begin{pmatrix} 0 & 1_n \\ -1_n & 0 \end{pmatrix} \rho = \begin{pmatrix} 0 & 1_n \\ -1_n & 0 \end{pmatrix}, \tag{1.3.2}$$

where 1_n is the $n \times n$ identity matrix and ρ^t is the transpose of ρ. Such a ρ must be of the form

$$\rho = \begin{pmatrix} C_a{}^b & D_{ab} \\ E^{ab} & F^a_{\ b} \end{pmatrix} \tag{1.3.3}$$

where C, D, E, F are $n \times n$ matrices such that

$$C^t F - E^t D = 1_n, \qquad C^t E = E^t C, \qquad D^t F = F^t D$$

(again the positioning of the indices is derived from the Einstein conventions); ρ acts on $V = \mathbb{R}^{2n} = \{(p_1, \ldots, p_n, q^1, \ldots, q^n)\}$ by

$$\begin{pmatrix} p \\ q \end{pmatrix} \mapsto \begin{pmatrix} C & D \\ E & F \end{pmatrix} \begin{pmatrix} p \\ q \end{pmatrix};$$

That is,

$$p_a \mapsto C_a{}^b p_b + D_{ab} q^b, \qquad q^a \mapsto E^{ab} p_b + F^a_{\ b} q^b. \tag{1.3.4}$$

Note that $SP(1, \mathbb{R}) = SL(2, \mathbb{R})$.

Since $SP(V, \omega)$ preserves the *Liouville form* $\omega^n = \omega \wedge \omega \wedge \cdots \wedge \omega$, ρ must have unit determinant. Consequently V has a natural orientation in which all symplectic frames are right-handed.

The group of complex matrices which satisfy (1.3.2) is $SP(n, \mathbb{C})$. It should be noted that $SP(n, \mathbb{R})$ is not the same as the compact group $SP(n) = SP(n, \mathbb{C}) \cap U(2n)$, which is also called the symplectic group.

1.4 Darboux's theorem

An important example of a symplectic manifold is the cotangent bundle $M = T^*Q$ of an n-dimensional manifold Q. This is the set of pairs (p, q) where $q \in Q$ and p is a covector at q. It is made into a manifold by using as coordinates the $2n$ functions $p_1, \ldots, p_n, q^1, \ldots, q^n$ where the qs are coordinates on Q and the ps are the corresponding components of covectors.

The symplectic structure on M is the *canonical* 2-*form*

$$\omega = \mathrm{d}p_a \wedge \mathrm{d}q^a.$$

There is an intrinsic construction for ω which gives a simple way of seeing that it is, in fact, independent of the choice of coordinates on Q. Let π denote the projection map $M \to Q : (p, q) \mapsto q$ and, for each $m = (p, q) \in M$, define $\theta(m) \in T_m^* M$ by

$$X \lrcorner \theta(m) = (\pi_* X) \lrcorner p; \qquad X \in T_m M.$$

As m varies, θ becomes a smooth 1-form on M, called the *canonical* 1-*form*. It is given in the coordinates p_a, q^b by

$$\theta = p_a \mathrm{d}q^a.$$

Thus ω is also determined by

$$\omega = \mathrm{d}\theta$$

without reference to any particular coordinate system on Q. Clearly ω is closed, since it is also exact, and nondegenerate.

In the typical applications, Q is the configuration space of some mechanical system, such as a collection of particles subject to holonomic constraints, and T^*Q is the phase space of the system.

The cotangent bundle is a fundamental example since all symplectic manifolds have this form locally, as follows from Darboux's theorem.

Theorem (1.4.1). Let (M, ω) be a $2n$-dimensional symplectic manifold and let $m \in M$. Then there is a neighbourhood U of m and a coordinate system $\{p_a, q^b\}$ $(a, b = 1, 2, \ldots, n)$ on U such that $\omega = \mathrm{d}p_a \wedge \mathrm{d}q^a$ in U.

Proof. The proof is taken from Moser (1965) and Weinstein (1971). It makes use of some properties of time-dependent vector fields and differential forms which are listed in appendix §A.1. The key step is the following.

Lemma (1.4.2). Let ω and ω' be symplectic structures on a manifold M and let $m \in M$. If $\omega(m) = \omega'(m)$, then there are neighbourhoods U and V of m and a diffeomorphism $\rho : U \to V$ such that $\rho(m) = m$ and $\rho^*(\omega') = \omega$.

Proof. Since $\mathrm{d}(\omega' - \omega) = 0$, there is a 1-form α on some neighbourhood W of m such that

$$\mathrm{d}\alpha = \omega' - \omega.$$

By adding the gradient of a scalar to α, we can ensure that $\alpha(m) = 0$.

Let Ω be the time-dependent 2-form

$$\Omega = \omega + t(\omega' - \omega)$$

on W: and let X be the time-dependent vector field on W determined by

$$X \lrcorner \Omega + \alpha = 0.$$

Note that $\Omega(m) = \omega(m)$, so $\Omega(m)$ is nondegenerate; by replacing W by a smaller neighbourhood, if necessary, we can make sure that Ω is nondegenerate throughout W. Then X is well defined and

$$\mathcal{L}_X \Omega = \mathrm{d}(X \lrcorner \Omega) + X \lrcorner \mathrm{d}\Omega + \partial_t \Omega = -\mathrm{d}\alpha + \omega - \omega' = 0,$$

where we have used $\mathrm{d}\Omega = 0$ and eqn (A.1.24).

It follows that if $\rho_{tt'}$ is the flow of X, then $\rho^*_{tt'}\Omega(t') = \Omega(t)$. But $\Omega(1) = \omega'$ and $\Omega(0) = \omega$; so, if we put $\rho = \rho_{01}$, then $\rho^*\omega' = \omega$. Moreover, $\rho(m) = m$ since $X(m) = 0$. We can also obtain ρ by allowing U to flow for unit time along $\tilde{X} \in V(M \times \mathbb{R})$, where $\tilde{X}$ is defined as in eqn (A.1.19). See Fig. 1.1.

The only difficulty is that ρ may not be defined on the whole of W since the integral curves of X may leave W. We know, however, that $\rho(m) = m$; so there must exist a neighbourhood $U \subset W$ of m such that ρ is defined on U and $\rho(U) \subset W$. The proof of the lemma is completed by taking $V = \rho(U)$. ∎

To complete the proof of the theorem, we have only to note that as a consequence of Proposition (1.2.5), it is possible to introduce a coordinate system $\{r_a, s^b\}$ in a neighbourhood of m such that $\omega = \mathrm{d}r_a \wedge \mathrm{d}s^a$ at m. So if we put $\omega' = \mathrm{d}r_a \wedge \mathrm{d}s^a$ and define p_a and q^a by

$$p_a = r_a \circ \rho, \qquad \text{and} \qquad q^a = s^a \circ \rho,$$

where ρ is constructed as in the lemma, then we shall have $\omega = \mathrm{d}p_a \wedge \mathrm{d}q^a$ in some neighbourhood of U. ∎

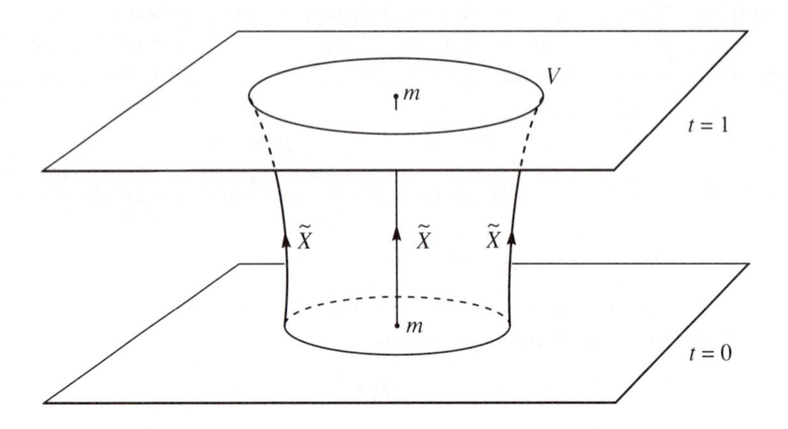

Fig. 1.1. The construction of ρ from the flow of $\tilde{X}$.

Coordinates $\{p_a, q^b\}$ in which $\omega = \mathrm{d}p_a \wedge \mathrm{d}q^a$ are called *canonical coordinates.* A symplectic manifold has a natural orientation in which all canonical coordinate systems are right-handed.

Canonical coordinates are far from unique. They can be changed by linear canonical transformations, as in eqn (1.3.4), and also by the general nonlinear canonical transformations that we shall consider in Chapter 4.

A (local) diffeomorphism $\rho : M_1 \to M_2$ between two symplectic manifolds (M_1, ω_1) and (M_2, ω_2) is said to be *canonical* whenever $\rho^*(\omega_2) = \omega_1$. As an immediate corollary of Darboux's theorem, we have the following.

Corollary (1.4.3). Let (M_1, ω_1) and (M_2, ω_2) be symplectic manifolds of the same dimension and let $m_1 \in M_1$ and $m_2 \in M_2$. Then there exist neighbourhoods $U_1 \subset M_1$ and $U_2 \subset M_2$ of m_1 and m_2 together with a canonical diffeomorphism $\rho : U_1 \to U_2$ such that $\rho(m_1) = m_2$.

In particular, all symplectic manifolds have the same local symplectic structure as a cotangent bundle of the appropriate dimension. However, in addition to its canonical symplectic structure, a cotangent bundle has two other natural structures which are not common to all symplectic manifolds.

The vertical foliation. A cotangent bundle is foliated by the individual cotangent spaces (the surfaces of constant q). This is an example of a polarization, about which more will be said later.

The canonical 1-form. On any manifold with symplectic structure ω, it is possible to find in some neighbourhood of each point a 1-form θ such that $\omega = \mathrm{d}\theta$. Such a 1-form is called a *symplectic potential.* In general θ exists only locally and is not unique. The canonical 1-form on a cotangent bundle is a natural global symplectic potential.

On a compact symplectic manifold, there cannot exist a global symplectic potential. For if $\omega = \mathrm{d}\theta$, then

$$\int_M \omega^n = \int_M \omega \wedge \omega \wedge \cdots \wedge \omega = \int_M \mathrm{d}(\theta \wedge \omega^{n-1}) = 0$$

by Stokes' theorem. But the integral on the left-hand side is nonzero, so we have a contradiction (Nagano 1968).

1.5 Hamiltonian vector fields

In the physical applications, (M, ω) is the phase space of a classical system and a smooth function $f \to \mathbb{R}$ represents a classical observable. When M is the cotangent bundle of a configuration space, f is simply a smooth function of position and momentum. We shall see later, however, that there are also symplectic phase spaces for systems with additional, internal degrees of freedom where f could represent a more general observable of a type not normally dealt with in classical mechanics, such as spin or hypercharge.

A classical observable plays two roles. First, as the term suggests, it is a measurable quantity that takes a definite value for any given state of the system. Second, it is the generator of a one parameter family of canonical transformations. For example, the Hamiltonian function generates the time evolution, and the components of the total linear and angular momenta generate spatial translations and rotations.

The geometric connection between the two roles is as follows. Let $f \in C^\infty(M)$ and let X_f be the vector field determined by

$$X_f \lrcorner\, \omega + \mathrm{d}f = 0. \tag{1.5.1}$$

Then X_f preserves ω in the sense that

$$\mathcal{L}_{X_f}\omega = X_f \lrcorner\, \mathrm{d}\omega + \mathrm{d}(X_f \lrcorner\, \omega) = -\mathrm{d}(\mathrm{d}f) = 0.$$

It follows that the flow ρ_t of X_f is a one-parameter family of local canonical diffeomorphisms of M (X_f does not generate global diffeomorphisms $M \to M$ unless it is complete). We shall call ρ_t and X_f the *canonical flow* and the *Hamiltonian vector field* generated by f.

In local canonical coordinates,

$$\omega = \mathrm{d}p_a \wedge \mathrm{d}q^a \qquad \text{and} \qquad \mathrm{d}f = \frac{\partial f}{\partial q^a}\mathrm{d}q^a + \frac{\partial f}{\partial p_a}\mathrm{d}p_a.$$

Hence,

$$X_f = \frac{\partial f}{\partial p_a}\frac{\partial}{\partial q^a} - \frac{\partial f}{\partial q^a}\frac{\partial}{\partial p_a}$$

and ρ_t is found by solving Hamilton's equations

$$\dot{q}^a = \frac{\partial f}{\partial p_a} \qquad \text{and} \qquad \dot{p}_a = -\frac{\partial f}{\partial q^a}.$$

Conversely, if X is a vector field on M such that

$$\mathcal{L}_X \omega = 0, \tag{1.5.2}$$

then $\mathrm{d}(X \lrcorner \omega) = 0$ since $\mathrm{d}\omega = 0$, so that, at least locally, there is a function f such that $X = X_f$; f is determined by X up to the addition of a constant.

A vector field X that satisfies (1.5.2) is said to be *locally Hamiltonian* and the set of all locally Hamiltonian vector fields is denoted by $V^{LH}(M)$. If, in addition, $X = X_f$ for some globally defined $f \in C^\infty(M)$, then f is said to be *Hamiltonian.* The set of Hamiltonian vector fields is denoted by $V^H(M)$. Clearly $V^H(M) \subset V^{LH}(M)$, but this is generally not an equality. For example, if $M = \mathbb{R}^2 - \{0\}$ and $\omega = r dr \wedge \mathrm{d}\theta$ in polar coordinates, then

$$X = \frac{1}{r}\frac{\partial}{\partial r}$$

is locally Hamiltonian, but not Hamiltonian. However, the Lie bracket of two locally Hamiltonian vector fields is always Hamiltonian, as is shown by the following proposition from Sternberg (1964).

Proposition (1.5.3). If $X, Y \in V^{LH}(M)$, then $[X, Y] = X_f$, where $f = 2\omega(X, Y)$.

Proof. Since $\mathcal{L}_X\omega = 0$ and $\mathrm{d}(Y \lrcorner \omega) = 0$,

$$\begin{aligned}
[X, Y] \lrcorner \omega &= \mathcal{L}_X(Y \lrcorner \omega) - Y \lrcorner \mathcal{L}_X \omega \\
&= X \lrcorner \mathrm{d}(Y \lrcorner \omega) + \mathrm{d}(X \lrcorner (Y \lrcorner \omega)) \\
&= -\mathrm{d}f.
\end{aligned}$$

∎

It follows that $V^{LH}(M)$ is a Lie algebra under Lie bracket and that $V^H(M)$ is an ideal.

Example. *The generators of the symplectic group.* Let (V, ω) be a symplectic vector space and let g be a quadratic form on V. Put $f(X) = \frac{1}{2}g(X, X)$. Then the canonical flow of f is a one parameter subgroup of $SP(V, \omega)$. In linear canonical coordinates,

$$f = -C_a{}^b q^a p_b - \tfrac{1}{2} D_{ab} q^a q^b + \tfrac{1}{2} E^{ab} p_a p_b,$$

where $D_{ab} = D_{(ab)}$ and $E^{ab} = E^{(ab)}$, and the subgroup is given by exponentiating

$$R : \begin{pmatrix} p \\ q \end{pmatrix} \mapsto \begin{pmatrix} C & D \\ E & -C^t \end{pmatrix} \begin{pmatrix} p \\ q \end{pmatrix},$$

where $C = (C_a{}^b)$, $D = (D_{ab})$, and $E = (E^{ab})$. The quadratic form is directly related to R by $g(X, Y) = 2\omega(X, RY)$. ∎

The definitions also make sense for complex functions and vector fields: for $f \in C^\infty_{\mathbb{C}}(M)$, we define $X_f \in V_{\mathbb{C}}(M)$ by eqn (1.5.1), as in the real case. Thus we also have the complex Lie algebras $V^H_{\mathbb{C}}(M)$ and $V^{LH}_{\mathbb{C}}(M)$ of complex Hamiltonian vector fields and complex locally Hamiltonian vector fields. In some contexts we shall also want to introduce complex symplectic potentials. That is, complex 1-forms θ such that $\mathrm{d}\theta = \omega$.

1.6 Poisson brackets

The *Poisson bracket* of $f, g \in C^\infty(M)$ is the function $[f, g] \in C^\infty(M)$ defined by

$$[f, g] = X_f(g). \tag{1.6.1}$$

In local canonical coordinates,

$$[f, g] = \frac{\partial f}{\partial p_a}\frac{\partial g}{\partial q^a} - \frac{\partial g}{\partial p_a}\frac{\partial f}{\partial q^a}. \tag{1.6.2}$$

Its properties are summarized in the following proposition.

Proposition (1.6.3). For any $f, g \in C^\infty(M)$, $[f, g] = -[g, f]$ and $[X_f, X_g] = X_{[f,g]}$. The *Jacobi identity*

$$[f, [g, h]] + [g, [h, f]] + [h, [f, g]] = 0$$

holds for any $f, g, h \in C^\infty(M)$.

Proof. The skew symmetry of the Poisson bracket follows from

$$X_f(g) = X_f \lrcorner \, \mathrm{d}g = -X_f \lrcorner \, (X_g \lrcorner \, \omega) = 2\omega(X_f, X_g), \tag{1.6.4}$$

or else directly from eqn (1.6.2). The second equation follows from Proposition (1.5.3). By using eqn (1.6.4),

$$X_f \lrcorner \, \mathrm{d}\big(\omega(X_g, X_h)\big) - \omega([X_f, X_g], X_h) = \tfrac{1}{2}[f, [g, h]] - \tfrac{1}{2}[[f, g], h].$$

The sum of the right-hand side over the three cyclic permutations of f, g, h is

$$[f, [g, h]] + [g, [h, f]] + [h, [f, g]],$$

while by (A.1.16) the sum of the left-hand side is $3\mathrm{d}\omega(X_f, X_g, X_h)$, which vanishes because ω is closed. Therefore the Jacobi identity holds. ∎

The definitions of X_f and $[f, g]$ do not make use of the closure of ω, so the calculation in the proof shows that the closure of ω is in fact equivalent to the Jacobi identity.

The Poisson bracket makes $C^\infty(M)$ into an infinite-dimensional Lie algebra (called the *algebra of classical observables*) and the map $f \mapsto X_f$ is a Lie algebra homomorphism of $C^\infty(M)$ onto $V^H(M)$. The kernel is the set of constant functions, so that, as Lie algebras,

$$V^H(M) = C^\infty(M)/\mathbb{R}.$$

Eqn (1.6.1) can also be used to define the Poisson bracket for complex-valued functions; $C^\infty_{\mathbb{C}}(M)$ then becomes a complex Lie algebra and we have $V^H_{\mathbb{C}}(M) = C^\infty_{\mathbb{C}}(M)/\mathbb{C}$.

1.7 Submanifolds and reduction

Before we turn to the mechanical applications, we shall look at a geometric construction which is interesting as a source of examples of symplectic manifolds and for the insight that it gives into Dirac's treatment of constraints in Lagrangian systems (Dirac 1958, 1964). This is the *reduction* of a submanifold of a symplectic manifold (Weinstein 1977).

We shall look at the construction first in a more general context. Let C be a smooth manifold and let σ be a 2-form on C such that the dimension of

$$K_m = \{X \mid X \lrcorner \sigma = 0\} \subset T_m C$$

is constant as m varies over C. Then K is a distribution on C. It is called the *characteristic distribution* of σ.

Let $X, Y \in V_K(C)$ (see §A.4: $V_K(C)$ denotes the set of vector fields tangent to K). Then for any $Z \in V(C)$,

$$3\mathrm{d}\sigma(X, Y, Z) = -\sigma([X, Y], Z)$$

by eqn (A.1.16). It follows that when σ is closed, $[X, Y] \lrcorner \sigma = 0$ and so $[X, Y] \in V_K(C)$. Therefore K is integrable.

We shall follow Souriau (1970) in calling (C, σ) a *presymplectic* manifold whenever σ is closed and of constant rank. A presymplectic manifold is *reducible* if its characteristic foliation is reducible (§A.4). In this case, the space of leaves $M' = C/K$ is a Hausdorff manifold and σ projects onto a symplectic struture ω' on M'. This is well defined since, if $X \in V_K(C)$, then

$$X \lrcorner \sigma = 0 = X \lrcorner \mathrm{d}\sigma.$$

Clearly ω' is also closed and nondegenerate. We shall call (M', ω') the *reduction* of (C, σ) or the *reduced phase space*. If $\Sigma \subset C$ is a section of K, then $(\Sigma, \sigma|_\Sigma)$ is also a symplectic manifold, which is canonically diffeomorphic to (M', ω'), by mapping $m \in \Sigma$ to the leaf of K through m.

Example. Let $C = \mathbb{R}^3 - \{0\}$ and $\sigma = r^{-3}(x\mathrm{d}y \wedge \mathrm{d}z + y\mathrm{d}z \wedge \mathrm{d}x + z\mathrm{d}x \wedge \mathrm{d}y)$, where $r^2 = x^2 + y^2 + z^2$. Then σ is closed and has rank 2. Reduction identifies (x, y, z) with $\lambda(x, y, z)$ for $\lambda > 0$. The reduced phase space is the same as the sphere $r = 1$ with the area element as its symplectic structure. ∎

We can apply the construction when C is a submanifold of a symplectic manifold (M, ω) and $\sigma = \omega|_C$ has constant rank: we think of C as a constraint in phase space. In this case, $K = TC \cap TC^\perp$, and when C is reducible, reduction is a nonlinear extension of the construction in (1.2.6).

Definition (1.7.1). A submanifold C of a symplectic manifold (M, ω) is *isotropic, coisotropic, Lagrangian* or *symplectic* if its tangent space is of the corresponding type as a subspace of T_mM at every $m \in C$.

As in the linear case, these categories are not exhaustive. Also T_mC can be a different type of subspace at different points of C. In Dirac's terminology, a coisotropic submanifold is a first class constraint (or, rather, a family of first class constraints), and a symplectic submanifold is a second class constraint.

In the symplectic and isotropic (and Lagrangian) cases, reduction is a trivial operation: in the symplectic case, because the characteristic distribution is zero-dimensional and $M' = C$; in the isotropic case, because the characteristic distribution is the whole tangent space at each point and M' is zero-dimensional. In the coisotropic case, $K = TC^\perp \subset TC$ and K has the maximum possible dimension given the dimension of C (the codimension of $C \subset M$). It is spanned by the Hamiltonian vector fields generated by functions on M that are constant on C. One can see this as follows. Let $m \in C$. Let $f \in C^\infty(M)$ be constant on C and let $X \in T_mC$. Then

$$2\omega(X, X_f) = X(f) = 0.$$

Hence $X_f(m) \in T_mC^\perp = K_m$. It is clear by counting dimensions that the vectors $X_f(m)$ also span K. Thus a first class constraint is given by the vanishing of k (the codimension) functions in involution—i.e. with vanishing Poisson brackets.

1.8 Time-dependent Hamiltonians

There is a straightforward extension of the construction of Hamiltonian vector fields and Poisson brackets to time-dependent functions on a symplectic manifold. In this section, we shall look at how this theory can be understood by reducing a constraint in an extended phase space in which time and the Hamiltonian appear as 'conjugate variables'. The terminology and notation are explained in §A.1.

Let (M, ω) be a symplectic manifold and let $h \in C^\infty(M \times \mathbb{R})$ be a time-dependent function on M. The time-dependent Hamiltonian vector field X_h is defined, as in the time-independent case, by

$$X_h \lrcorner \omega + dh = 0,$$

or, equivalently, by

$$X_h = \frac{\partial h}{\partial p_a}\frac{\partial}{\partial q^a} - \frac{\partial h}{\partial q^a}\frac{\partial}{\partial p_a}.$$

in local canonical coordinates. As before, $\mathcal{L}_{X_h}\omega = 0$, so the flow of X_h preserves ω.

The Poisson bracket of two time-dependent functions f, g is defined by $[f, g] = 2\omega(X_f, X_g)$; or equivalently by (1.6.2). Thus

$$\frac{\mathrm{d}g}{\mathrm{d}t} = [f, g] + \partial_t g,$$

where $\mathrm{d}/\mathrm{d}t$ is the derivative along the integral curves of X_f.

The extended phase space is the symplectic manifold $\tilde{M} = M \times \mathbb{R}^2$ with the symplectic structure

$$\tilde{\omega} = \omega + \mathrm{d}s \wedge \mathrm{d}t,$$

where $(s, t) \in \mathbb{R}^2$. Put $\tilde{h} = h + s$ and let C to be the submanifold of $\tilde{M}$ on which $\tilde{h} = 0$. This is coisotropic and its characteristic distribution is spanned by

$$X_{\tilde{h}} = \frac{\partial h}{\partial p_a}\frac{\partial}{\partial q^a} - \frac{\partial h}{\partial q^a}\frac{\partial}{\partial p_a} - \frac{\partial h}{\partial t}\frac{\partial}{\partial s} + \frac{\partial}{\partial t}.$$

Let (M', ω') be the reduction of C. Each point of M' is an integral curve of $X_{\tilde{h}}$ in C (see Fig. 1.2).

For each t, we can embed M in C by

$$\iota_t : M \to C : m \mapsto (m, -h(m, t), t) \in \tilde{M}$$

When X_h is complete, the image of $\iota_t(M)$ intersects all the integral curves of $X_{\tilde{h}}$ in C. By mapping these curves to their intersections with $\iota_t(M)$, we obtain a canonical diffeomorphism $\tau_t : M' \to M$. However, τ_t changes with t; in fact the flow of X_h on M is given by $\rho_{tt'} = \tau_{t'} \circ \tau_t^{-1}$. Thus one can generate the dynamical evolution either by treating h as a time-dependent Hamiltonian on M or by treating $\tilde{h}$ as a time-independent Hamiltonian on the extended phase space $(\tilde{M}, \tilde{\omega})$.

The restriction of $\tilde{\omega}$ to C is

$$\mathrm{d}p_a \wedge \mathrm{d}q^a + \mathrm{d}t \wedge \mathrm{d}h.$$

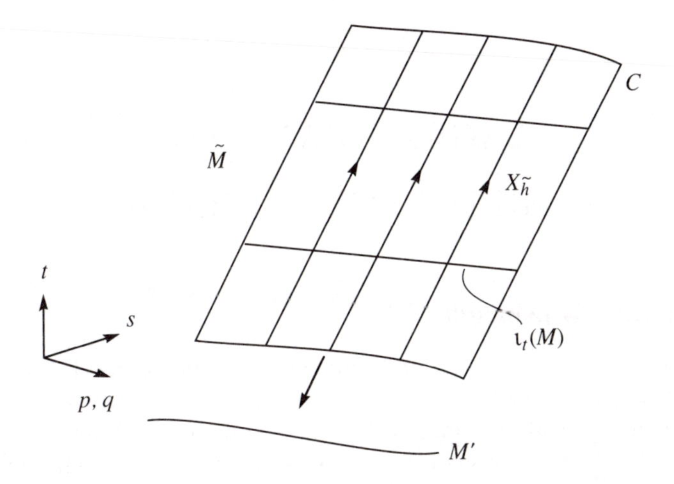

Fig. 1.2. The constraint surface $C \subset \tilde{M}$ and the construction of M'.

From the point of view of symplectic geometry, this is what is meant by the statement that time and the Hamiltonian are 'canonically conjugate variables'. (Two observables f, g on a phase space are canonically conjugate if $f = p_a$ and $g = q^a$ for the same value of a in some local canonical coordinate system. As it stands, this is a rather weak condition since it implies no more than that $[f, g] = 1$; but in many cases the coordinate system is singled out in some more or less natural way by the system in question and the relationship between f and g is more special. Thus for a particle moving in space, the three components of the momentum are canonically conjugate to the corresponding Cartesian coordinates. Although C is not a symplectic manifold, the meaning of the terminology in this case is clear.)

2

LAGRANGIAN AND HAMILTONIAN MECHANICS

2.1 Hamilton's principle

THE advantage of the Lagrangian formulation of classical mechanics is that it leads to a simple, direct procedure for writing down the dynamical equations of a mechanical system in arbitrary coordinates. When constraints are involved, this is of considerable practical importance since it allows one to carry out a systematic elimination of the constrained degrees of freedom, and of the constraint forces, in the initial stages of the analysis of a dynamical problem. The procedure is based on the invariance of Lagrange's equations under general coordinate transformations in configuration space, a fact that is easy to demonstrate by straightforward, but less than transparent calculation.[1]

There are two ways of understanding the underlying structure responsible for the invariance. One, which leads to Hamilton's principle, uses the observation that Lagrange's equations are precisely the Euler-Lagrange equations for the action functional. The other, which is the basis of the canonical formalism, involves, as a first step, the reduction of Lagrange's equations to Hamilton's equations by replacing the velocity coordinates by the corresponding generalized momenta, and then observing that coordinate transformations in configuration space can be regarded as special examples of canonical coordinate transformations in phase space.

In this chapter, we shall examine the symplectic geometry on which the second approach is based, and we shall explore the geometric relationship between Hamilton's principle and the canonical formalism.[2]

The simplest case to deal with is that of a conservative dynamical system subject only to holonomic constraints. The set of allowed configurations is then represented by a smooth, n-dimensional manifold Q (n is the number of degrees of freedom) and the set of kinematically possible states of motion by the tangent bundle TQ; Q is called the *configuration space* and TQ the *velocity phase space*—or simply, the *phase space*, when the qualification is unnecessary. The points of TQ are pairs (q, v) where $q \in Q$ and v is an element of T_qQ, representing a generalized velocity.

As coordinates on TQ, we can use the $2n$ functions $q^1, \ldots, q^n, v^1, \ldots, v^n$ where $q^1, \ldots, q^n$ are coordinates on Q and $v^1, \ldots, v^n$ are the components of the vs. Strictly, of course, we should distinguish between q^a, which is a function on Q, and $q^a \circ \pi$ (where $\pi : TQ \to Q$ is the projection), which is a function on TQ. However, it is generally simpler not to do this except in cases of possible confusion.

According to Hamilton's principle, the dynamical behaviour of the system is determined by a function $L = L(q, v) \in C^\infty(TQ)$, called the *Lagrangian*, which contains all the necessary information about the distribution of mass within the system and the external forces that act on it. Typically, $L = T - V$ where T is the kinetic energy and V is the potential of the external forces. The principle states that the dynamical trajectories are the solutions of the variational equation $\delta I = 0$, where I is the action functional

$$I = \int_{t_1}^{t_2} L(q, \dot{q})\, \mathrm{d}t \tag{2.1.1}$$

and the variations are over curves $t \mapsto q(t)$ in Q with fixed endpoints $q(t_1)$ and $q(t_2)$; the parameter t is time and the dot denotes differentiation with respect to t, so that $\dot{q}$ is the tangent vector to the curve. The variational equation is equivalent to the statement that, in TQ, the orbits[3] are solutions of Lagrange's equations

$$\frac{\mathrm{d}}{\mathrm{d}t}\left(\frac{\partial L}{\partial v^a}\right) - \frac{\partial L}{\partial q^a} = 0 \quad \text{and} \quad \dot{q^a} = v^a. \tag{2.1.2}$$

This, in turn, implies that the tangent vector X to any orbit in TQ satisfies

$$X \lrcorner\, \omega_L + \mathrm{d}h = 0, \tag{2.1.3}$$

where ω_L is the closed 2-form

$$\omega_L = \frac{\partial^2 L}{\partial q^a \partial v^b} \mathrm{d}q^a \wedge \mathrm{d}q^b + \frac{\partial^2 L}{\partial v^a \partial v^b} \mathrm{d}v^a \wedge \mathrm{d}q^b \tag{2.1.4}$$

and h is the *Hamiltonian*

$$h = v^a \frac{\partial L}{\partial v^a} - L. \tag{2.1.5}$$

We can check directly that (2.1.3) follows from (2.1.2) by writing out (2.1.3) in coordinates. Thus if $q^a = q^a(t)$ is a solution of (2.1.2), then the orbit in TQ is $(q^a, v^b) = \big(q^a(t), \dot{q}^b(t)\big)$ and

$$X = \dot{q}^a \frac{\partial}{\partial q^a} + \ddot{q}^a \frac{\partial}{\partial v^a}.$$

Hence

$$\begin{aligned} X \lrcorner\, \omega_L &= 2\frac{\partial^2 L}{\partial q^{[a} \partial v^{b]}} \dot{q}^a \mathrm{d}q^b + \frac{\partial^2 L}{\partial v^a \partial v^b}(\ddot{q}^a \mathrm{d}q^b - \dot{q}^a \mathrm{d}v^b) \\ \mathrm{d}h &= \frac{\partial^2 L}{\partial v^a \partial v^b} v^a \mathrm{d}v^b + \frac{\partial^2 L}{\partial v^a \partial q^b} v^a \mathrm{d}q^b - \frac{\partial L}{\partial q^a}\mathrm{d}q^a, \end{aligned}$$

and therefore

$$\begin{aligned} X \lrcorner\, \omega_L + \mathrm{d}h &= \left[\dot{q}^b \frac{\partial^2 L}{\partial v^a \partial q^b} + \ddot{q}^b \frac{\partial^2 L}{\partial v^a \partial v^b} - \frac{\partial L}{\partial q^a} \right] \mathrm{d}q^a \\ &= \left[\frac{\mathrm{d}}{\mathrm{d}t} \left(\frac{\partial L}{\partial v^a} \right) - \frac{\partial L}{\partial q^a} \right] \mathrm{d}q^a \\ &= 0. \end{aligned}$$

It is easy to check by direct calculation that both ω_L and h are independent of the choice of coordinates on Q. In the case of ω_L, this can also be deduced from $\omega_L = \mathrm{d}\theta_L$, where

$$\theta_L = \frac{\partial L}{\partial v^a} \mathrm{d}q^a, \tag{2.1.6}$$

together with the fact that the 1-form θ_L is characterized by the following coordinate-free construction. Let $(q, v) \in TQ$. The tangent space to the fibre of TQ through (q, v) can be identified with $T_q Q$. Thus a vector $u \in T_q Q$ can be lifted to a vector $F_u \in T_{(q,v)}(TQ)$: F_u is the tangent at $s = 0$ to the curve $s \mapsto (q, v + su)$. It is vertical in the sense that its projection onto Q vanishes. In coordinates, if

$$u = u^a \frac{\partial}{\partial q^a}, \quad \text{then} \quad F_u = u^a \frac{\partial}{\partial v^a}.$$

The value of θ_L at (q, v) is given invariantly by $Y \lrcorner\, \theta_L = F_{\pi_* Y} \lrcorner\, \mathrm{d}L$.

Note that the Hamiltonian can also be characterized invariantly: its value at (q, v) is $h = F_v \lrcorner\, \mathrm{d}L - L$.

Definition (2.1.7). A Lagrangian L is *regular* or *nondegenerate* whenever ω_L is everywhere nondegenerate; equivalently, whenever

$$\det \left[\frac{\partial^2 L}{\partial v^a \partial v^b} \right] \neq 0$$

at every point of TQ.

When L is regular, w_L is a symplectic structure on TQ and (2.1.3) is equivalent to the pair of equations (2.1.2); see §2.5. The time evolution is the canonical flow generated by h and every integral curve of X_h is of the form $(q^a, v^b) = \big(q^a(t), \dot{q}^b(t)\big)$ for some orbit of the system in Q.

However, things are not always so straightforward. Sometimes the equations in (2.1.2) are either inconsistent—for example, when $L = q^1$—or reduce to trivial identities—for example, when $L = v^1$ (Dirac 1964).

Before looking in more detail at this problem, we shall look at two other ways of understanding the transition from the Lagrangian equations (2.1.1) to the Hamiltonian equation (2.1.3). We shall assume, for the moment, that L is nondegenerate.

2.2 Momentum phase space

Given a nondegenerate Lagrangian $L(q, v)$, we have constructed a symplectic structure ω_L on TQ and a Hamiltonian h which generates the time evolution of the system. By Darboux's theorem, therefore, we can find canonical coordinates p_a, q^b on TQ in which the equations of motion take the Hamiltonian form

$$\dot{q}^a = \frac{\partial h}{\partial p_a}, \qquad \dot{p}_a = -\frac{\partial h}{\partial q^a}.$$

If we think of (TQ, ω_L) simply as a symplectic manifold, then there is no reason to prefer one canonical coordinate system over another. However, the Lagrangian also picks out a preferred class of canonical coordinates on TQ in which the qs are coordinates on Q (pulled back to TQ) and the ps are defined by

$$p_a = \frac{\partial L}{\partial v^a}. \tag{2.2.1}$$

They are canonical since $\theta_L = p_a \mathrm{d}q^a$, and so $\omega_L = \mathrm{d}p_a \wedge \mathrm{d}q^a$. The coordinate p_a is the *generalized momentum* conjugate to q^a.

There is another way of looking at conjugate momenta, which is a special case of a construction that we shall look at in a more general context in §4.7. This involves the *Legendre transformation*, which is the map

$$\rho : TQ \to T^*Q : (q, v) \mapsto (q, p)$$

where $p \in T_q^*Q$ is defined invariantly by

$$u \lrcorner p = F_u \lrcorner \mathrm{d}L; \qquad u \in T_qQ, \tag{2.2.2}$$

or in coordinates by

$$p = \frac{\partial L}{\partial v^a} dq^a.$$

It follows from (2.2.2) that

$$\rho^*\theta = \theta_L \text{ and hence that } \rho^*\omega = \omega_L,$$

where θ and ω are the canonical 1- and 2-forms on T^*Q.

We can use ρ to transfer our configuration and momentum coordinates to T^*Q. They then coincide with the coordinates on T^*Q introduced at the beginning of §1.4.

The main advantage of going over to T^*Q, which is called the *momentum phase space*, is that we only have to deal with one symplectic structure, whatever the Lagrangian. However, ρ need not be a global diffeomorphism, and some information may be lost in the transfer. The Legendre transformation is still well defined when L is degenerate, but it is then not even a local diffeomorphism, as it always is in the regular case. In the Dirac-Bergmann terminology, the image $\rho(TQ) \subset T^*Q$ is the *primary constraint* associated with L (see §2.5).

2.3 The space of motions

The *space of motions* M of a Lagrangian $L(q, v)$ is the set of solutions $q^a = q^a(t)$ of the equations of motion (2.1.2) (Souriau 1970). It is given a topology and made into a manifold by using as coordinates the values of q^a and $\dot{q}^a$ at some particular time t_0. Note that if $t \mapsto q^a(t)$ is a solution, then so is $t \mapsto q^a(t+k)$, where k is a constant. These are generally distinct solutions, and are regarded as distinct points of M, even though the two orbits occupy the same point set in Q.

When L is regular and X_h is complete, the map

$$\tau_{t_0} : M \to TQ : q^a(.) \mapsto (q^a, v^b) = (q^a(t_0), \dot{q}^a(t_0))$$

is a diffeomorphism; but when X_h is incomplete, as in the example of the Kepler system below, τ_{t_0} is not well defined and M need not be diffeomorphic to TQ; it may even be non-Hausdorff. When L is degenerate, τ_{t_0} need not be surjective.

We shall use the 'integration by parts' calculation, by which one derives the Euler-Lagrange equations from the first variation of action, to introduce a closed 2-form ω on M.

Let t_0 and t_1 be fixed values of the time. Then

$$I_{01} = \int_{t_0}^{t_1} L(q, \dot{q})\, \mathrm{d}t$$

is a smooth function on M. A tangent vector U to M at a solution $q = q(t)$ is represented by a solution $u = u(t)$ of the linearized equations of motion

$$\frac{\mathrm{d}}{\mathrm{d}t}\left(\frac{\partial^2 L}{\partial v^a \partial v^b}\dot{u}^b + \frac{\partial^2 L}{\partial v^a \partial q^b} u^b\right) - \frac{\partial^2 L}{\partial q^a \partial v^b}\dot{u}^b - \frac{\partial^2 L}{\partial q^a \partial q^b} u^b = 0. \qquad (2.3.1)$$

The derivative of I_{01} along U is given by

$$\begin{aligned} U \lrcorner\, \mathrm{d}I_{01} &= \int_{t_0}^{t_1} \left(\frac{\partial L}{\partial q^a} u^a + \frac{\partial L}{\partial v^a}\dot{u}^a\right) \mathrm{d}t \\ &= \left[\frac{\partial L}{\partial v^a} u^a\right]_{t_0}^{t_1} + \int_{t_0}^{t_1} \left[\frac{\partial L}{\partial q^a} - \frac{\mathrm{d}}{\mathrm{d}t}\left(\frac{\partial L}{\partial v^a}\right)\right] u^a\, \mathrm{d}t. \qquad (2.3.2) \end{aligned}$$

Now the integral vanishes because $q = q(t)$ is a solution of the equations of motion. Thus if we define for each t a 1-form θ_t on M by

$$U \lrcorner \theta_t = u^a(t) \frac{\partial L}{\partial v^a},$$

with the partial derivative on the right-hand side evaluated at $(q, v) = \big(q(t), \dot{q}(t)\big)$, then eqn (2.3.2) implies that

$$dI_{01} = \theta_{t_1} - \theta_{t_0}. \tag{2.3.3}$$

So the difference between the 1-forms θ_t for two different values of t is exact. It follows that the closed 2-form $\omega = \mathrm{d}\theta_t$ does not depend on t, and is therefore a natural structure on M. We note also that h is constant along the orbits and so is well defined as a function on M.

There is one part of this that needs more care: it may not be possible to define θ_t at every point of M, because, for example, some of the solutions may run into singularities in the potential before t, as happens in the Kepler system. Thus the 1-forms may only be defined on subsets of M, and eqn (2.3.3) may only hold on the intersections of these subsets. Nevertheless, the 2-form ω is well defined globally; it will always be closed, but may not be exact.

In the most straightforward case, in which L is regular and X_h is complete, ω is a symplectic structure on M, and, for any t, $\tau_t^*(\omega_L) = \omega$, so τ_t is a canonical diffeomorphism. On M, the Hamiltonian h generates the flow

$$q^a(\cdot) \mapsto q^a(\cdot + s). \tag{2.3.4}$$

In other words, the operation of moving a constant parameter distance s along an integral curve of X_h takes the solution $q^a = q^a(t)$ to the solution $q^a = q^a(t+s)$. In one sense, therefore, M is more or less the same as the phase spaces TQ and T^*Q. There is, however, a change in point of view involved in passing from TQ to M: in phase space, the system evolves by moving along an integral curve of X_h. In the space of motions, on the other hand, the state of the system is represented by a fixed point of M which does not change with t, while the physical observables become time-dependent functions. For example the velocities are represented by the time-dependent functions $v^a \circ \tau_t$ on M. There is an analogous distinction in quantum mechanics between the Schrödinger and Heisenberg pictures.

Example. Let Q be a Riemannian manifold with metric g_{ab} and let $L = \frac{1}{2} g_{ab} v^a v^b$. Then M is the space of affinely parametrized geodesics on Q. Eqn (2.3.1) is the Jacobi equation

$$D^2 u^a + R^a{}_{bcd} v^b u^c v^d = 0$$

where $D = v^a \nabla_a$ (Milnor 1963, Kobayashi and Nomizu 1969). The symplectic form is $\omega(U, U') = \frac{1}{2}(u'_a D u^a - u_a D u'^a)$, which is constant along geodesics. See Fig. 2.1. ∎

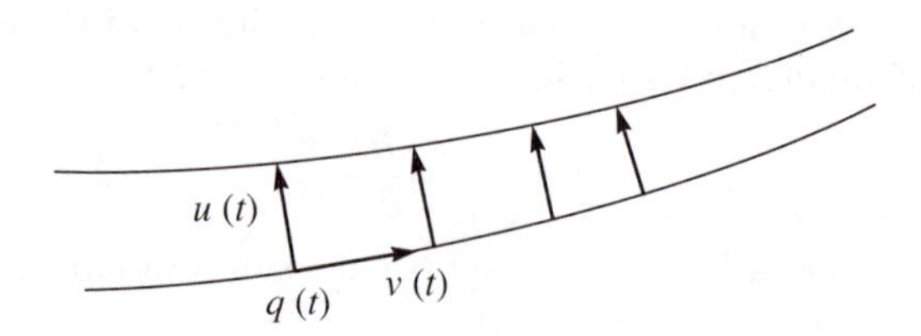

Fig. 2.1. A Jacobi field is a tangent vector to the space of geodesics.

Example. *The Kepler system.* Consider the motion of a particle in the x, y plane under the influence of an inverse-square-law central force. The system has two degrees of freedom and Lagrangian

$$L = \tfrac{1}{2}(\dot{x}^2 + \dot{y}^2) + \frac{1}{r}$$

where $r^2 = x^2 + y^2$. The orbits in $Q = \mathbb{R}^2 - \{0\}$ are the Keplerian conics, each of which has a focus at the origin.

The space M of negative energy motions is not Hausdorff because a sequence of orbits with nonvanishing angular momentum can have more than one radial orbit as its limit. Consider, for example, a family of orbits with the same energy

$$E = \tfrac{1}{2}(\dot{x}^2 + \dot{y}^2) - \frac{1}{r} < 0.$$

These all have period $\tau = 2\pi(-2E)^{-3/2}$, so that $t \mapsto \big(x(t), y(t)\big)$ and $t \mapsto \big(x(t+n\tau), y(t+n\tau)\big)$ define the same solution in the family and the same point of M for any integer n. In the limit of vanishing angular momentum, the orbits become radial: they collide with the centre of force and are no longer periodic. The radial orbits

$$t \mapsto \big(x(t), y(t)\big) \quad \text{and} \quad t \mapsto \big(x(t + n\tau), y(t + n\tau)\big)$$

with energy E are distinct, but cannot be separated by open subsets of M. See Fig. 2.2.

The space of motions is 'regularized' by making the radial motions periodic, allowing the particle to bounce off the centre and reverse its trajectory. The space of negative energy orbits then becomes a Hausdorff manifold, which we shall denote by $\hat{M}$. We can see the structure of $\hat{M}$ by using a 'spinor' transformation, which maps the orbits of the Kepler system onto those of a two-dimensional isotropic oscillator, as follows.[4]

Define new real variables α, β by

$$(\alpha + \mathrm{i}\beta)^2 = 2(x + \mathrm{i}y).$$

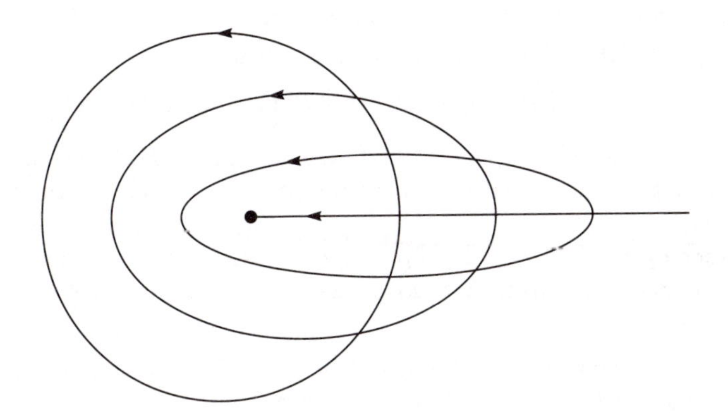

Fig. 2.2. A sequence of elliptical orbits with the same $E < 0$. The limiting radial orbit is not periodic.

Then $r = \frac{1}{2}(\alpha^2 + \beta^2)$ and

$$L = r(\dot{\alpha}^2 + \dot{\beta}^2) + \frac{1}{r}.$$

The corresponding Lagrangian equations are

$$\frac{\mathrm{d}}{\mathrm{d}t}\left(2r\dot{\alpha}\right) - \frac{E\alpha}{r} = 0, \quad \frac{\mathrm{d}}{\mathrm{d}t}\left(2r\dot{\beta}\right) - \frac{E\beta}{r} = 0,$$

where $E = r(\dot{\alpha}^2 + \dot{\beta}^2) - r^{-1}$ is the energy, which is constant on the orbits. Hence if we reparametrize the orbits by a non-physical time parameter s, defined by $\dot{s} = r^{-1}$, $s(0) = 0$, then

$$\frac{\mathrm{d}^2\alpha}{\mathrm{d}s^2} - \frac{E\alpha}{2} = 0, \qquad \frac{\mathrm{d}^2\beta}{\mathrm{d}s^2} - \frac{E\beta}{2} = 0.$$

For $E < 0$, these are the equations of motion of a two-dimensional isotropic oscillator with angular frequency $(-\frac{1}{2}E)^{\frac{1}{2}}$. Note that s is continuous, but not differentiable, at the points where the radial orbits bounce off the centre of force.

Each point of $\hat{M}$ corresponds to two oscillator orbits since (α, β) and $(-\alpha, -\beta)$ are mapped to the same point in the x, y plane. For example, $\alpha = \sin\frac{1}{2}s$, $\beta = 0$ and $\alpha = -\sin\frac{1}{2}s$, $\beta = 0$ are both mapped to the $E = -\frac{1}{2}$ radial orbit on the x-axis which bounces off the centre at $t = 0, 2\pi, 4\pi, \ldots$.

The oscillator orbits are labelled by five parameters: the energy $E < 0$, together with the values of

$$\alpha,\ \beta,\ \mu = \frac{\mathrm{d}\alpha}{\mathrm{d}s} = r\dot{\alpha}, \quad \text{and} \quad \nu = \frac{\mathrm{d}\beta}{\mathrm{d}s} = r\dot{\beta}$$

at, say, $s = 0$. The orbits that correspond to solutions of the Kepler problem are those for which $\mu^2 + \nu^2 - 1 = rE$, where $r = \frac{1}{2}(\alpha^2 + \beta^2)$. Thus we can identify $\hat{M}$ with the quotient of the hypersurface

$$N = \{\mu^2 + \nu^2 - \tfrac{1}{2}E(\alpha^2 + \beta^2) = 1, E < 0\} \subset \mathbb{R}^5$$

by the $\mathbb{Z}_2$ action $(\alpha, \beta, \mu, \nu, E) \mapsto (-\alpha, -\beta, -\mu, -\nu, E)$. The subset of $\hat{M}$ on which E takes a fixed negative value is diffeomorphic to the projective space $\mathbb{RP}_3$.

Since $\theta_L = r\dot{\alpha}\mathrm{d}\alpha + r\dot{\beta}\mathrm{d}\beta = \mu\mathrm{d}\alpha + \nu\mathrm{d}\beta$, the symplectic structure $\hat{\omega}_L$ on $\hat{M}$ is the projection from N to $\hat{M}$ of the 2-form $\mathrm{d}\mu \wedge \mathrm{d}\alpha + \mathrm{d}\nu \wedge \mathrm{d}\beta$, restricted to N. ∎

The construction of M and ω can be generalized in various ways. First, there is no need to assume that the Lagrangian is time independent (§2.4). Second, we can allow L to depend on higher derivatives of the qs. We can see how this works without introducing too much extra notation by considering the Lagrange equations

$$\frac{\mathrm{d}^2}{\mathrm{d}t^2}\left(\frac{\partial L}{\partial \ddot{q}^a}\right) - \frac{\mathrm{d}}{\mathrm{d}t}\left(\frac{\partial L}{\partial \dot{q}^a}\right) + \frac{\partial L}{\partial q^a} = 0, \tag{2.3.5}$$

generated by a Lagrangian $L = L(q, \dot{q}, \ddot{q})$ (in the traditional abuse of notation in which the $\dot{q}$s and $\ddot{q}$s are regarded either as functions of t or as independent coordinates, according to context). The space of motions is now the $4n$-dimensional manifold M of solutions of (2.3.5), on which we can use as coordinates the $4n$ values of the qs, $\dot{q}$s, $\ddot{q}$s, and $\dddot{q}$s at a particular time.

If U is a tangent vector to M represented by a solution $u(t)$ of the linearized form of (2.3.5), then eqn (2.3.2) must now be replaced by

$$\begin{aligned} U \lrcorner\, \mathrm{d}I_{01} &= \left[u^a \frac{\partial L}{\partial \dot{q}^a} - u^a \frac{\mathrm{d}}{\mathrm{d}t}\left(\frac{\partial L}{\partial \ddot{q}^a}\right) + \dot{u}^a \frac{\partial L}{\partial \ddot{q}^a}\right]_{t_0}^{t_1} \\ &\quad + \int_{t_0}^{t_1} u^a \left[\frac{\partial L}{\partial q^a} - \frac{\mathrm{d}}{\mathrm{d}t}\left(\frac{\partial L}{\partial \dot{q}^a}\right) + \frac{\mathrm{d}^2}{\mathrm{d}t^2}\left(\frac{\partial L}{\partial \ddot{q}^a}\right)\right] \mathrm{d}t. \end{aligned}$$

Eqn (2.3.3) still holds if we now define θ_t by

$$U \lrcorner\, \theta_t = u^a \frac{\partial L}{\partial \dot{q}^a} - u^a \frac{\mathrm{d}}{\mathrm{d}t}\left(\frac{\partial L}{\partial \ddot{q}^a}\right) + \dot{u}^a \frac{\partial L}{\partial \ddot{q}^a}$$

with the right-hand side evaluated at t. Once again, therefore, $\omega = \mathrm{d}\theta_t$ is a well defined closed 2-form on M. It is a symplectic structure when L satisfies the regularity condition

$$\det\left[\frac{\partial^2 L}{\partial \ddot{q}^a \partial \ddot{q}^b}\right] \neq 0.$$

Example. If $L = \frac{1}{2}\ddot{q}^2$, then $\omega = \mathrm{d}\ddot{q} \wedge \mathrm{d}\dot{q} - \mathrm{d}\dddot{q} \wedge \mathrm{d}q$. ∎

2.4 Time-dependent Lagrangians

Hamilton's principle also determines the dynamical behaviour of a system with a time-dependent Lagrangian $L(q, v, t) \in C^\infty(TQ \times \mathbb{R})$. As before, it states that the trajectories are the solutions of the variational equation $\delta I = 0$, where I is the action functional

$$I = \int_{t_0}^{t_1} L(q, \dot{q}, t)\,\mathrm{d}t$$

and the variations are over curves $t \mapsto q^a(t)$ in Q with fixed endpoints $q^a(t_0)$ and $q^a(t_1)$. Again it is equivalent[5] to the Lagrange equations (2.1.2).

The construction of the 2-form ω on the space of motions M—the manifold of solutions of (2.1.2)—also goes through without change. If we fix t_0 and use the initial data $q^a = q^a(t_0)$, $v^a = \dot{q}^a(t_0)$ as coordinates on M, then

$$\omega = \frac{\partial^2 L}{\partial q^a \partial v^b}\mathrm{d}q^a \wedge \mathrm{d}q^b + \frac{\partial^2 L}{\partial v^a \partial v^b}\mathrm{d}v^a \wedge \mathrm{d}q^b, \tag{2.4.1}$$

as in the time-independent case (the right-hand side is evaluated at $t = t_0$). It is a symplectic structure on M provided that L satisfies the regularity condition (2.1.7).

In the traditional approach to the canonical formalism, $L(q, v, t)$ is handled in the same way as a time-independent Lagrangian. One again derives Hamilton's equations from the variational principle, with the Hamiltonian and the momenta defined by

$$h = v^a \frac{\partial L}{\partial v^a} - L \quad \text{and} \quad p_a = \frac{\partial L}{\partial v^a}.$$

The only difference is that h now depends explicitly on time.

Another way to understand the transition to the time-dependent Hamiltonian formalism is to replace L by a 'homogeneous' Lagrangian in which t is treated as a configuration coordinate, on the same footing as the qs. Such an approach seems very natural from the point of view of relativistic mechanics.

Let $\tilde{Q} = Q \times \mathbb{R}$ be the *space of events*: each $(q, t) \in \tilde{Q}$ represents a particular configuration at a particular time. The tangent bundle $T\tilde{Q} = TQ \times \mathbb{R}^2$ is the *extended (velocity) phase space.*

As coordinates on $T\tilde{Q}$, we shall use the $2n+2$ functions q^a, t, v^a, w where the qs are coordinates Q, the vs are corresponding velocity coordinates on TQ, and w is the velocity coordinate associated with t.

The homogeneous Lagrangian $\tilde{L} : T\tilde{Q} \to \mathbb{R}$ is defined by

$$\tilde{L}(q, t, v, w) = wL(q, w^{-1}v, t).$$

Its action functional is

$$\tilde{I}(\gamma) = \int_{\tilde{t}_0}^{\tilde{t}_1} \tilde{L}(q, t, q', t')\, \mathrm{d}\tilde{t}$$

where $\gamma : \tilde{t} \mapsto \big(q(\tilde{t}), t(\tilde{t})\big)$ is a path in $\tilde{Q}$, with fixed endpoints $(q_0, t_0) = \big(q(\tilde{t}_0), t(\tilde{t}_0)\big)$ and $(q_1, t_1) = \big(q(\tilde{t}_1), t(\tilde{t}_1)\big)$, and the prime denotes differentiation with respect to $\tilde{t}$. This generates the equations of motion

$$\frac{\mathrm{d}}{\mathrm{d}\tilde{t}}\left(\frac{\partial \tilde{L}}{\partial v^a}\right) - \frac{\partial \tilde{L}}{\partial q^a} = 0, \quad \frac{\mathrm{d}}{\mathrm{d}\tilde{t}}\left(\frac{\partial \tilde{L}}{\partial w}\right) - \frac{\partial \tilde{L}}{\partial t} = 0,$$

in which we substitute $q^{a\prime} = \mathrm{d}q^a/\mathrm{d}\tilde{t}$ for v^a, and $t' = \mathrm{d}t/\mathrm{d}\tilde{t}$ for w.

We shall show that $\tilde{L}$ is essentially equivalent to L. The qualification is necessary because $\tilde{I}$ is invariant under reparametrization of γ, so the variational equation $\delta\tilde{I} = 0$ does not determine the parametrization of the orbits.

In fact if $t \mapsto q(t)$ is any path in Q with endpoints $q(t_0) = q_0$ and $q(t_1) = q_1$, and if t is expressed as any increasing function of $\tilde{t}$, then $\gamma : \tilde{t} \mapsto \big(q(t(\tilde{t})), t(\tilde{t})\big)$ is a path in $\tilde{Q}$ with endpoints (q_0, t_0) and (q_1, t_1), and

$$\tilde{I}(\gamma) = \int_{t_0}^{t_1} L(q, \dot{q}, t)\, \mathrm{d}t.$$

It follows that if $t \mapsto q(t)$ is a solution of the variational problem $\delta I = 0$, then γ is a solution of the problem $\delta\tilde{I} = 0$. Hence if $t \mapsto q(t)$ satisfies the equations of motion generated by L, then $\tilde{t} \mapsto \big(q(t(\tilde{t})), t(\tilde{t})\big)$ satisfies the equations of motion generated by $\tilde{L}$, whatever the choice of the function $t(\tilde{t})$. By applying the time-independent theory to $\tilde{L}$, with t treated as a configuration coordinate and $\tilde{t}$ as the 'time', we conclude that if $q = q(t)$ is a solution of the Lagrangian equations generated by L and if

$$X = q^{a\prime}\frac{\partial}{\partial q^a} + q^{a\prime\prime}\frac{\partial}{\partial v^a} + t'\frac{\partial}{\partial t} + t''\frac{\partial}{\partial w}$$

is the tangent to the corresponding orbit $\tilde{t} \mapsto (q, t, q', t')$ in $T\tilde{Q}$, then

$$X \lrcorner\, \omega_{\tilde{L}} + \mathrm{d}\tilde{h} = 0 \qquad (2.4.2)$$

where

$$\tilde{h} = v^a\frac{\partial \tilde{L}}{\partial v^a} + w\frac{\partial \tilde{L}}{\partial w} - \tilde{L} \quad \text{and} \quad \omega_{\tilde{L}} = \mathrm{d}\left(\frac{\partial \tilde{L}}{\partial v^a}\mathrm{d}q^a + \frac{\partial \tilde{L}}{\partial w}\mathrm{d}t\right).$$

However, $\tilde{L}$ is homogeneous of degree one in the velocity coordinates v^a and w, so $\tilde{h} = 0$ and $\omega_{\tilde{L}}$ is degenerate.[6]

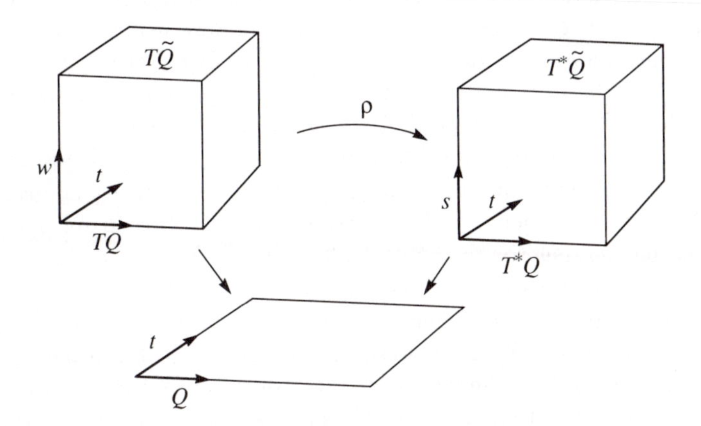

Fig. 2.3. The Legendre transformation from $T\tilde{Q}$ to $T^*\tilde{Q}$.

We can make the special choice $t = \tilde{t}$. Then $t' = 1$ and $t'' = 0$. So in this case the orbit in $T\tilde{Q}$ lies in the hypersurface $C = \{w = 1\}$ and its tangent X satisfies

$$X \lrcorner\, \omega_{\tilde{L}} = 0, \quad X(w) = 0, \quad \text{and} \quad X \lrcorner\, \mathrm{d}t = 1, \tag{2.4.3}$$

the first equation following from (2.4.2).

When L satisfies the regularity condition (2.1.7), the restriction of $\omega_{\tilde{L}}$ to C is a presymplectic form with a one-dimensional characteristic distribution. The three equations (2.4.3) determine a vector field X on C whose integral curves are the dynamical trajectories in $T\tilde{Q}$, parametrized by t. Moreover, X spans the characteristic distribution on C. There is just one orbit through each point of C and the orbits in C are the leaves of the characteristic foliation. It follows that we can identify M with the reduction of C. The identification is canonical (i.e. it preserves symplectic structures) because the restriction of $\omega_{\tilde{L}}$ to $\{w = 1, t = t_0\}$ coincides with (2.4.1).

We can get a clearer picture by using the Legendre transformation of $\tilde{L}$ to transfer the dynamics to the extended momentum phase space $T^*\tilde{Q} = T^*Q \times \mathbb{R}^2$. The Legendre transformation is the map $\rho : T\tilde{Q} \to T^*\tilde{Q}$ defined by

$$p_a = \frac{\partial \tilde{L}}{\partial v^a} = \frac{\partial L}{\partial v^a}, \quad s = \frac{\partial \tilde{L}}{\partial w} = L - w^{-1} v^a \frac{\partial L}{\partial v^a},$$

where p_a and s are the momenta conjugate to q^a and t, and L and its partial derivatives are evaluated at $q, w^{-1}v, t$. We shall assume that, in addition to regularity, L satisfies the stronger condition that the first of these equations can be solved to express the vs as a functions of p and t.

Then the restriction of ρ to C is a diffeomorphism from C to the hypersurface $\tilde{h} = 0$ in $T^*\tilde{Q}$, where $\tilde{h} = h + s$. We are here using h—in the usual abuse of notation—also to denote the function $h(p, q, t)$ constructed from $h(q, v, t)$ by substituting $v^a = v^a(p, q, t)$; $h(p, q, t)$ can be regarded either as a time-dependent function on T^*Q or as a function on $T^*\tilde{Q}$.

From the time-independent theory, $\rho^*(\tilde{\omega}) = \omega_{\tilde{L}}$ where $\tilde{\omega}$ is the canonical 2-form on $T^*\tilde{Q}$. It follows that the image under ρ_* of the dynamical vector field X on C is tangent to the characteristic distribution on $\tilde{h} = 0$ and so must be parallel to $X_{\tilde{h}}$. But $[s, t] = 1$, so $X_{\tilde{h}} \lrcorner \, \mathrm{d}t = 1$. Therefore $\rho_*(X) = X_{\tilde{h}}$ since $X \lrcorner \, \mathrm{d}t = 1$. See Fig. 2.3.

We conclude that the dynamical trajectories in $T^*\tilde{Q} = T^*Q \times \mathbb{R}^2$ are the integral curves of $X_{\tilde{h}}$ in the hypersurface $h + s = 0$. From the discussion in §1.8, it follows that the orbits in T^*Q are generated by treating h as a time-dependent function on T^*Q. Thus the relationship between the space of motions and the traditional Hamiltonian formalism is essentially a special case of the theory described in §1.8. As in the time-independent case, however, there is a shift in point of view between M, where the state is represented by a fixed point and the physical observables are time dependent, and T^*Q, where the state of the system evolves under the Hamiltonian flow generated by h.

2.5 Constraints

Three general types of constraint arise in Lagrangian mechanics. First there are constraints that are applied *a priori* and that represent some physical restriction on the system. For example, a particle confined to a smooth surface in space or a rigid sphere rolling without slipping on a rough plane. Second, there are constraints that are imposed after the equations of motion have been found and that pick out some subset of the allowed motions. For example, a restriction on the total energy or angular momentum of the system. Finally, there is the type of constraint that arises when the Lagrangian is irregular and dynamical trajectories exist in only a restricted portion of phase space. There are interesting examples in field theory (electromagnetism, gravity, Yang-Mills theory), but it is also possible to construct artificial examples in finite dimensions by using more coordinates than are necessary to fix the state of the system.

Holonomic and nonholonomic constraints

We shall begin by considering a constraint of the first type which restricts the motion to a submanifold $C \subset TQ$ of codimension r. We shall suppose that C is a *linear constraint* in the sense that it is given locally by the vanishing of r functions $f_1, \ldots, f_r$ of the form

$$f_i(q, v) = v \lrcorner \, \beta_i(q) + k_i(q)$$

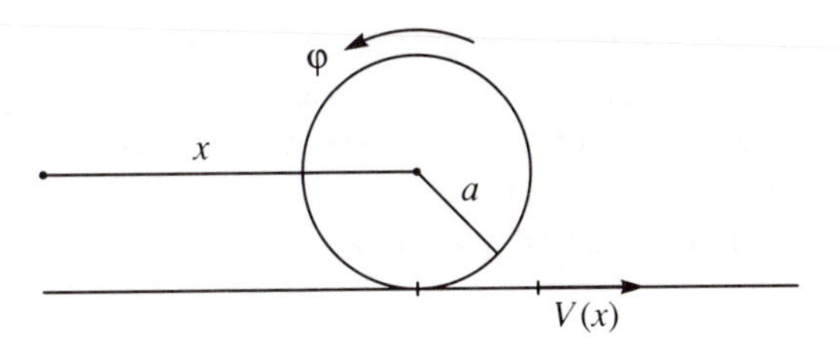

Fig. 2.4. A nonholonomic constraint.

where $\beta_i \in \Omega^1(Q)$, $k_i \in C^\infty(Q)$ $(i = 1, \ldots, r)$.

Example. A particle constrained to move on the surface of the sphere $x^2 + y^2 + z^2 = 1$. Here $r = 2$ and we can take

$$\beta_1 = 0, \quad k_1 = x^2 + y^2 + z^2 - 1, \quad \beta_2 = 2x\mathrm{d}x + 2y\mathrm{d}y + 2z\mathrm{d}z, \quad k_2 = 0. \quad \blacksquare$$

Example. A vertical wheel rolls on a rough horizontal rail. The rail is forced to move parallel to itself with a velocity $V(x)$ which depends on x, where x is the horizontal displacement of the centre of the wheel from some fixed origin. In this case Q is 2-dimensional and $r = 1$. Dropping the subscript, we can take

$$\beta = \mathrm{d}x + a\mathrm{d}\phi, \qquad k(x, \phi) = -V(x),$$

where ϕ is the angle turned through by the wheel and a is its radius. See Fig. 2.4. ■

Example. The constraint $w = 1$ on the orbits of a time-dependent system in $T\tilde{Q}$ is also of this form, with $r = 1$ and $\beta = \mathrm{d}t$, $k = -1$. ■

In a time-independent system in which the constraint is maintained by 'workless' forces, the motion is determined by a Lagrangian $L(q, v)$ according to a modified form of Hamilton's principle. Let $t \mapsto q(t)$ be a path in Q which is consistent with the constraint in the sense that $\big(q(t), \dot{q}(t)\big) \in C$ for each t. An infinitesimal variation of the path is represented by a vector field $u(t)$ along $q(t)$ (the varied path is $q^a = q^a(t) + \epsilon u^a(t)$ for some small value of a parameter ϵ).

The modified form of Hamilton's principle is that the orbits of the system satisfy

$$\delta I = \epsilon \int_{t_0}^{t_1} \left(u^a \frac{\partial L}{\partial q^a} + \dot{u}^a \frac{\partial L}{\partial v^a} \right) \mathrm{d}t = 0 \tag{2.5.1}$$

for all variations such that $u(t_0) = u(t_1) = 0$ and, for $t \in [t_0, t_1]$,

$$u \lrcorner \beta_i = 0 \quad i = 1, 2, \ldots, n. \tag{2.5.2}$$

These are the variations that are 'consistent with the instantaneous constraints' (Whittaker 1904). In the first example, (2.5.2) implies that u is tangent to the sphere $x^2 + y^2 + z^2 = 1$. In general, however, (2.5.2) is *not* the condition that the varied path should also satisfy the constraint, since it does *not* imply that the lifted vector field

$$u^a \frac{\partial}{\partial q^a} + \dot{u}^a \frac{\partial}{\partial v^a}$$

along $(q, v) = \big(q(t), \dot{q}(t)\big)$ is tangent to C. This is illustrated by the second example, where the condition on u has the following interpretation. Freeze the motion of the rail and the wheel, and then displace the wheel by rolling it a small distance along the rail, so that $\delta x + a\delta\phi = 0$. By doing this at each value of t, we obtain a variation of the path which satisfies $u \lrcorner (\mathrm{d}x + a\mathrm{d}\phi) = 0$ (condition 2.5.2). The variation is consistent with the instantaneous constraint (i.e. the rolling condition at the point of contact between the rail and the wheel), although the varied path will not itself satisfy the rolling constraint unless the function $V(x)$ has been specially chosen.

The variational equation (2.5.1) implies

$$\int_{t_0}^{t_1} u^a(t) \left[\frac{\partial L}{\partial q^a} - \frac{\mathrm{d}}{\mathrm{d}t}\left(\frac{\partial L}{\partial v^a}\right)\right] \mathrm{d}t = 0,$$

and so the orbits satisfy

$$\frac{\mathrm{d}}{\mathrm{d}t}\left(\frac{\partial L}{\partial v^a}\right) - \frac{\partial L}{\partial q^a} = \lambda_1 \beta_{1a} + \cdots + \lambda_r \beta_{ra}, \quad \dot{q}^a = v^a,$$

where the λs are functions of t (they are called *Lagrange multipliers*). It follows that if X is tangent to an orbit in $C \subset TQ$, then

$$X \lrcorner \omega_L + \mathrm{d}h = \lambda_1 \pi^*(\beta_1) + \cdots + \lambda_r \pi^*(\beta_r), \tag{2.5.3}$$

where ω_L and h are defined by (2.1.4) and (2.1.5), and $\pi : TQ \to Q$ is the projection.

When the system is in the configuration $q \in Q$, a force is represented by a covector $\mu \in T_q^*Q$. If $u \in T_qQ$, then $\epsilon u \lrcorner \mu$ is the work done by the force during the small displacement $q^a \mapsto q^a + \epsilon u^a$ in the configuration. In particular, the covector $\mu = \lambda_1\beta_1 + \cdots + \lambda_r\beta_r$ represents the total constraint force.

The physical condition under which the modified form of Hamilton's principle holds is that the forces maintaining the constraint should do no work under any small displacement in the configuration consistent with the instantaneous constraints (i.e. $u \lrcorner \mu = 0$ whenever $u \lrcorner \beta_i = 0$ for each i).

In the second example, as in most rolling problems, this holds because the constraint force—the friction at the point of contact—does no work when the motion of the rail is frozen and the wheel is rolled a small distance along it. Note, however, that it is not required that the constraint forces should do no work during the actual motion of the system. In fact μ works during the motion at the rate $X \lrcorner \mathrm{d}h$, which is generally nonzero.

It is possible to write down constraints that are inconsistent (for example, $q^1 = 0$, $v^1 = 1$), for which there are no solutions of the equations of motion. When the constraints are consistent and there are solutions, the system will generally not be Hamiltonian and the construction of the 2-form ω on the space of motions M will fail. This is because a tangent vector to the space of motions will be represented by a solution $u^a(t)$ of the linearized form of the equations of motion which also satisfies the linearized constraint, which is the condition that

$$u^a \frac{\partial}{\partial q^a} + \dot{u}^a \frac{\partial}{\partial v^a} \tag{2.5.4}$$

should be tangent to C. But this does *not* imply (2.5.2), so the second integral in (2.3.2) need not vanish.

Things are much simpler when C is a *holonomic*, which means that $C = TQ' \subset TQ$ for some submanifold $Q' \subset Q$ of codimension s. Then $r = 2s$ and we can take

$$f_1 = k_1, \ldots, f_s = k_s, \; f_{s+1} = v \lrcorner \mathrm{d}k_1, \ldots, f_r = v \lrcorner \mathrm{d}k_s.$$

where $k_1, \ldots, k_s \in C^\infty(Q)$ are independent functions that vanish on Q'. In this case the 1-forms $\pi^*\beta_i$ vanish on restriction to C and (2.5.3) implies

$$(X \lrcorner \omega_L + \mathrm{d}h)|_C = 0. \tag{2.5.5}$$

The 1-form $\theta_L|_C$ can be computed either by treating L as a Lagrangian on TQ, finding θ_L and restricting the result to $C = TQ'$, or by restricting L to $C = TQ'$ and treating L as a Lagrangian on TQ'. The result is the same whichever route is followed. The same must be true of $\omega_L|_C = \mathrm{d}\theta_L|_C$ and $h|_C$. In the holonomic case, therefore, it makes no difference whether the constraint is imposed before or after calculating ω_L and h. Moreover, the 2-form ω on the space of motions is well defined since, with C holonomic, (2.5.2) holds whenever (2.5.4) is tangent to C.

Constraints in Hamiltonian systems

Consider a system in which the phase space is a symplectic manifold (M, ω) and the dynamical evolution is generated by a Hamiltonian $h \in C^\infty(M)$.

Definition (2.5.6). A constraint compatible with h is a submanifold $C \subset M$ tangent to X_h.

Let C be such a constraint. Locally, C is determined by fixing the values of a number of constants of the motion—i.e. functions that are constant along the integral curves of X_h.

Put $\sigma = \omega|_C$, and let K be the characteristic distribution on C. We shall suppose that C is reducible, so that $M' = C/K$ is a Hausdorff manifold, with a symplectic structure ω'. If $Y \in V_K(C)$, then

$$Y \lrcorner \, \mathrm{d}h = -2\omega(X_h, Y) = -2\sigma(X_h, Y) = 0.$$

It follows that $h|_C$ is the pullback of a function h' on M'.

Each integral curve of the Hamiltonian vector field generated by h' in M' is the projection into M' of a family of integral curves of X_h, all of which lie in C. In many cases, it is enough to know the orbits in M' and the distinction between the members of the corresponding family in C is irrelevant. A typical example would be a system of particles in which the total linear momentum is conserved—e.g. two particles interacting through their mutual gravitational attraction. The constraint would be given by fixing the values of the three components of momentum, and the points of M' would represent the various possible motions relative to the centre of mass: the reduction would throw away the overall translational degrees of freedom.

Irregular Lagrangians

The third type of constraint arises when the Lagrangian $L(q, v)$ fails to satisfy the regularity condition (2.1.7). In the regular case, there is a unique vector field X on TQ such that

$$X \lrcorner \, \omega_L + \mathrm{d}h = 0. \tag{2.5.7}$$

In fact if we write

$$X = u^a \frac{\partial}{\partial q^a} + \alpha^a \frac{\partial}{\partial v^a},$$

then

$$\begin{aligned} X \lrcorner \, \omega_L + \mathrm{d}h &= \left[u^b \frac{\partial^2 L}{\partial v^a \partial q^b} + \alpha^b \frac{\partial^2 L}{\partial v^a \partial v^b} - \frac{\partial L}{\partial q^a} \right] \mathrm{d}q^a + \\ &\quad (v^a - u^a) \left[\frac{\partial^2 L}{\partial v^a \partial v^b} \mathrm{d}v^b + \frac{\partial^2 L}{\partial v^a \partial q^b} \mathrm{d}q^b \right]. \end{aligned}$$

The vanishing of the dv terms forces $u^a = v^a$ and the vanishing of the first square bracket determines the accelerations α^a in terms of q and v. The integral curves of X are given by

$$\dot{q}^a = v^a, \quad v^b \frac{\partial^2 L}{\partial v^a \partial q^b} + \dot{v}^b \frac{\partial^2 L}{\partial v^a \partial v^b} - \frac{\partial L}{\partial q^a} = 0.$$

They are precisely the curves of the form $(q, v) = \big(q(t), \dot{q}(t)\big)$ where $q = q(t)$ is a solution of Lagrange's equation.

The irregular case is more complicated for three reasons. First because it may not be possible to solve (2.5.7) for X at every point of TQ. Second, because even when we can solve (2.5.7), we may not be able to find a solution that also satisfies $u^a = v^a$—this is no longer an automatic consequence of (2.5.7). Third because the existence of a solution of (2.5.7) at some point of TQ does not guarantee the existence of a dynamical trajectory through the point. However,

(1) every solution $q = q(t)$ of Lagrange's equations determines a path $(q, v) = \big(q(t), \dot{q}(t)\big)$ in TQ whose tangent vector X satisfies both (2.5.7) and the condition $u^a = v^a$; these paths in TQ are the dynamical trajectories;

(2) if the Lagrangian is an inhomogeneous quadratic of the form[7]

$$L = \tfrac{1}{2} g_{ab} v^a v^b + A_a v^a - V,$$

where g_{ab} is constant, then every path $(q, v) = \big(q(t), v(t)\big)$ in TQ with a tangent vector that satisfies (2.5.7) is of the form $t \mapsto \big(q(t), \dot{q}(t) + w(t)\big)$, where $q(t)$ is a solution of Lagrange's equations and

$$w^a(t) \frac{\partial^2 L}{\partial v^a \partial v^b} = w^a(t) g_{ab} = 0.$$

Let C denote the set of all points in TQ that lie on dynamical trajectories (solutions of Lagrange's equations); C is called the *constraint* associated with L.

The construction of the closed 2-form ω on the space of motions M is the same as in the regular case, and by taking the values of q and $\dot{q}$ at some time t_0, we can identify (M, ω) with $(C, \omega_L|_C)$. In general ω is degenerate. In the field theory examples, the points of the reduction of (M, ω) are the physical states, and making identifications along the characteristic foliation of M is interpreted as the removal of gauge freedom.

Let C_1 denote the subset of TQ on which it is possible to solve (2.5.7) for X and let C_2 denote the subset of C_1 on which, in addition, it is possible to satisfy $u^a = v^a$. Clearly $C \subset C_2 \subset C_1$.

Example. Let $Q = \mathbb{R}^4$ and $L = \frac{1}{2}(v^1)^2 + q^2 v^1 + q^4 v^2 - \frac{1}{2}(q^3)^2$. Then $h = \frac{1}{2}\left((v^1)^2 + (q^3)^2\right)$ and

$$dh = v^1 \mathrm{d}v^1 + q^3 \mathrm{d}q^3, \quad \omega_L = \mathrm{d}v^1 \wedge \mathrm{d}q^1 + \mathrm{d}q^2 \wedge \mathrm{d}q^1 + \mathrm{d}q^4 \wedge \mathrm{d}q^2.$$

The general solution of Lagrange's equations is

$$q^1 = at + b, \quad q^2 = c, \quad q^3 = 0, \quad q^4 = at + d,$$

where a, b, c, d are constants. Thus $C = \{q^3 = v^2 = v^3 = 0,\ v^1 = v^4\}$. However, $C_1 = \{q^3 = 0\}$, at points of which the solution of (2.5.7) is

$$X = v^1\left(\frac{\partial}{\partial q^1} + \frac{\partial}{\partial q^4}\right) + u^3\frac{\partial}{\partial q^3} + \alpha^2\frac{\partial}{\partial v^2} + \alpha^3\frac{\partial}{\partial v^3} + \alpha^4\frac{\partial}{\partial v^4},$$

with $u^3, \alpha^2, \alpha^3, \alpha^4$ arbitrary. If we also impose $u^a = v^a$, then we must have $v^1 = v^4,\ v^2 = 0$. Thus $C_2 = \{q^3 = v^2 = 0,\ v^1 = v^4\}$. In this example, C, C_1, and C_2 are all distinct. There are paths in C_1 with tangent vectors that satisfy (2.5.7), but which do not come from solutions of Lagrange's equations; for example, $q^a = 0, v^1 = v^2 = v^3 = 0, v^4 = kt$. This cannot happen in the regular case. ∎

Dirac (1950, 1958) and Bergmann (1956) introduced a systematic procedure for identifying the constraints and finding the evolution of the system in momentum phase space T^*Q. Gotay, Nester and Hinds (1978) restated the construction in a geometric language. See also Simms (1980b). It goes as follows.

Let $M_1 \subset T^*Q$ be the image of the Legendre transformation

$$\rho : TQ \to T^*Q : (q, v) \mapsto (p, q); \quad p_a = \frac{\partial L}{\partial v^a};$$

and let ω_1 be the restriction to M_1 of the canonical 2-form on T^*Q; M_1 is called the *primary constraint.*

We shall assume for simplicity that $\rho^{-1}(m)$ is connected for each $m \in M_1$—although it is easy to construct examples where this is not the case; for example, $L = (v^1)^3$. The tangent vectors to the surfaces $\rho^{-1}(m)$ in TQ are of the form

$$Y = y^a\frac{\partial}{\partial v^a} \quad \text{where} \quad Y \lrcorner\, \omega_L = y^a \frac{\partial^2 L}{\partial v^a \partial v^b}\mathrm{d}q^b = 0$$

and they satisfy $Y \lrcorner\, \mathrm{d}h = 0$. It follows that $h = h_1 \circ \rho$ for some $h_1 \in C^\infty(M_1)$. In the example, M_1 is given by $p_2 - q^4 = p_3 = p_4 = 0$,

$$\omega_1 = \mathrm{d}p_1 \wedge \mathrm{d}q^1 + \mathrm{d}q^4 \wedge \mathrm{d}q^2, \quad \text{and} \quad h_1 = \tfrac{1}{2}\left((p_1 - q^2)^2 + (q^3)^2\right).$$

We now look for vectors X_1 tangent to M_1 such that $X_1 \lrcorner\, \omega_1 + \mathrm{d}h_1 = 0$; these will only exist on a subset $M_2 = \rho(C_1) = \rho(C_2)$, called the *secondary constraint*. Moreover, even where X_1 exists, it may not be possible to make it tangent to M_2; so we let ω_2 and h_2 denote the restrictions of ω_1 and h_1 to M_2, and look for vectors X_2 tangent to M_2 such that $X_2 \lrcorner\, \omega_2 + \mathrm{d}h_2 = 0$, and so on. We assume that each M_i is a submanifold. The process terminates when $M_k = M_{k-1}$, at which stage we can find a vector field X_k on M_k such that $X_k \lrcorner\, \omega_k + \mathrm{d}h_k = 0$. However, X_k need not be unique. The final constraint $M_k \subset T^*Q$ is the image of C under ρ and the integral curves of X_k are the images of the dynamical trajectories in TQ. The reduction (M', ω') of (M_k, ω_k) is the same as the reduction of the space of motions, with the usual shift in point of view between the 'Schrödinger' and 'Heisenberg' interpretations. The points of M' represent the 'physical states' and the Hamiltonian h_k descends to M' where it generates the time evolution.

In the example, we can use p_1, q^1, q^2, q^3, q^4 as coordinates on M_1. The secondary constraint is $M_2 = \{q^3 = 0\} \subset M_1$ on which

$$X_1 = (p_1 - q^2)\left(\frac{\partial}{\partial q^1} + \frac{\partial}{\partial q^4}\right) + x^3 \frac{\partial}{\partial q^3}$$

with x^3 arbitrary. Since we can make X_1 tangent to M_2 by choosing $x^3 = 0$, the construction ends with the secondary constraint.

2.6 Transformations of the Lagrangian

The Lagrangian is not uniquely determined by the classical equations of motion. For example, in any time-independent system with Lagrangian $L(q, v)$, we can replace L by

$$L'(q, v) = L(q, v) + v^a \beta_a(q) + K \tag{2.6.1}$$

where $\beta = \beta_a \mathrm{d}q^a$ is a closed 1-form on Q and K is a constant. Then

$$\theta_{L'} = \theta_L + \pi^* \beta, \quad \omega_{L'} = \omega_L, \quad h' = h - K,$$

and the equations of motion are unaltered. The change simply adds a constant to the action functional.

Similarly, in the time-dependent case, we can replace L by

$$L'(q, v, t) = L(q, v, t) + v^a \beta_a(q, t) + K(q, t) \tag{2.6.2}$$

where $\beta_a \mathrm{d}q^a + K\mathrm{d}t$ is a closed 1-form on $Q \times \mathbb{R}$.

Example. *Motion in a magnetic field.* A particle of unit mass and charge moves in a magnetic field $\mathbf{B}$ in Euclidean space $\mathbb{E}$. The field satisfies $\operatorname{div}\mathbf{B} = 0$ or, equivalently $\mathrm{d}F = 0$ where

$$F = 2\left(B_1 \mathrm{d}y \wedge \mathrm{d}z + B_2 \mathrm{d}z \wedge \mathrm{d}x + B_3 \mathrm{d}x \wedge \mathrm{d}y\right). \tag{2.6.3}$$

The equation of motion is $\dot{\mathbf{v}} = \mathbf{v} \wedge \mathbf{B}$, which is generated by the Lagrangian $L = \frac{1}{2}\mathbf{v} \cdot \mathbf{v} + \mathbf{A} \cdot \mathbf{v}$, where $\mathbf{A}$ is a vector potential. That is, $\mathbf{B} = \operatorname{curl} \mathbf{A}$; or, equivalently, $F = 2\mathrm{d}\alpha$, where $\alpha = \mathbf{A} \cdot \mathrm{d}\mathbf{r}$.

The vector potential is not unique: it can be replaced by $\alpha + \beta$, where $\mathrm{d}\beta = 0$. Such a gauge transformation induces a transformation of the Lagrangian of the form (2.6.1), with $K = 0$.

Both the Lagrangian and the Legendre transformation depend on the choice of potential; but the corresponding Hamiltonian system does not because, in velocity phase space, neither h nor ω_L depend on α. When we transfer h to momentum phase space by the Legendre transformation, we get $h = \frac{1}{2}(\mathbf{p} - \mathbf{A}) \cdot (\mathbf{p} - \mathbf{A})$ and the dependence on the choice of potential reappears.

There is, however, a gauge-invariant way of transferring the Hamiltonian system to momentum phase space, which is to use the free particle Hamiltonian $\frac{1}{2}\mathbf{p}\cdot\mathbf{p}$, but to replace the canonical 2-form ω on $T^*\mathbb{E}$ by the 'charged' symplectic structure $\omega + \frac{1}{2}\pi^*(F)$, where π is the projection $T^*\mathbb{E} \to \mathbb{E}$ (Souriau 1970, Torrence and Tulczyjew 1973, Sommers 1973b, Śniatycki 1974, Guillemin and Sternberg 1984). This is related to the Hamiltonian description in velocity phase space by the Legendre transformation of the free-particle Lagrangian $\frac{1}{2}\mathbf{v} \cdot \mathbf{v}$. ∎

Example. *Galilean transformations.* A free particle of unit mass in Euclidean space $\mathbb{E}$ has Lagrangian $L = \frac{1}{2}\mathbf{v} \cdot \mathbf{v}$. We shall treat this as a time-dependent system by regarding L as an element of $C^\infty(T\mathbb{E} \times \mathbb{R})$.

The dynamical equation $\dot{\mathbf{v}} = 0$ is invariant under Galilean transformations of $T\mathbb{E} \times \mathbb{R}$ of the form

$$(\mathbf{q}, \mathbf{v}, t) \mapsto (\mathbf{q} + \mathbf{u}t, \mathbf{v} + \mathbf{u}, t)$$

where $\mathbf{q}$ is the position vector in $\mathbb{E}$ and $\mathbf{u}$ is a constant velocity: the transformation relates the coordinate systems of two inertial frames. The Lagrangian is not invariant, however, and transforms according to

$$L \mapsto L' = \tfrac{1}{2}\mathbf{v} \cdot \mathbf{v} + \mathbf{v} \cdot \mathbf{u} + \tfrac{1}{2}\mathbf{u} \cdot \mathbf{u}$$

which is a special case of (2.6.2). ∎

The indeterminacy of the Lagrangian is the first point at which we come across an ambiguity in constructing quantum mechanics from classical mechanics. It is not hard to cope with the existence of a family of Lagrangians related by transformations like (2.6.1) or (2.6.2), but there are other less trivial ways of transforming the Lagrangian without affecting the equations of motion. For example, if the Lagrangian $L(q, v)$ is independent of time and is a homogeneous quadratic in the velocities, then it is a constant of the motion and any function of L with non-vanishing derivative generates the same dynamical equations. It is not always obvious how to select the 'correct' Lagrangian, and therefore the correct symplectic form, as a starting point for quantization.

We end this chapter by noting that it is also possible under certain conditions to recover the Lagrangian formalism from the canonical formalism. If h is a Hamiltonian on a cotangent bundle T^*Q, then

$$\tau : (p, q) \mapsto (q, v) = \big(q, \pi_*(X_h)\big)$$

is a map from T^*Q to TQ. It is given in coordinates by

$$v^a = \frac{\partial h}{\partial p_a}.$$

When τ is a diffeomorphism, $L = (X_h \lrcorner\, \theta - h) \circ \tau^{-1}$, where θ is the canonical 1-form, is a Lagrangian on TQ that generates the same dynamics as h. That is, the Legendre transformation of L is τ^{-1} and the dynamical trajectories of L in TQ are mapped by τ^{-1} onto the integral curves of X_h in T^*Q.

3

SYMMETRY

3.1 Symmetry in quantum physics

SYMMETRY principles play an important part in theoretical physics, both as a means of understanding conservation laws and as computational devices providing quick and elegant solutions to problems that would otherwise be tedious to solve. In quantum theory, their role is especially significant since they not only clarify basic concepts and simplify the formal development of the subject, but they also provide detailed information about physical systems for which no explicit dynamical models exist. However, the methods used to investigate symmetry in quantum systems are often based on abstract group-theoretical techniques, which bypass the canonical formalism. They include no general rules for going over to the classical limit and for directly relating the symmetries of the quantum system to those of a corresponding classical system.

In this chapter, we shall look at some methods for studying symmetry in classical systems that closely mirror techniques commonly used in quantum theory. The quantum notion of a unitary representation—an action of a group on a quantum phase space that preserves transition amplitudes—is replaced by its classical analogue—an action by canonical transformations on a classical phase space. The idea is to study the symmetries of quantum systems by quantizing the corresponding classical systems. The classical limit is then built in from the start and the relationship between the symmetries at the classical and quantum levels is explicit.

We shall begin by looking at the moment map, which enables one to relate the geometry of a symplectic manifold with symmetry to the structure of its symmetry group.[1]

3.2 The moment map

Let (M, ω) be a symplectic manifold and let $\mathcal{G}$ be a real Lie algebra that acts on M on the right by infinitesimal canonical transformations. That is, there is a linear map

$$\mathcal{G} \to V^{LH}(M) : A \mapsto X_A$$

such that $X_{[A,B]} = [X_A, X_B]$ for every $A, B \in \mathcal{G}$. We shall consider in the next few sections whether there is a natural way to assign an observable $h_A \in C^\infty(M)$ to each $A \in \mathcal{G}$ so that h_A generates X_A. For example, when $\mathcal{G}$ is the Lie algebra of the rotation group, the problem is to identify the components of angular momentum.

Clearly it is necessary that each of the individual vector fields X_A should be (globally) Hamiltonian. This will be the case if, for example, M is simply connected or if $[\mathcal{G},\mathcal{G}]=\mathcal{G}$, where $[\mathcal{G},\mathcal{G}]$ is the derived algebra.[2] Each h_A is then determined up to the addition of a constant; but even if we require that $A \mapsto h_A$ should be linear, which involves no additional restrictions on M or $\mathcal{G}$, there is still considerable freedom since we can replace h_A by $h_A + f(A)$ where f is any element of the dual space $\mathcal{G}^*$.

In certain circumstances the remaining freedom can be further reduced or eliminated altogether by imposing the condition

$$h_{[A,B]} = [h_A, h_B] \qquad \forall A, B \in \mathcal{G}. \tag{3.2.1}$$

When this can be satisfied, the action of $\mathcal{G}$ on M is said to be *Hamiltonian*.[3]

Definition (3.2.2). When (3.2.1) holds the map $A \mapsto h_A$ is called a *Hamiltonian* and the dual map

$$\mu : M \to \mathcal{G}^* : m \mapsto f_m,$$

where $f_m(A) = h_A(m)$, is called a *moment* for the action.

'Moment' because such a map generalizes the association of linear momentum with translation and angular momentum with rotation. There are similar definitions at the level of group actions.

Definition (3.2.3). Let G be a Lie group with Lie algebra $\mathcal{G}$ and suppose that G acts on M on the right by canonical transformations, so that the elements g of G determine diffeomorphisms $g : M \to M$ such that

$$g^*\omega = \omega \quad \text{and} \quad (mg)g' = m(gg').$$

We say that the action is *Hamiltonian* whenever the corresponding infinitesimal action of $\mathcal{G}$ is Hamiltonian; and, when (3.2.1) holds, we call $\mu : M \to \mathcal{G}^*$ a moment for the action of G.

The distinction between a Hamiltonian action and a general action by canonical transformations is analogous to the distinction at the quantum level between unitary and projective representations; and the conditions for the existence of a moment map are similar to those under which it is possible to derive a nonprojective representation from a given projective representation.

3.3 The cocycle condition

Definition (3.3.1). A *cocycle* (of degree 2) on a Lie algebra $\mathcal{G}$ is a skew-symmetric bilinear form $\alpha \in \mathcal{G}^* \wedge \mathcal{G}^*$ such that

$$\alpha([A, B], C) + \alpha([B, C], A) + \alpha([C, A], B) = 0$$

for every $A, B, C \in \mathcal{G}$.

If $\mathcal{G}$ is a Lie algebra, then each $f \in \mathcal{G}^*$ determines a cocycle δf by

$$\delta f(A, B) = \tfrac{1}{2} f([A, B]),$$

with the cocycle condition (3.3.1) following from the Jacobi identity

$$[[A, B], C] + [[B, C], A] + [[C, A], B] = 0$$

in $\mathcal{G}$. Cocycles of the form δf are called *coboundaries* and two cocycles that differ by a coboundary are said to be equivalent (or *cohomologous*). The set of equivalence classes forms a group under addition. It is called the second cohomology group of $\mathcal{G}$, and is denoted by $H^2\mathcal{G}$. (The first cohomology group $H^1\mathcal{G}$ is the dual of the quotient space $\mathcal{G}/[\mathcal{G}, \mathcal{G}]$. There are also higher degree cohomology groups.[4] The general definitions are given in Chevalley and Eilenberg (1948).)

Proposition (3.3.2). An action $A \mapsto X_A$ of a Lie algebra $\mathcal{G}$ by infinitesimal canonical transformations on a connected symplectic manifold (M, ω) determines an element $[\Omega]$ of $H^2\mathcal{G}$. If there is a moment, then $[\Omega] = 0$. Conversely, if $[\Omega] = 0$ and if each of the individual vector fields X_A is Hamiltonian, then the action of $\mathcal{G}$ on M is Hamiltonian.

The usefulness of this is that $H^2\mathcal{G} = 0$ for many of the Lie algebras that arise in simple physical systems, so it can be used to establish the existence of a moment in a straightforward way.

Proof. The definition of $[\Omega]$ is as follows. Pick $m \in M$ and put $\Omega(A, B) = \omega_m(X_A, X_B)$, where ω_m is the value of ω at m. Since X_A and X_B are locally Hamiltonian,

$$\omega([X_A, X_B], X_C) = -X_C(\omega(X_A, X_B)),$$

for any $A, B, C \in \mathcal{G}$, as a consequence of (1.5.3). Therefore,

$$2\omega([X_A, X_B], X_C) = \omega([X_A, X_B], X_C) - X_C(\omega(X_A, X_B)).$$

By (A.1.16), the sum of the right-hand side over the three cyclic permutations of A, B, C is $3\mathrm{d}\omega(X_A, X_B, X_C)$. But $\mathrm{d}\omega = 0$, so Ω satisfies the cocycle condition.

A different choice of m gives an equivalent cocycle. For if m' is some other point of M, then

$$\omega_m(X_A, X_B) - \omega_{m'}(X_A, X_B) = \tfrac{1}{2}\int_m^{m'} [X_A, X_B] \lrcorner\, \omega = \tfrac{1}{2}\int_m^{m'} X_{[A,B]} \lrcorner\, \omega$$

for $A, B \in \mathcal{G}$, where the integral is along any path in M from m to m', again by using (1.5.3). The second integral is a linear function of $[A, B]$, and is therefore a coboundary.

If there exists a moment map, then

$$2\omega(X_A, X_B) = [h_A, h_B] = h_{[A,B]},$$

where h is a Hamiltonian. In this case $2\Omega(A, B) = h_{[A,B]}(m)$, and so $\Omega = \delta\Theta$ where $\Theta(A) = h_A(m)$.

Suppose, conversely, that the Xs are Hamiltonian and that $\Omega = \delta\Theta$ for some $\Theta \in \mathcal{G}^*$. Then each X_A is generated by some $h_A \in C^\infty(M)$, which is determined by X_A up to the addition of a constant. If we fix the constants by imposing $h_A(m) = \Theta(A)$, then h_A depends linearly on A and

$$[h_A, h_B](m) - h_{[A,B]}(m) = 2\Omega(A, B) - \Theta([A, B]) = 0.$$

since $[h_A, h_B] = 2\omega(X_A, X_B)$. But $[h_A, h_B] - h_{[A,B]}$ is constant since both $[h_A, h_B]$ and $h_{[A,B]}$ generate $X_{[A,B]}$ (§1.6). Therefore $[h_A, h_B] = h_{[A,B]}$ everywhere and so there exists a moment map. ∎

3.4 Examples

Invariant potentials

A moment exists whenever there is an invariant symplectic potential. This can be seen as follows. Suppose that $\mathcal{G}$ acts on (M, ω) by infinitesimal canonical transformations and that there is a 1-form θ on M such that $\mathrm{d}\theta = \omega$ and

$$\mathcal{L}_{X_A}\theta = 0$$

for every $A \in \mathcal{G}$. For each $A \in \mathcal{G}$, put $h_A = X_A \lrcorner\, \theta$. Then

$$0 = \mathcal{L}_{X_A}\theta = X_A \lrcorner\, \mathrm{d}\theta + \mathrm{d}(X_A \lrcorner\, \theta) = X_A \lrcorner\, \omega + \mathrm{d}h_A,$$

so that h_A generates X_A. Also,

$$[h_A, h_B] = X_A(X_B \lrcorner\, \theta) = X_B \lrcorner\, \mathcal{L}_{X_A}\theta + [X_A, X_B] \lrcorner\, \theta = X_{[A,B]} \lrcorner\, \theta = h_{[A,B]}.$$

Therefore h is a Hamiltonian.

An important special case arises when $\mathcal{G}$ acts by *point transformations.* That is when M is the momentum phase space or the velocity phase space of some system with a configuration space Q and each X_A is the lift[5] to T^*Q or TQ of a vector field on Y_A on Q. In the case of momentum phase space, the flow of X_A preserves the canonical 1-form, and $h_A(p,q) = Y_A(q) \lrcorner\, p$. In the case of velocity phase space, the point transformations must be also symmetries of the Lagrangian $L(q,v)$. That is, $X_A(L) = 0$ for each $A \in \mathcal{G}$. Then $\mathcal{L}_{X_A}\theta_L = 0$ and the moment is given by $h_A = X_A \lrcorner\, \theta_L$. If the Lagrangian is irregular, then h_A descends to the reduced phase space, as in the next example.

Reduction

Invariant global symplectic potentials exist only in special cases. But it is often possible to construct a moment in much the same way by using reduction.

Proposition (3.4.1). Let (M', ω') be a reducible presymplectic manifold, with $\omega' = \mathrm{d}\theta'$, and let $A \mapsto X'_A$ be an action of $\mathcal{G}$ on M' such that $\mathcal{L}_{X'_A}\theta' = 0$ for every $A \in \mathcal{G}$. Let (M, ω) be the reduced phase space. For each $A \in \mathcal{G}$, put $h'_A = X'_A \lrcorner\, \theta'$. Then under the reduction map $M' \to M$

(1) each X'_A projects onto a Hamiltonian vector field X_A on M;

(2) each h'_A is the pullback of a function $h_A \in C^\infty(M)$; and

(3) $A \mapsto h_A$ is a Hamiltonian for the action $A \mapsto X_A$ of $\mathcal{G}$ on M.

Proof. Let Y' be tangent to the characteristic distribution on M'. Then for each $A \in \mathcal{G}$,

$$[X'_A, Y'] \lrcorner\, \omega' = \mathcal{L}_{X'_A}(Y' \lrcorner\, \omega') - Y' \lrcorner\, \mathcal{L}_{X'_A}\omega' = 0$$

since $\mathcal{L}_{X'_A}\omega' = \mathrm{d}(\mathcal{L}_{X'_A}\theta')$. Therefore $[X'_A, Y']$ is also tangent to the characteristic foliation, so X'_A projects onto a well-defined vector field $X_A \in V(M)$. The flow of X_A preserves ω since the flow of X'_A preserves ω'. Therefore X_A is locally Hamiltonian. Further, $Y'(h'_A) = Y' \lrcorner\, \mathrm{d}(X'_A \lrcorner\, \theta') = 0$ since $Y' \lrcorner\, \omega' = 0$ and

$$\mathrm{d}(X'_A \lrcorner\, \theta') = \mathcal{L}_{X'_A}\theta' - X'_A \lrcorner\, \mathrm{d}\theta' = -X'_A \lrcorner\, \omega'.$$

Hence h'_A is the pullback of some $h_A \in C^\infty(M)$. By the same calculation as in the previous example, h is a Hamiltonian. ∎

Semisimple Lie algebras

Let $\mathcal{G}$ be a real Lie algebra and, for each $A \in \mathcal{G}$, define $\mathrm{ad}_A : \mathcal{G} \to \mathcal{G}$ by

$$\mathrm{ad}_A(B) = [A, B].$$

Then $\mathrm{ad}_A \in \mathrm{gl}(\mathcal{G})$ (the Lie algebra of infinitesimal linear transformations of $\mathcal{G}$) and

$$\mathrm{ad}_{[A,B]} = \mathrm{ad}_A \mathrm{ad}_B - \mathrm{ad}_B \mathrm{ad}_A$$

for each A, B. Thus $A \mapsto \mathrm{ad}_A$ is a representation of $\mathcal{G}$ on its underlying vector space. It is called the *adjoint representation.*

Definition (3.4.2). The *Killing form* is the symmetric bilinear form

$$k(A, B) = \mathrm{tr}(\mathrm{ad}_A \mathrm{ad}_B); \qquad A, B \in \mathcal{G}.$$

If k is nondegenerate, then $\mathcal{G}$ is said to be *semisimple.*

For example, the Lie algebra of the orthogonal group $O(n)$ is the set of skew-symmetric $n \times n$ matrices with the Killing form

$$k(A, B) = (n - 2)\mathrm{tr}(AB),$$

which is nondegenerate for $n \geq 3$. On the other hand, the unitary group $U(n)$ has as its Lie algebra the set of $n \times n$ Hermitian matrices,[6] with Lie bracket $[A, B] = \mathrm{i}(AB - BA)$ and Killing form

$$k(A, B) = -2n\mathrm{tr}(AB) + 2\mathrm{tr}(A)\mathrm{tr}(B),$$

which is degenerate whatever the value of n (in the adjoint representation, the multiples of the identity matrix in the Lie algebra are mapped to zero). However, the Lie algebra of $SU(n)$, which is the set of trace-free Hermitian matrices, *is* semisimple: its Killing form is $k(A, B) = -2n\,\mathrm{tr}(AB)$. So are the Lie algebras of $O(p, q)$ $(p + q > 2)$ and $SU(p, q)$.

The Lie algebra of the symplectic group $SP(n, \mathbb{R})$ is the set of $2n \times 2n$ real matrices A such that

$$A\begin{pmatrix} 0 & 1_n \\ -1_n & 0 \end{pmatrix} = \begin{pmatrix} 0 & -1_n \\ 1_n & 0 \end{pmatrix} A^t,$$

where 1_n is the $n \times n$ identity matrix. It is also semisimple, with Killing form $k(A, B) = 2(n + 1)\mathrm{tr}(AB)$.

Proposition (3.4.3). For any semisimple Lie algebra $\mathcal{G}$, $[\mathcal{G}, \mathcal{G}] = \mathcal{G}$ and $H^2\mathcal{G} = 0$.

These are the Whitehead lemmas (Whitehead 1937; see also Hochschild 1942, Chevalley and Eilenberg 1948). Their importance in the present context is that they imply that every canonical action of a semisimple Lie algebra is Hamiltonian, and has a unique moment: $[\mathcal{G},\mathcal{G}]=\mathcal{G}$ implies that each individual generator is a globally Hamiltonian vector field (Proposition 1.5.3); $H^2\mathcal{G}=0$ then implies the existence of a moment, which is unique since $[\mathcal{G},\mathcal{G}]=\mathcal{G}$ and since the corresponding Hamiltonian is determined on $[\mathcal{G},\mathcal{G}]$ by $h_{[A,B]}=2\omega(X_A,X_B)$. In the case of a semisimple symmetry group, therefore, there is a natural and unambiguous way to associate an observable with each generator of the group. In particular, this is true of the rotation group $SO(3)$ and the Lorentz group $O(1,3)$.

Proof. Introduce a basis in $\mathcal{G}$. Then for $X,Y\in\mathcal{G}$,

$$[X,Y]^a = C^a{}_{bc}X^bY^c,$$

where the $C^a{}_{bc}$s are the structure constants and $a,b,\ldots=1,2,\ldots,\dim\mathcal{G}$, with summation and range conventions. The Killing form has components

$$k_{ab}=C^c{}_{da}C^d{}_{cb}. \tag{3.4.4}$$

Since k is nondegenerate, we can use it to raise and lower indices. By the skew-symmetry of $[\cdot,\cdot]$, $C_{abc}=-C_{acb}$. Also, for $X,Y,Z\in\mathcal{G}$,

$$\begin{aligned} C_{abc}X^aY^bZ^c &= k(X,[Y,Z]) \\ &= \mathrm{tr}\,\{\mathrm{ad}_X(\mathrm{ad}_Y\mathrm{ad}_Z-\mathrm{ad}_Z\mathrm{ad}_Y)\} \\ &= \mathrm{tr}\,\{\mathrm{ad}_Y(\mathrm{ad}_Z\mathrm{ad}_X-\mathrm{ad}_X\mathrm{ad}_Z)\} \\ &= k(Y,[Z,X]) \\ &= C_{abc}Y^aZ^bX^c. \end{aligned}$$

Therefore $C_{abc}=C_{[abc]}$. In terms of the structure constants, the Jacobi identity is

$$C_{a[bc}C^a{}_{d]e}=0. \tag{3.4.5}$$

To prove the first part of the proposition, let $X\in[\mathcal{G},\mathcal{G}]^\perp$, the orthogonal complement of $[\mathcal{G},\mathcal{G}]$ with respect to k. Then $X^aC_{abc}=0$. By multiplying eqn (3.4.4) by X^b and contracting over b, we obtain $X_a=0$. Therefore $[\mathcal{G},\mathcal{G}]^\perp=0$ and so $[\mathcal{G},\mathcal{G}]=\mathcal{G}$.

To prove the second part, we need a lemma.

Lemma (3.4.6). Suppose that $\alpha_{a[b}C^a{}_{cd]}=0$, $\alpha_{ab}=-\alpha_{ba}$, and $\alpha_{ab}C^{ab}{}_c=0$. Then $\alpha_{ab}=0$.

Proof. It follows from what we are given that $C^{bc}{}_e\alpha_{a[b}C^a{}_{cd]} = 0$. By expanding the skew-symmetrizer, relabelling the dummy indices, and substituting from eqn (3.4.4), we obtain

$$\alpha_a{}^b(C^a{}_{cd}C^c{}_{eb} + C^a{}_{ce}C^c{}_{bd}) - \alpha_{ed} = 0.$$

By using the Jacobi identity, therefore,

$$\alpha_{ed} = -\alpha_a{}^b C^a{}_{cb}C^c{}_{de} = 0$$

since $\alpha_{ab}C^{ab}{}_c = 0$. ■

Let us return now to the proof that $H^2\mathcal{G} = 0$. Suppose that $\Omega \in \mathcal{G}^* \wedge \mathcal{G}^*$ satisfies the cocycle condition, so that $\Omega_{ab} = -\Omega_{ba}$ and $\Omega_{a[b}C^a{}_{cd]} = 0$. Put $\Theta_a = -2C_{abc}\Omega^{bc}$ and $\alpha_{ab} = \Omega_{ab} - \frac{1}{2}\Theta_c C^c{}_{ab}$; that is, $\alpha = \Omega - \delta\Theta$. Then $\alpha_{ab} = -\alpha_{ba}$, $\alpha_{a[b}C^a{}_{cd]} = 0$, and

$$\alpha_{bc}C^{bc}{}_a = (\Omega_{bc} - \tfrac{1}{2}\Theta_d C^d{}_{bc})C^{bc}{}_a = -\tfrac{1}{2}\Theta_a + \tfrac{1}{2}\Theta_a = 0.$$

Therefore, by the lemma, $\alpha = 0$ and $\Omega = \delta\Theta$. It follows that $H^2\mathcal{G} = 0$. ■

Abelian Lie algebras

At the other extreme, there are the abelian Lie algebras, in which $[A, B] = 0$ for every $A, B \in \mathcal{G}$. Here $[\mathcal{G}, \mathcal{G}] = 0$ and $H^2\mathcal{G} = \mathcal{G}^* \wedge \mathcal{G}^*$. There is no guarantee that the individual generators are Hamiltonian, and when they are, there is no guarantee that there is a moment; and when there is a moment, it is not unique since we can add to it any element of $\mathcal{G}^*$.

For example, there is no moment for the action by translation of the abelian Lie algebra of constant vector fields on the phase space $\mathbb{R}^{2n}$ with the symplectic form $\omega = \mathrm{d}p_a \wedge \mathrm{d}q^a$. The corresponding element of $H^2\mathcal{G}$ is $[\omega]$.

The translations of Minkowski space-time form an abelian Lie algebra. While we can always associate 'angular momenta' with the generators of a canonical action of the Lorentz group, we may not be able to associate 'linear momenta' with the generators of the group of translations—although, as the next example shows, there is a moment whenever both Lorentz rotations and translations act with the correct Lie brackets.

Inhomogeneous orthogonal groups

Let g be a definite or indefinite inner product on $\mathbb{R}^n$ ($n > 2$) and let $IO(g)$ be the corresponding inhomogeneous orthogonal group—the group of affine transformations of $\mathbb{R}^n$ that preserve the metric determined by g. The principal examples are the Poincaré group (where $n = 4$ and g has signature $+ - - -$) and the isometry group of Euclidean space.

The elements of the Lie algebra io(g) of $IO(g)$ can be represented as formal sums $A+X$, where $X \in \mathbb{R}^n$ is a column vector with entries X^a ($a = 1, 2, \ldots, n$) and A is a skew-symmetric $n \times n$ matrix with entries A^{ab}. The Xs generate translations, and the As generate (pseudo) orthogonal transformations; they act on $\mathbb{R}^n$ by $A : X \mapsto A(X)$, where

$$A(X)^a = A^{ab} X_b \tag{3.4.7}$$

(the indices are raised and lowered by g and its inverse; note that they now run from 1 to n, and not to the dimension of the Lie algebra as in the last example). The Lie brackets are

$$[A, B]^{ab} = A^{ac} B_c{}^b - B^{ac} A_c{}^b, \qquad [A, X] = A(X), \qquad [X, Y] = 0. \tag{3.4.8}$$

It is clear that $[\mathrm{io}(g), \mathrm{io}(g)] = \mathrm{io}(g)$, although io($g$) is not semisimple. Nonetheless, we have the following.

Proposition (3.4.9). Every action of io(g) by infinitesimal canonical transformations is Hamiltonian.

Proof. We have only to show that $H^2\mathrm{io}(g) = 0$, which we shall do by following Bargmann (1954). Let Ω be a cocycle. Since the (pseudo) orthogonal transformations of $\mathbb{R}^n$ form a semisimple group, we can assume, without loss of generality, that $\Omega(A, B) = 0$ whenever A and B are generators of orthogonal transformations. Then

$$\Omega(A + X, B + Y) = K_{abc}(A^{ab} Y^c - B^{ab} X^c) + T_{ab} X^a Y^b$$

where the Ks and Ts are constants such that $K_{abc} = -K_{bac}$ and $T_{ab} = -T_{ba}$. By the cocycle condition,

$$\begin{aligned} 0 &= \Omega([A, B], X) + \Omega(B, A(X)) - \Omega(A, B(X)) \\ &= 2K^a{}_{cd} A_{ab} B^{bc} X^d + K_{ec}{}^a B^{ec} A_{ad} X^d - K^{ab}{}_e A_{ab} B^{ec} X_c \end{aligned}$$

$$\begin{aligned} 0 &= \Omega(A(X), Y) - \Omega(A(Y), X) \\ &= T^a{}_c A_{ab} (X^b Y^c - Y^b X^c). \end{aligned}$$

Since these hold for any skew-symmetric A, B and for any X, Y,

$$2\delta^{[a}_{[c} K^{b]}{}_{e]d} + K_{ec}{}^{[a} \delta^{b]}_d - K^{ab}{}_{[e} g_{c]d} = 0 \tag{3.4.10}$$

$$T^{[a}{}_{[c} \delta^{b]}_{d]} = 0. \tag{3.4.11}$$

By contracting over b, d in (3.4.11), $T_{ab} = 0$ (since $n > 2$); and, by contracting over b, d in (3.4.10),

$$(n-1)K_{ec}{}^{a} = 2\delta^{a}_{[c}K_{e]b}{}^{b}$$

It follows that $\Omega(A, X) = \frac{1}{2}\Theta(A(X))$, where $\Theta(X) = 4K_{ab}{}^{b}X^{a}/(n-1)$, and hence that $\Omega = \delta\Theta$. ∎

The Galilean group

The Galilean group is the symmetry group of space-time in Newtonian mechanics. It is the set of transformations of the form

$$\begin{pmatrix} t \\ x \\ y \\ z \end{pmatrix} \mapsto \left(\begin{array}{c|c} 1 & 0 \\ \hline v_1 & \\ v_2 & H \\ v_3 & \end{array}\right) \begin{pmatrix} t \\ x \\ y \\ z \end{pmatrix} + \begin{pmatrix} X^0 \\ X^1 \\ X^2 \\ X^3 \end{pmatrix} \tag{3.4.12}$$

where $H \in SO(3)$, $(v_1, v_2, v_3) = \mathbf{v}$ is the velocity of the transformation, and the Xs are constants.

The group acts on the phase space of a free particle of mass m, which is $\mathbb{R}^6$, with symplectic form $\mathrm{d}p_a \wedge \mathrm{d}q^a$. When $\mathbf{v} = 0 = X^0$, the transformations are isometries of Euclidean space. They act in the obvious way by rotating the position and momentum vectors $\mathbf{q} = (q^1, q^2, q^3)$ and $\mathbf{p} = (p_1, p_2, p_3)$, and by translating $\mathbf{q}$. The time translation $t \mapsto t + X^0$ acts by $\mathbf{q} \mapsto \mathbf{q} + m^{-1}X^0\mathbf{p}$, leaving $\mathbf{p}$ fixed; and the boosts ($H = 1$, $X = 0$) act by translating the momentum. That is, by $\mathbf{q} \mapsto \mathbf{q}$ and $\mathbf{p} \mapsto \mathbf{p} + m\mathbf{v}$. All the translations of $\mathbb{R}^6$ are included. Because there is no moment for the action of the translation group on $\mathbb{R}^6$, there is no moment for the action of the whole Galilean group. This is consistent with the observation (§2.6) that the Lagrangian of a free particle in Euclidean space is not invariant under Galilean transformations. (Souriau 1970, Guillemin and Sternberg 1977; the cohomology of the Galilean group is discussed in Bargmann 1954.)

Compact phase spaces

Let (M, ω) be a compact symplectic manifold and let $h, h' \in C^\infty(M)$. Then, by following Guillemin and Sternberg (1984),

$$[h, h']\omega^n = X_h(h')\omega^n = \mathcal{L}_{X_h}(h'\omega^n) = \mathrm{d}(h'X_h \lrcorner\, \omega^n)$$

since $\mathcal{L}_{X_h} w^n = 0$. Therefore

$$\int_M [h, h']\,\omega^n = 0.$$

It follows that if $\mathcal{G}$ acts on M by infinitesimal canonical transformations and if each X_A is globally Hamiltonian, then we can construct a Hamiltonian for the action of $\mathcal{G}$ by imposing

$$\int_M h_A \, \omega^n = 0$$

to fix the arbitrary constants in the generators. In particular, every canonical action of a Lie algebra on a compact simply-connected symplectic manifold is Hamiltonian.

Circle and torus actions

The circle group T is the group of complex numbers of unit modulus under multiplication. The n-torus is the n-fold product $T^n = T \times \cdots \times T$. It is abelian, and has Lie algebra $\mathbb{R}^n$.

Suppose that T^n has a Hamiltonian action on a compact connected symplectic manifold (M, ω). Then the moment

$$\mu : M \to \mathbb{R}^n : m \mapsto \big(h_1(m), \ldots, h_n(m)\big)$$

has a striking convexity property (Atiyah 1982, Guillemin and Sternberg 1982a): its image is a convex polytope in $\mathbb{R}^n$. It is the convex hull of the images of the fixed points of T^n.

The first step in Atiyah's proof is to use Morse theory to show that if $h = x_1 h_1 + \cdots + x_n h_n$, where the xs are constant, then the level surfaces of h are connected. The essential point is that because X_h is either periodic or quasiperiodic (its orbits are either closed or wind around a torus), the critical points of h must have even index. That is, at points where $\mathrm{d}h = 0$, the matrix of second partial derivatives of h must have an even number of negative eigenvalues. It is not hard to see why this should be so when the critical points are nondegenerate. At a critical point m, the canonical flow of h induces a family of linear canonical transformations in the tangent space $T_m M$. These are generated by the second partial derivatives of h, which can be regarded as the components of a quadratic form g on $T_m M$ (p. 10). If g is nondegenerate, then its determinant must be positive. Otherwise there would exist a real nonzero solution of the characteristic equation $\det(g - 2\lambda\omega) = 0$. The orbit of the corresponding eigenvector $X \in T_m M$ would be of the form $t \mapsto \mathrm{e}^{\lambda t} X$, which would be inconsistent with periodicity or quasiperiodicity.

The next step is to show by induction on n that $\mu^{-1}(x)$ is either empty or connected for each $x \in \mathbb{R}^n$; and finally to deduce the convexity result for T^n from the connectivity result for T^{n-1}.

One consequence of these results of Atiyah, and Guillemin and Sternberg is that if $h \in C^\infty(M)$ generates a circle action on a compact phase space, then h has a unique local maximum (and a unique local minimum). This is relevant to the general theory of coherent states (§9.2).

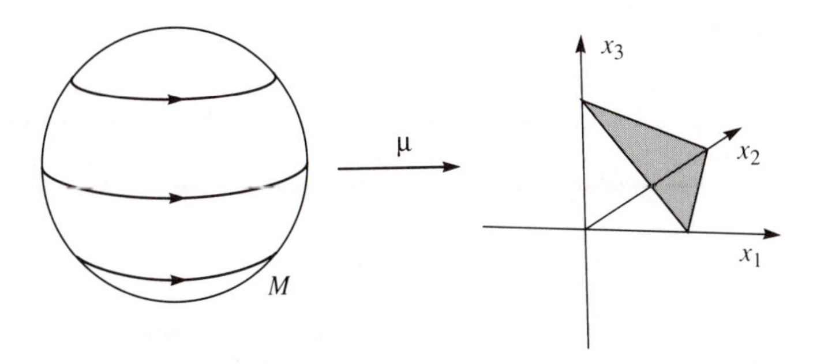

Fig. 3.1. The moment of a torus action on M.

Example. Let (M, ω) be the reduction of the constraint[7]

$$\tfrac{1}{2}\left((p_1)^2 + (p_2)^2 + (p_3)^2 + (q^1)^2 + (q^2)^2 + (q^3)^2\right) = 1$$

in $\mathbb{R}^6$ with its standard symplectic structure $\mathrm{d}p_a \wedge \mathrm{d}q^a$. Then M is compact and connected (it is diffeomorphic to $\mathbb{CP}_2$). Put $h_a = \frac{1}{2}((p_a)^2 + (q^a)^2)$ $(a = 1, 2, 3)$. The hs generate an action of T^3 on $\mathbb{R}^6$ and on M. The image in $\mathbb{R}^3$ of the moment map from M is

$$\{(x_1, x_2, x_3) \mid x_1 + x_2 + x_3 = 1,\ x_1 > 0,\ x_2 > 0,\ x_3 > 0\},$$

which is the region of the plane $x_1 + x_2 + x_3 = 1$ shown in Fig. 3.1.

3.5 Elementary classical systems

In quantum theory, a system with a symmetry group G is *elementary* if G acts irreducibly on the Hilbert space of states. At the classical level, the analogous condition is transitivity: a classical system with symmetry is elementary if the symmetry group acts transitively on its phase space, so that any state can be transformed into any other by an element of the group. In the case of the Galilean group, for example, a single particle is an elementary system, but a pair of particles is not since their relative speed cannot be changed by a Galilean transformation.

The problem of determining all the elementary quantum systems with a given symmetry group can be very hard; but the analogous 'classical' problem has a straightforward solution in an elegant construction which, under certain conditions, gives all the symplectic manifolds on which a given Lie group has a transitive action (Kirillov 1976, Souriau 1970, and Kostant 1970a; see also Guillemin and Sternberg 1984). We shall look at the construction in this section.

Invariant vector fields

We shall need some facts about the relationship between Lie groups and Lie algebras. Let G be a Lie group and let $\mathcal{G}$ be its Lie algebra: $\mathcal{G}$ is the tangent space at $e \in G$. Each $g \in G$ determines two diffeomorphisms $G \to G$: the right translation ρ_g and the left translation λ_g, defined by

$$\rho_g : g' \mapsto g'g \quad \text{and} \quad \lambda_g : g' \mapsto gg'.$$

By acting on $A \in \mathcal{G}$ with right and left translations, we obtain two vector fields on G: the right-invariant vector field R_A and the left-invariant vector field L_A, defined by

$$R_A(e) = A = L_A(e), \quad \rho_{g*}(R_A) = R_A, \quad \lambda_{g*}(L_A) = L_A$$

for every $g \in G$.

The flow of the right-invariant vector field R_A is the one-parameter group of left translations $(g, t) \mapsto \mathrm{e}^{tA}g$ and the flow of the left-invariant vector field L_A is the one-parameter group of right translations $(g, t) \mapsto g\mathrm{e}^{tA}$, where $t \in \mathbb{R}$ and $A \mapsto \mathrm{e}^A$ is the exponential map (note the potentially confusing interchange of 'left' and 'right'). It follows, by using the approximation

$$\mathrm{e}^{sA}\mathrm{e}^{tB} \sim \mathrm{e}^{st[A,B]}\mathrm{e}^{tB}\mathrm{e}^{sA},$$

which holds to the second order in s and t, that the invariant vector fields generated by $A, B \in \mathcal{G}$ have Lie brackets

$$[R_A, R_B] = -R_{[A,B]}, \quad [L_A, L_B] = L_{[A,B]}, \quad \text{and} \quad [R_A, L_B] = 0.$$

(Fig. 3.2). Because of the second of these identities, it is not uncommon to define $\mathcal{G}$ to be the Lie algebra of left-invariant vector fields.

By combining left and right translations, we obtain for each $g \in G$ a map

$$\iota_g : G \to G : g' \mapsto gg'g^{-1} = \lambda_g \rho_{g^{-1}}(g')$$

which fixes e. Its derivative at e is therefore a linear transformation of $\mathcal{G}$, which is usually denoted by Ad_g, although we shall use the shorthand gA for $\mathrm{Ad}_g A$. The construction gives an action of G on $\mathcal{G}$ (on the left) by Lie algebra homomorphisms. It is called the *adjoint action*.

The adjoint action induces the *coadjoint action* on the dual space $\mathcal{G}^*$ by[8]

$$g : \mathcal{G}^* \to \mathcal{G}^* : f \mapsto fg \quad \text{where} \quad fg(A) = f(gA) \quad \text{for} \quad A \in \mathcal{G}.$$

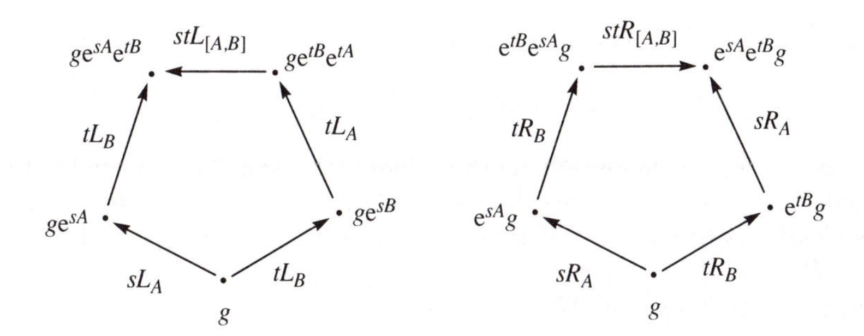

Fig. 3.2. The Lie brackets of invariant vector fields.

This is an action on the right; that is $(fg)g' = f(gg')$. The infinitesimal form is

$$fe^{tA}(B) = f(B) + tf([A,B]) + O(t^2), \tag{3.5.1}$$

for $A, B \in \mathcal{G}$.

By acting on $f \in \mathcal{G}^*$ by right translations, we can construct a right-invariant 1-form θ_f on G, which is characterized by

$$\theta_f(e) = f \quad \text{and} \quad \rho_g^*(\theta_f) = \theta_f \quad \text{for every } g \in G,$$

or alternatively by $R_A \lrcorner\, \theta_f = f(A)$ for every $A \in \mathcal{G}$. The exterior derivative $\omega_f = \mathrm{d}\theta_f$ is a right-invariant 2-form[9] such that

$$\omega_f(R_A, R_B) = \tfrac{1}{2} f([A,B]) \tag{3.5.2}$$

for every $A, B \in \mathcal{G}$.

Coadjoint orbits

Suppose that G is connected. Under the coadjoint action, $\mathcal{G}^*$ breaks up into orbits of the form

$$M = \{fg \mid g \in G\},$$

on each of which G acts transitively. The orbits are connected because G is connected. The first result of Kirillov, Souriau, and Kostant is that they are all symplectic manifolds.

It follows from (3.5.1) that the action on $\mathcal{G}^*$ of the one-parameter subgroup generated by $A \in \mathcal{G}$ is the flow of the vector field $X_A \in V(\mathcal{G}^*)$, where the value of X_A at f is $f' \in T_f\mathcal{G}^* = \mathcal{G}^*$, where

$$f'(B) = f([A,B]). \tag{3.5.3}$$

Let $f \in \mathcal{G}^*$ and let M be the orbit through f. The vector fields X_A span the $T_f M$; and if X_A and $X_{A'}$ have the same value at f, then

$$f([A-A',B]) = f([A,B]) - f([A',B]) = 0$$

for any $B \in \mathcal{G}$. It follows that if we put

$$\omega(X_A, X_B) = \tfrac{1}{2} f([A,B]), \qquad (3.5.4)$$

then $\omega(X_A, X_B)$ depends only on the values of X_A and X_B at f, and not on A and B themselves. Therefore ω is a well-defined bilinear form on $T_f M$. It is clearly antisymmetric. It is also nondegenerate since $f([A,B]) = 0$ for every $B \in \mathcal{G}$ only if X_A vanishes at f. As f varies over M, ω becomes a nondegenerate 2-form on M.

Proposition (3.5.5). The 2-form ω is a symplectic structure on M and it is invariant under the coadjoint action.

Proof. We have to show that ω is closed and that $g^*(\omega) = \omega$ for every $g \in G$. For $A \in \mathcal{G}$, define $h_A \in C^\infty(M)$ by $h_A(f) = f(A)$. Then, for $A, B \in \mathcal{G}$,

$$(X_A \lrcorner \, \mathrm{d}h_B)(f) = \frac{\mathrm{d}}{\mathrm{d}t}\Big(f(B + t[A,B]) \Big) = f([A,B]).$$

It follows that $X_A \lrcorner \, \mathrm{d}h_B = h_{[A,B]}$ and hence that

$$X_A \lrcorner \, (X_B \lrcorner \, \omega + \mathrm{d}h_B) = 0 \qquad (3.5.6)$$

since $X_A \lrcorner \, (X_B \lrcorner \, \omega) = 2\omega(X_B, X_A) = h_{[B,A]}$. Equation (3.5.6) holds for any $A \in \mathcal{G}$, so

$$X_B \lrcorner \, \omega + \mathrm{d}h_B = 0$$

for every $B \in \mathcal{G}$. Hence for any $A, B \in \mathcal{G}$,

$$\begin{aligned} \mathcal{L}_{X_A}(X_B \lrcorner \, \omega + \mathrm{d}h_B) &= X_B \lrcorner \, \mathcal{L}_{X_A}\omega + [X_A, X_B] \lrcorner \, \omega + \mathrm{d}(X_A \lrcorner \, \mathrm{d}h_B) \\ &= X_B \lrcorner \, \mathcal{L}_{X_A}\omega + X_{[A,B]} \lrcorner \, \omega + \mathrm{d}h_{[A,B]} \\ &= X_B \lrcorner \, \mathcal{L}_{X_A}\omega. \end{aligned}$$

But the left-hand side vanishes. Therefore $\mathcal{L}_{X_A}\omega = 0$, and so $g^*(\omega) = \omega$ for every $g \in G$ since G is connected. Finally, for any $A \in \mathcal{G}$,

$$0 = \mathcal{L}_{X_A}\omega = \mathrm{d}(X_A \lrcorner \, \omega) + X_A \lrcorner \, \mathrm{d}\omega = \mathrm{d}(-\mathrm{d}h_A) + X_A \lrcorner \, \mathrm{d}\omega = X_A \lrcorner \, \mathrm{d}\omega.$$

Therefore $\mathrm{d}\omega = 0$. ∎

The proof establishes more than the proposition: it shows that $A \mapsto h_A$ is a Hamiltonian, and hence that the identity map $M \to \mathcal{G}^*$ is a moment for the action of G on M.

Two connected groups with the same Lie algebra have the same orbits with the same symplectic structures, and the whole construction can in fact be expressed in terms of $\mathcal{G}$ rather than G.

Reduction of the group manifold

There is another way of looking at the orbits in $\mathcal{G}^*$, which is sometimes helpful, particularly when it comes to quantization. Let $f \in \mathcal{G}^*$ and let M be the orbit through f. Then $\pi : g \mapsto fg$ maps G onto M, with $\pi(e) = f$. It follows from (3.5.2) and (3.5.4), together with the invariance of ω_f and ω, that $\pi^*(w) = \omega_f$.

The alternative starting point is to treat (G, ω_f) as a presymplectic manifold. Its characteristic foliation is spanned by the vectors R_A for which $A \in \mathcal{G}_f$, where

$$\mathcal{G}_f = \{A \in \mathcal{G} \mid X_A(f) = 0\}.$$

The reduction is a symplectic manifold M_f on which G acts transitively on the right. When M is simply-connected, $M = M_f$, but in general M and M_f are not quite the same because[10] $M = G/G_f$, where

$$G_f = \{g \in G \mid fg = f\}$$

is the stabilizer of f, while $M_f = G/(G_f)_0$, where $(G_f)_0$ is the connected subgroup generated by $\mathcal{G}_f$. That is, $(G_f)_0$ is the identity component of G_f.

The advantage of working on G is that there is an invariant potential, since $\omega_f = \mathrm{d}\theta_f$. In fact, the map $M_f \mapsto M$ induced by the inclusion $(G_f)_0 \subset G_f$ is the moment constructed from θ_f by reduction (3.4.1).

Example. The Lie algebra of $SL(2, \mathbb{R})$ is the set of real matrices

$$A = \begin{pmatrix} z & x+y \\ x-y & -z \end{pmatrix}.$$

If $f(A) = z$, then G_f is the diagonal subgroup, which has two components. The identity component $(G_f)_0$ consists of diagonal matrices with positive entries. ■

The second result is that any symplectic manifold on which G has a transitive Hamiltonian action is a covering space of an orbit in $\mathcal{G}^*$. In principle, therefore, all elementary classical sytems with symmetry group G which admit a moment can be found by analysing the orbits in $\mathcal{G}^*$.

Proposition (3.5.7). Let (M', ω') be a symplectic manifold on which G acts transitively on the right by canonical transformations and let $\mu : M' \to \mathcal{G}^*$ be a moment. Then μ is a local canonical diffeomorphism from (M', ω') onto an orbit $(M, \omega) \subset \mathcal{G}^*$.

Proof. Let X'_A denote the Hamiltonian vector field on M' generated by $A \in \mathcal{G}$ and let h'_A be the Hamiltonian constructed from μ, so that $h'_A = h_A \circ \mu$.

Since μ is a moment, $X'_A \lrcorner \, \mathrm{d}h'_B = h'_{[A,B]}$ for every $A, B \in \mathcal{G}$. Therefore $X'_A \lrcorner \, \mathrm{d}\mu = X_A \circ \mu$. To understand this formula, note that μ is a function on M' with values in the vector space $\mathcal{G}^*$, so its exterior derivative is a 1-form with values in $\mathcal{G}^*$. The expression on the right-hand side, $X_A \circ \mu$, is a function on M' with values in $\mathcal{G}^*$.

It follows that

$$\mu_*(X'_A) = X_A \tag{3.5.8}$$

for every $A \in \mathcal{G}$ and hence, since G is connected, $\mu(m'g) = \mu(m')g$ for every $m' \in M'$, $g \in G$. These properties of the moment do not depend on transitivity of the action on M'.

Since the action *is* transitive, $M = \mu(M')$ is an orbit in $\mathcal{G}^*$. Moreover, for every $A, B \in \mathcal{G}$,

$$2\omega'(X'_A, X'_B) = h'_{[A,B]} = h_{[A,B]} \circ \mu = 2\omega(X_A, X_B) \circ \mu$$

Therefore $\mu^*(\omega) = \omega'$ and so $\mu : M' \to M$ is a local canonical diffeomorphism. ∎

Example. *The rotation group and* $SU(2)$. The Lie algebra of $SO(3)$ is the set of 3×3 skew symmetric matrices. It can be identified with $\mathbb{R}^3$ by writing $A = a_1 X + a_2 Y + a_3 Z$, where

$$X = \begin{pmatrix} 0 & 0 & 0 \\ 0 & 0 & 1 \\ 0 & -1 & 0 \end{pmatrix}, \quad Y = \begin{pmatrix} 0 & 0 & -1 \\ 0 & 0 & 0 \\ 1 & 0 & 0 \end{pmatrix}, \quad Z = \begin{pmatrix} 0 & 1 & 0 \\ -1 & 0 & 0 \\ 0 & 0 & 0 \end{pmatrix}.$$

Note that $[X, Y] = -Z$, and so on.[11]

The dual space is also $\mathbb{R}^3$, with $f(A) = f_1 a_1 + f_2 a_2 + f_3 a_3$, and the coadjoint orbits are the spheres centred on the origin. The sphere of radius s is the classical phase space for the rotational degrees of freedom of an elementary particle with spin s. Its symplectic form is the area element, divided by s.

The group $SU(2)$ is the double cover of $SO(3)$, and has the same Lie algebra. Each $g \in SU(2)$ is a matrix

$$g = \begin{pmatrix} z^0 & z^1 \\ -\bar{z}^1 & \bar{z}^0 \end{pmatrix}$$

where $z^0\bar{z}^0 + z^1\bar{z}^1 = 1$. Thus $SU(2)$ is diffeomorphic to the unit 3-sphere $S^3 \subset \mathbb{C}^2$. The Lie algebra is the set of 2×2 trace-free hermitian matrices, with Lie bracket $[A, B] = \mathrm{i}(AB - BA)$. It is identified with the Lie algebra of $SO(3)$ by mapping X,Y, and Z to $\frac{1}{2}\sigma_1$, $\frac{1}{2}\sigma_2$, and $\frac{1}{2}\sigma_3$, where the σs are the Pauli matrices

$$\sigma_1 = \begin{pmatrix} 0 & 1 \\ 1 & 0 \end{pmatrix}, \quad \sigma_2 = \begin{pmatrix} 0 & -\mathrm{i} \\ \mathrm{i} & 0 \end{pmatrix}, \quad \sigma_3 = \begin{pmatrix} 1 & 0 \\ 0 & -1 \end{pmatrix}.$$

The right-invariant forms on $SU(2)$ determined by $f = (0, 0, s)$ are

$$\theta_f = \mathrm{i}s(z^0\mathrm{d}\bar{z}^0 + z^1\mathrm{d}\bar{z}^1 - \bar{z}^0\mathrm{d}z^0 - \bar{z}^1\mathrm{d}z^1)$$

and

$$\omega_f = 2\mathrm{i}s(\mathrm{d}z^0 \wedge \mathrm{d}\bar{z}^0 + \mathrm{d}z^1 \wedge \mathrm{d}\bar{z}^1),$$

and the reduction map from $SU(2)$ onto the coadjoint orbit of f (the sphere of radius s) is the Hopf map

$$(z^0, z^1) \mapsto s(z^1\bar{z}^0 + z^0\bar{z}^1, \mathrm{i}z^1\bar{z}^0 - \mathrm{i}z^0\bar{z}^1, z^0\bar{z}^0 - z^1\bar{z}^1). \qquad \blacksquare$$

Example. Let (M, ω) be a coadjoint orbit in $\mathcal{G}^*$ for some Lie group. In spite of its simplicity, it is not always easy to translate (3.5.4) into an expression for ω in local coordinates. One trick that we shall use later on to obtain an explicit representation of ω is to look for a manifold C on which $\mathcal{G}$ acts by $A \mapsto X'_A \in V(C)$, together with a surjection $\pi : C \to M$ and a 1-form $\theta' \in \Omega^1(C)$, such that

(C1) $\pi^{-1}(m)$ is connected for each $m \in M$;

(C2) $\pi_* X'_A = X_A$;

(C3) for each $m' \in C$, $X'_A \lrcorner\, \theta'(m') = f(A)$, where $f = \pi(m')$.

Then θ' is invariant under the action of $\mathcal{G}$ on C, and, by (3.5.4), $\mathrm{d}\theta' = \pi^*\omega$. It follows that (M, ω) is the reduction of $(C, \mathrm{d}\theta')$. One possibility, of course, is to take $C = G$, $\theta' = \theta_f$. But other choices are often useful.

Non-Hamiltonian actions

When the action is not Hamiltonian, Proposition (3.5.7) can be modified as follows (Souriau 1970, Marle 1976). Suppose that (M', ω'), X'_A, and h'_A are as in the proof, but that it is no longer true that $[h'_A, h'_B] = h'_{[A,B]}$, although h'_A is still linear in A. Put $\Omega(A, B) = \frac{1}{2}[h'_A, h'_B] - \frac{1}{2}h'_{[A,B]}$, for $A, B \in \mathcal{G}$. Then Ω is constant, and so is an element of $\mathcal{G}^* \wedge \mathcal{G}^*$. It satisfies the cocycle condition and its class in $H^2\mathcal{G}$ is the class determined by the action of $\mathcal{G}$ on M'.

The idea is to use Ω to modify the action of G on $\mathcal{G}^*$. The linear coadjoint transformations are replaced by the affine transformations generated by the vector fields

$$\tilde{X}_A = X_A + \Omega_A, \tag{3.5.9}$$

where $\Omega_A \in \mathcal{G}^*$ is defined by $\Omega_A(B) = 2\Omega(A, B)$. The symplectic form on an orbit under the modified action is defined by replacing (3.5.3) by

$$\tilde{\omega}(\tilde{X}_A, \tilde{X}_B) = \tfrac{1}{2} f([A, B]) + \Omega(A, B).$$

With these new definitions, $M' \to \mathcal{G}^* : m' \mapsto f'_m$, where $f_{m'}(A) = h'_A(m')$, is a local canonical diffeomorphism from M' onto an orbit in $\mathcal{G}^*$.

3.6 Discrete transformations

We shall also be interested in how classical systems behave under discrete transformations such as time reversal and spatial reflection, as well as under the action of a connected symmetry group such as the proper orthochronous Lorentz group. To allow for the possibility of representing discrete symmetries at the quantum level by anti-unitary operators, we shall allow their classical counterparts to act by anticanonical transformations. That is, by diffeomeorphisms $g : M \to M$ such that $g^*\omega = -\omega$ (Rawnsley 1972).

We shall say that a symplectic manifold (M, ω) is *elementary* with respect to a general Lie group G whenever G acts transitively on M on the right by canonical or anticanonical transformations. That is, for each $g \in G$, $g^*\omega = \epsilon_g \omega$ where $\epsilon_g = \pm 1$. Then ϵ is constant on the connected components of G and is a homomorphism $G \to \mathbb{Z}_2$. The identity component G_0 of G acts by canonical transformations, so the connected components of M are elementary with respect to G_0, according to the definition for connected groups. If we make the simplifying assumptions that the coadjoint orbits of G_0 are simply-connected, that $\mathcal{G} = [\mathcal{G}, \mathcal{G}]$, and that $H^2\mathcal{G} = 0$, then it is straightforward to describe all the elementary symplectic manifolds, as follows.

Suppose that (M, ω) is elementary. There is then a unique moment $\mu : M \to \mathcal{G}^*$ for the action of G_0. The restriction of μ to each connected component of M is a canonical diffeomorphism onto a coadjoint orbit of G_0 in $\mathcal{G}^*$.

Let h be the Hamiltonian constructed from μ and, for each $A \in \mathcal{G}$, let X_A be the Hamiltonian vector field on M generated by h_A. Then for each $A \in \mathcal{G}$ and $g \in G$, $g_* X_{gA} = X_A$ and $h_A \circ g$ generates $\epsilon_g X_{gA}$. Moreover

$$[h_A \circ g, h_B \circ g] = \epsilon_g h_{[A,B]} \circ g.$$

Therefore

$$A \mapsto \epsilon_g h_{g^{-1}A} \circ g$$

is also a Hamiltonian. However, under our assumptions about G, the moment and Hamiltonian are unique. Therefore $h_A \circ g = \epsilon_g h_{gA}$ and so

$$\mu(mg) = \epsilon_g(\mu(m))g$$

for every $m \in G$.

Thus M consists of a collection of copies of (not necessarily distinct) coadjoint orbits of G_0, and the action of G on M is given by combining the coadjoint action of G on the $\mathcal{G}^*$ with $\mathcal{G}^* \to \mathcal{G}^* : f \mapsto \epsilon_g f$, where ϵ is a homomorphism from G to $\mathbb{Z}_2$.

Example. If $G = O(3)$, then $G_0 = SO(3)$ and there are two choices for ϵ: the identity homomorphism and the determinant. The orbits of $SO(3)$ are the spheres centred on the origin in $\mathbb{R}^3$. We can make a single orbit M into an elementary symplectic manifold for $O(3)$ in two different ways, either by making $-I$ act on M as the identity, or by making it act by reflection in the origin (i.e. $f \mapsto -f$, which reverses orientation and is therefore anticanonical). We can also take two copies of M which are interchanged by $-I$, either with or without a reflection. These are the only possibilities. ∎

3.7 Marsden-Weinstein reduction

When a connected Lie group G has a Hamiltonian action on a symplectic manifold (M, ω), the image of M under the moment map $\mu : M \to \mathcal{G}^*$ is a union of coadjoint orbits, whether or not the action is transitive. This is because $\mu \circ g = g \circ \mu$ for every $g \in G$ (see 3.5.8).

Example. *Composite systems.* Suppose that (M_1, ω_1) and (M_2, ω_2) are homogeneous symplectic manifolds of the Lie group G. Then G acts on the *composite* symplectic manifold $M = M_1 \times M_2$. In general, this action is not transitive.

If μ_1 and μ_2 are moments for the actions on M_1 and M_2, then $\mu = \mu_1 + \mu_2$ is a moment for the action on M. The image of M consists of all $f_1 + f_2 \in \mathcal{G}^*$ such that $f_1 \in \mu_1(M_1)$ and $f_2 \in \mu_2(M_2)$. For example, if $G = SO(3)$ and M_1 and M_2 are the orbits corresponding to two values s_1 and s_2 of the spin, then $\mu(M)$ is the union of the orbits on which

$$s_1 + s_2 \geq s \geq |s_1 - s_2|,$$

which is a classical analogue of the triangle rule. ∎

There is an analogy between the decomposition of $\mu(M)$ into orbits and the decomposition of a reducible representation of G into a sum of irreducibles. The 'multiplicity' with which the orbit of f appears in the decomposition is measured by the volume of Marsden-Weinstein reduction

MW_f of M with respect to f, which is defined below. In certain circumstances, this suggestive analogy becomes precise on quantization (Guillemin and Sternberg 1982b). Quantization associates a reducible representation with the action of G on M and an irreducible representation with the orbit of f. In the semiclassical limit, the multiplicity of the irreducible representation in the reducible representation is given by the volume[12] of MW_f.

In §2.5, we saw how to reduce the degrees of freedom in a Hamiltonian system by fixing the values of constants of the motion. If the corresponding canonical flows generate a Hamiltonian action of a Lie group G, then the following lemma identifies the characteristic foliation on the constraint manifold in terms of the coadjoint action of G. It generalizes the abelian case, in which the constants of motion are in involution—the constraint is then coisotropic, and the characteristic foliation is spanned by the Hamiltonian vector fields generated by the constants of the motion.

Lemma (3.7.1). Let G be a connected Lie group which has a Hamiltonian action on a symplectic manifold (M, ω), let $\mu : M \to \mathcal{G}^*$ be a moment, and for each $A \in \mathcal{G}$, let X_A be the Hamiltonian vector field on M generated by A. Let $f \in \mathcal{G}^*$ be a regular value of μ and suppose that $C = \mu^{-1}(f)$ is a presymplectic submanifold of M. Then the characteristic foliation of C is spanned by the vector fields X_A where $A \in \mathcal{G}_f$.

Recall that $\mathcal{G}_f = \{A \in \mathcal{G} \,|\, f([A, \cdot]) = 0\}$ is the Lie algebra of the stabilizer of f under the coadjoint action. That C is presymplectic means that the rank of $\omega|_C$ is constant: this is not, in fact, an independent condition (Libermann and Marle 1987, Lemma 6.2).

Proof. We have to show that if $m \in C$, then

$$T_m C \cap T_m C^\perp = \{X_A(m) \,|\, A \in \mathcal{G}_f\}. \tag{3.7.2}$$

Let h be the Hamiltonian constructed from μ. Then C is the intersection over all $A \in \mathcal{G}$ of the subsets of M on which $h_A = f(A)$. Therefore

$$T_m C = \bigcap_{\mathcal{G}} T_A \quad \text{where} \quad T_A = \{X \in T_m M \,|\, X \lrcorner \, \mathrm{d}h_A = 0\}.$$

It follows that $T_m C^\perp$ is the linear span of the subspaces $T_A^\perp \subset T_m M$. Now $T_A^\perp$ is spanned by $X_A(m)$ (if $X_A(m) = 0$, then $T_A^\perp = 0$). Hence

$$T_m C^\perp = \{X_A(m) \,|\, A \in \mathcal{G}\}.$$

For any $g \in G$, $\mu(mg) = \mu(m)g^\cdot$. Therefore the only elements of G that map C to itself are those in the stabilizer of f. Consequently the only elements of $\mathcal{G}$ for which $X_A(m)$ is tangent to C are those in $\mathcal{G}_f$. Hence we have (3.7.2). ∎

If C is reducible, then the reduced phase space is called the Marsden-Weinstein reduction of M with respect to f, and is denoted by MW_f (Marsden and Weinstein 1974, Meyer 1973, Marle 1976). It is obtained from C by factoring out the action of $(G_f)_0$ (the identity component of the stabilizer of f). An important special case is when $f = 0$. Then the characteristic foliation is spanned by all the Hamiltonian vector fields generated by $\mathcal{G}$ and reduction involves factoring out the action of G itself.

Guillemin and Sternberg (1984) give more details. They also give a different construction of MW_f that emphasizes the fact that a different choice of f on the same orbit in $\mathcal{G}^*$ gives an equivalent reduced phase space. Libermann and Marle (1987) generalize to the non-Hamiltonian case, in which $\mathcal{G}_f$ is replaced by the Lie algebra of the stabilizer of f under the affine action. (3.5.9)

4

HAMILTON-JACOBI THEORY

4.1 The Hamilton-Jacobi equation

IN Hamiltonian mechanics, the position and momentum coordinates are treated symmetrically and one is free to make canonical transformations in phase space which mix up the two sets of coordinates without destroying the basic shape of the equations of motion. The idea in Hamilton-Jacobi theory is to use this freedom to bring the equations into an analytically trivial form. In the case of a time-independent system, this can be done by introducing a canonical coordinate system p'_a, q'^b in which the Hamiltonian depends only on the q's.

In practical terms, the procedure is this: starting with a coordinate system q^a on the configuration space Q, and the corresponding canonical coordinate system p_a, q^b on phase space, one looks for a family of solutions $S(q, q')$ of the partial differential equation

$$h\left(\frac{\partial S}{\partial q^a}, q^b\right) = \text{constant}, \tag{4.1.1}$$

depending on n parameters $q'^1, \ldots, q'^n$. Here, $n = \dim Q$ and h is the Hamiltonian; the partial derivatives replace the ps in the local coordinate expression for h and the constant on the right-hand side is allowed to change with the parameters. The transformation to the new coordinates is then found by solving

$$p_a = \frac{\partial S}{\partial q^a}, \qquad p'_b = -\frac{\partial S}{\partial q'^b}$$

for p' and q' as functions of q and p.

Eqn (4.1.1) is the *Hamilton-Jacobi equation* (more accurately, it is the time-independent form of the Hamilton-Jacobi equation). Although it was originally introduced as a device for obtaining analytic solutions to mechanical problems, it is, in fact, of very little use in that context since the only generally applicable method for solving such first-order partial differential equations involves the reverse construction and depends, as a first step, on the integration of Hamilton's equations (for example, see Courant and Hilbert 1962). Apart from this the only other practical method is to separate the variables, and this can only be done for a very special class of Hamiltonians.

The equation is, however, of central importance in the context of quantization since it provides a link between classical and quantum dynamics

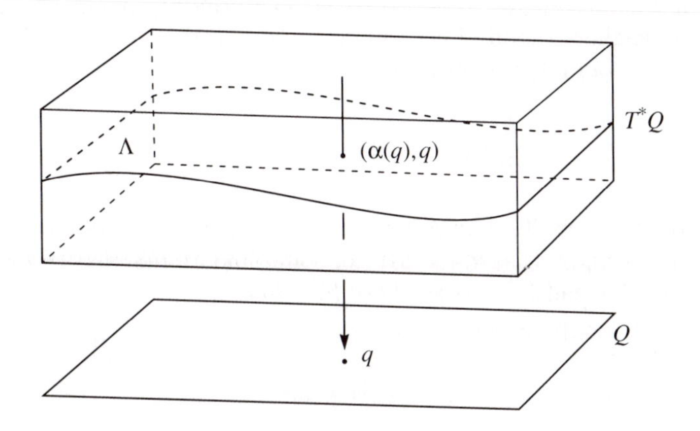

Fig. 4.1. The graph of a 1-form.

by appearing as the first order approximation when Schrödinger's equation is solved asymptotically as $\hbar \to 0$.

In this chapter, we shall look at some geometric ideas that are suggested by the Hamilton-Jacobi equation and that help to clarify the workings of the classical theory. The most important of these is the concept of a *real polarization*, which arises in the global and geometric interpretation of complete integrals of the Hamilton-Jacobi equation.

We shall begin with some facts about generating functions.[1]

4.2 Lagrangian submanifolds

Let Q be an n-dimensional manifold (the configuration space of some classical system) and let $\alpha \in \Omega^1(Q)$. The *graph* of α is the n-dimensional submanifold

$$\Lambda = \{(p, q) \mid p = \alpha(q)\}$$

of T^*Q (Fig. 4.1).

Proposition (4.2.1). Let Λ be the graph of a 1-form α. Then Λ is Lagrangian if and only if $\mathrm{d}\alpha = 0$.

Proof. Let ω and θ be the canonical 1-form and 2-form on T^*Q. Then

$$\alpha = \alpha^*(\theta),$$

where on the right-hand side, α is interpreted as the map $q \mapsto (\alpha(q), q)$, and on the left-hand side as a 1-form on Q. Therefore $\mathrm{d}\alpha = \alpha^*(\omega)$ and so $\mathrm{d}\alpha = 0$ if and only if $\omega|_\Lambda = 0$. ∎

When Λ is Lagrangian, α is closed and so we can find a smooth function S in some neighbourhood of each point of Q such that $\alpha = \mathrm{d}S$. We call S a local *generating function* of Λ. In coordinates,

$$\alpha = \frac{\partial S}{\partial q^a}\mathrm{d}q^a \quad \text{and} \quad \Lambda = \left\{p_a = \frac{\partial S}{\partial q^a}\right\}.$$

The Lagrangian condition on Λ also implies that the restriction of θ to Λ is closed, and therefore that $\theta|_\Lambda = \mathrm{d}W$ for some smooth function W on Λ—at least locally. We call W a local *phase function* on Λ.

If W is a local phase function, then

$$S = W \circ \alpha, \tag{4.2.2}$$

is a local generating function, and conversely; here α is again interpreted as a map from Q to $\Lambda \subset T^*Q$. Note that W and S are determined by Λ up to the addition of a constant.

Another consequence of the Lagrangian condition is that $\theta - \pi^*(\alpha)$ is a symplectic potential on T^*Q, where π is the projection onto Q. Thus Λ is the zero set of a symplectic potential. Conversely, any submanifold of a symplectic manifold given by $\theta' = 0$, where θ' is a symplectic potential, is isotropic, because the vanishing of θ' implies the vanishing of the restriction of $\mathrm{d}\theta'$ to the submanifold. But it need not be Lagrangian since it can have dimension less that $\frac{1}{2}\dim M$. For example, $\frac{1}{2}(p_a\mathrm{d}q^a - q^a\mathrm{d}p_a)$ vanishes only at the origin in $\mathbb{R}^{2n}$, but is a potential for the standard symplectic structure $\mathrm{d}p_a \wedge \mathrm{d}q^a$.

4.3 Canonical transformations

The generating function of a canonical diffeomorphism can be understood as a special example of the construction in §4.2. Let Q and Q' be n-dimensional manifolds and let $\rho : T^*Q \to T^*Q'$. Define $\overline{\rho} : T^*Q \to T^*Q'$ by

$$\overline{\rho}(p, q) = (-p', q') \quad \text{where} \quad (p', q') = \rho(p, q),$$

and let $\Lambda_\rho \subset T^*(Q \times Q') = T^*Q \times T^*Q'$ be the graph of $\overline{\rho}$. That is,

$$\Lambda_\rho = \{(p, p', q, q') \mid (p, q) \in T^*Q,\ (p', q') \in T^*Q',\ \rho(p, q) = (-p', q')\}$$

We want to show that ρ is canonical if and only if Λ_ρ is Lagrangian. This is a consequence of the following proposition, together with the fact that ρ is canonical if and only if $\overline{\rho}$ is anticanonical (that is, $\rho^*\omega' = -\omega$, where ω and ω' are the canonical 2-forms on T^*Q and T^*Q').

Proposition (4.3.1). Let (M, ω) and (M', ω') be symplectic manifolds of the same dimension. Let $\phi : M \to M'$ and let $\Lambda = \{(m, \phi(m))\} \subset M \times M'$ be the graph of ϕ. Then ϕ is anti-canonical if and only if Λ is Lagrangian.

Proof. Note that $\dim \Lambda = \frac{1}{2}\dim(M \times M')$. Let Ω be the symplectic structure on $M \times M'$ (the sum of ω and ω'), and let $\mathrm{pr} : \Lambda \to M$ and $\mathrm{pr}' : \Lambda \to M'$ be the projections onto M and M'. Then $\Omega|_\Lambda = \mathrm{pr}^*\omega + \mathrm{pr}'^*\omega'$. But $\mathrm{pr}' = \phi \circ \mathrm{pr}$. Therefore

$$\Omega|_\Lambda = \mathrm{pr}^*(\omega + \phi^*\omega'),$$

which vanishes if and only if $\omega = -\phi^*\omega'$. ∎

A local *generating function* of a canonical diffeomorphism $\rho : T^*Q \to T^*Q'$ is a local generating function $S \in C^\infty(Q \times Q')$ of the corresponding Lagrangian submanifold $\Lambda_\rho \subset T^*(Q \times Q')$. It is determined by ρ up to the addition of a constant.

We can recover ρ from S by introducing coordinates. Let q^a and q'^a be coordinates on Q and Q' and let p_a, q^b and p'_a, q'^b be the corresponding canonical coordinate systems on T^*Q and T^*Q'. Then S is a function of q and q', and ρ is given by solving

$$p_a = \frac{\partial S}{\partial q^a}, \qquad p'_a = -\frac{\partial S}{\partial q'^a} \tag{4.3.2}$$

for p' and q' as functions of p and q; the minus sign in the second equation comes from the minus sign in front of p' in the definition of Λ_ρ. Any S generates a Lagrangian submanifold of $T^*(Q \times Q')$ and hence a local canonical diffeomorphism. Conversely, almost every canonical diffeomorphism can be recovered from a local generating function. The exceptions are those for which Λ_ρ fails to be a section of $T^*(Q \times Q')$—for example the identity transformation $p'_a = p_a$, $q'^a = q^a$.

We can also interpret (4.3.2) as a passive transformation between two canonical coordinate systems on the same symplectic manifold.

Example. A linear canonical transformation is of the form

$$p'_a = C_a{}^b p_b + D_{ab}q^b, \qquad q'^a = E^{ab}p_b + F^a_b q^b$$

where C, D, E, F are $n \times n$ matrices as on p. 5. It is generated by

$$S = \tfrac{1}{2}K_{ab}q^a q^b + L_{ab}q^a q'^b + \tfrac{1}{2}M_{ab}q'^a q'^b,$$

where $K = -E^{-1}F$, $L = E^{-1}$, and $M = -CE^{-1}$, provided that E is nonsingular (note that K and M are symmetric). ∎

4.4 The two-point characteristic function

A function $S \in C^\infty(Q \times Q)$ generates a canonical diffeomorphism $T^*Q \to T^*Q$. We shall now consider how to relate this construction to the generating functions of canonical flows that were introduced in §1.5.

Let $h \in C^\infty(T^*Q)$ and suppose that X_h is complete. Then the flow $\rho_t : T^*Q \to T^*Q$ of X_h is a one-parameter family of canonical diffeomorphisms.

Let θ be the canonical 1-form on T^*Q. Then

$$\mathcal{L}_{X_h}\theta = X_h \lrcorner\, \omega + \mathrm{d}(X_h \lrcorner\, \theta) = \mathrm{d}L, \tag{4.4.1}$$

where $L = X_h \lrcorner\, \theta - h$ is the Lagrangian (see p. 37). Therefore, for any $t \in \mathbb{R}$,

$$\theta - \rho_t^*\theta = \mathrm{d}A \quad \text{where} \quad A(m) = -\int_0^t L\,\mathrm{d}t',$$

with the integration along the integral curve of X_h from $m \in T^*Q$ to $\rho_t(m)$.

For fixed t, $\rho_t : T^*Q \to T^*Q$ is a canonical diffeomorphism, and so determines a Lagrangian submanifold $\Lambda = \Lambda_{\rho_t} \subset T^*(Q \times Q) = T^*Q \times T^*Q$. We shall construct its generating function by transferring A from T^*Q to $Q \times Q$.

Let $\mathrm{pr} : T^*(Q \times Q) \to T^*Q$ denote the projection onto the first factor. Then $\mathrm{pr}^*\big(\theta - \rho_t^*(\theta)\big) = \mathrm{d}(A \circ \mathrm{pr})$. Therefore $A \circ \mathrm{pr}$ is a local phase function on Λ. It follows that

$$S(q, q', t) = -\int_0^t L\,\mathrm{d}t'$$

is a local generating function of ρ_t. It is called Hamilton's *two-point characteristic function*. The integration is along the integral curve $t' \mapsto \big(p(t'), q(t')\big)$ of X_h such that $q(0) = q$ and $q(t) = q'$. There is a unique curve with this property provided that Λ is a section of $T^*(Q \times Q)$.

This last condition generally fails in the large, and as a consequence S is often many valued and badly behaved for certain pairs of points (q, q'). An illustration is provided by the geodesics of a Riemannian metric g on Q, which are the trajectories of the Hamiltonian $h = \frac{1}{2}g^{ab}p_a p_b$. Provided that q is close to q', there is a unique geodesic from q to q' which minimizes the distance, and S is well defined. However, S is singular when q and q' are conjugate along a geodesic, and is many valued when there is more than one geodesic joining q and q'.

Equation (4.4.1) also holds when h is time-dependent: L is then a time-dependent function on T^*Q and the flow of X_h is a family of maps $\rho_{tt'} : T^*Q \to T^*Q$ labelled by $t, t' \in \mathbb{R}$. A similar argument establishes that, for fixed t, t', $\rho_{tt'}$ is generated by

$$S(q, q', t, t') = -\int_t^{t'} L\,\mathrm{d}t'', \tag{4.4.2}$$

where the integration is along the integral curve $t'' \mapsto \big(p(t''), q(t'')\big)$ of X_h such that $q(t) = q$ and $q(t') = q'$.

4.5 Real polarizations

Definition (4.5.1). A real polarization of a symplectic manifold (M, ω) is a foliation of M by Lagrangian submanifolds; that is, a smooth distribution P which is

(1) *integrable*: if $X, Y \in V_P(M)$, then $[X, Y] \in V_P(M)$; and

(2) *Lagrangian*: for each $m \in M$, P_m is a Lagrangian subspace of $T_m M$.

As with more general foliations, P is said to be *reducible* if the space of leaves M/P is a Hausdorff manifold.

Example. The vertical foliation (§1.4) of a cotangent bundle T^*Q is a real polarization. The leaves are the cotangent spaces T_q^*Q, which are isotropic since $\omega = \mathrm{d}p_a \wedge \mathrm{d}q^a$ vanishes on restriction to surfaces of constant q. ∎

Example. Again with $M = T^*Q$, let F be a closed 2-form on Q and let $\omega_F = \omega + \frac{1}{2}\pi^*(F)$ be the charged symplectic structure on M (see §2.6; π is the projection $T^*Q \to Q$). Then

$$\omega_F = \mathrm{d}p_a \wedge \mathrm{d}q^a + \tfrac{1}{2}F_{ab}\mathrm{d}q^a \wedge \mathrm{d}q^b$$

and the vertical foliation is a polarization. ∎

Example. Let $\alpha_1, \ldots, \alpha_k$ be linearly independent closed 1-forms on Q and let $\sim$ be the equivalence relation

$$(p, q) \sim (p + n_1\alpha_1 + \cdots + n_k\alpha_k, q)$$

whenever $n_1, \ldots, n_k \in \mathbb{Z}$. Then ω is a well-defined symplectic structure on the quotient manifold $T^*Q/\sim$ and the vertical foliation is again a polarization, only its leaves now have topology $T^k \times \mathbb{R}^{n-k}$, where T^k is the k-torus.

The definition of $\sim$ does not actually require that the individual αs should be defined globally since it is enough to be given $\{\alpha_1, \ldots, \alpha_k\}$ up to permutation and sign change. Nevertheless, the construction requires strong topological restrictions on Q; in particular, it is necessary that Q should admit a field of $(n - k)$-planes (see Steenrod 1951). ∎

We shall see that locally, and also globally under quite general conditions, combinations of these are the only possible examples. The proof makes use of a natural flat affine connection on the leaves of a real polarization. The construction is due to Weinstein (1971) and is an application of a general theory of Bott's (1972). We shall look at the connection in the next section in a more general context which will be useful later on, and which involves no extra complication.

A real polarization is 'half a canonical coordinate system': locally, the leaves are the surfaces of constant q^a, where $q^1, \ldots, q^n$ are functions in involution (i.e. with vanishing Poisson brackets). The Hamiltonian vector fields generated by the qs are tangent to the leaves and are parallel with respect to the flat connection.

4.6 Coisotropic foliations

Let E be a coisotropic foliation of a symplectic manifold (M, ω). That is, $m \mapsto E_m$ is an integrable distribution; and, for each $m \in M$, E_m is a coisotropic subspace of $T_m M$. Associated with E, there is a second distribution $D = E^\perp$, defined by

$$m \mapsto D_m = E_m^\perp \subset E_m.$$

Proposition (4.6.1). For any coisotropic foliation E, $D = E^\perp$ is an isotropic foliation.

Proof. Clearly D is isotropic. We have to show that it is also integrable. Let $X, Y \in V_D(M)$ and let $Z \in V_E(M)$. Then $\omega(X,Y) = \omega(Y,Z) = \omega(Z,X) = 0$. Also $[Y,Z]$ and $[Z,X]$ are both tangent to E since E is integrable and $D \subset E$. Therefore, $\omega([Y,Z],X) = \omega([Z,X],Y) = 0$. Hence

$$\omega([X,Y],Z) = -3\mathrm{d}\omega(X,Y,Z) = 0$$

by (A.1.16). Since this holds for any $Z \in V_E(M)$, it follows that $[X,Y] \in V_D(M)$. ∎

Note that this does not hold the other way round: if D is a given isotropic foliation, then $E = D^\perp$ is a coisotropic distribution, but it is not necessarily integrable. For example, if E has codimension one, then it is automatically coisotropic, and D, being one-dimensional, is automatically integrable; but not every distribution of codimension one is integrable.

Definition (4.6.2). Let E be a coisotropic foliation and let $D = E^{\perp}$. The partial connection

$$\nabla : V_E(M) \times V_D(M) \to V_D(M) : (X, Y) \mapsto \nabla_X Y$$

is defined by

$$(\nabla_X Y) \lrcorner \omega = \mathcal{L}_X(Y \lrcorner \omega) = X \lrcorner \mathrm{d}(Y \lrcorner \omega).$$

The partial connection enables us to differentiate $Y \in V_D(M)$ along $X \in V_E(M)$. To see that $\nabla_X Y \in V_D(M)$, let $X' \in V_E(M)$. Then

$$\omega(\nabla_X Y, X') = \tfrac{1}{2} X' \lrcorner \mathcal{L}_X(Y \lrcorner \omega) = X(\omega(Y, X')) - \omega(Y, [X, X']) = 0$$

since X' and $[X, X']$ are tangent to E.

If h is constant on the leaves of E, then X_h is tangent to D and $\nabla_Y X_h$ vanishes for every $Y \in V_E(M)$.

Since D and E are integrable, there are local coordinates

$$x^1, \dots, x^d, y^1, \dots, y^{2(n-d)}, z^1 \dots, z^d$$

($d = \dim D$) such that the leaves of E are the surfaces on which the zs are constant and the leaves of D are the surfaces on which both the zs and the ys are constant. The xs are coordinates on the leaves of D. Put

$$\alpha_{ij} = 2\omega\left(\frac{\partial}{\partial x^i}, \frac{\partial}{\partial z^j}\right).$$

Then $\det(\alpha_{ij}) \neq 0$, so we can define 'Christoffel symbols' Γ^i_{jk} by

$$\Gamma^i_{jk} = \alpha^{il} \frac{\partial \alpha_{kl}}{\partial x^j} \quad \text{where} \quad \alpha^{il} \alpha_{jl} = \delta^i_j$$

(with range and summation conventions for $i, j, \dots = 1, \dots, d$). If $X, Y \in V_D(M)$ then

$$X = X^i \frac{\partial}{\partial x^i}, \quad Y = Y^i \frac{\partial}{\partial x^i}, \quad \text{and} \quad \nabla_X Y = X^i \left(\frac{\partial Y^j}{\partial x^i} + \Gamma^j_{ik} Y^k\right) \frac{\partial}{\partial x^j}.$$

Thus ∇ restricts to an affine connection on each leaf of D, with the Γs as connection coefficients.

The following proposition establishes that ∇ has all the properties of a flat torsion-free affine connection, except that it is defined only for a restricted class of vector fields. In particular, the connection on a leaf of D is flat and torsion-free.

Proposition (4.6.3). For every $X, X' \in V_E(M)$, $Y, Y' \in V_D(M)$ and $f \in C^\infty(M)$,

(a) $\nabla_X(fY) = f\nabla_X Y + X(f)Y$;

(b) $\nabla_Y Y' - \nabla_{Y'} Y = [Y, Y']$;

(c) $\nabla_X \nabla_{X'} Y - \nabla_{X'} \nabla_X Y = \nabla_{[X,X']} Y$.

Proof. Part (a) follows directly from the definition. Part (b) follows from

$$\begin{aligned}(\nabla_Y Y' - \nabla_{Y'} Y) \lrcorner\, \omega &= Y \lrcorner\, \mathrm{d}(Y' \lrcorner\, \omega) - Y' \lrcorner\, \mathrm{d}(Y \lrcorner\, \omega) \\ &= Y \lrcorner\, \mathcal{L}_{Y'}\omega - Y' \lrcorner\, \mathcal{L}_Y \omega \\ &= [Y, Y'] \lrcorner\, \omega\end{aligned}$$

by using (A.1.15). Part (c) follows from

$$\begin{aligned}(\nabla_X \nabla_{X'} Y - \nabla_{X'} \nabla_X Y) \lrcorner\, \omega &= (\mathcal{L}_X \mathcal{L}_{X'} - \mathcal{L}_{X'} \mathcal{L}_X)(Y \lrcorner\, \omega) \\ &= \mathcal{L}_{[X,X']}(Y \lrcorner\, \omega) \\ &= (\nabla_{[X,X']} Y) \lrcorner\, \omega.\end{aligned}$$

∎

The expressions in (4.6.2) can be used to define $\nabla_X Y$ when $X \in V_D(M)$, $Y \in V_E(M)$, so there is also a 'dual' partial connection with which vectors tangent to E can be differentiated along D.

4.7 Geometry of real polarizations

A real polarization is a special case of a coisotropic foliation, with $P = E = D$. The partial connection determines a flat affine connection on each leaf, with respect to which the Hamiltonian vector fields tangent to the leaves are covariantly constant.

When M is the cotangent bundle of some manifold Q and P is the vertical foliation, the flat connection on the leaves coincides with the one given by their vector space structures. For if q^a is a coordinate system on Q and q^a, p_b is the corresponding canonical system on T^*Q, then $q^a \in C_P^\infty(M)$ and

$$X_{q^a} = -\frac{\partial}{\partial p_a},$$

which restricts to a constant vector on each T_q^*Q.

The following proposition shows that when the leaves of P are complete and have the topology of $\mathbb{R}^n$, all that is needed to recover a cotangent bundle structure from a general real polarization is the location of the zero section.[2]

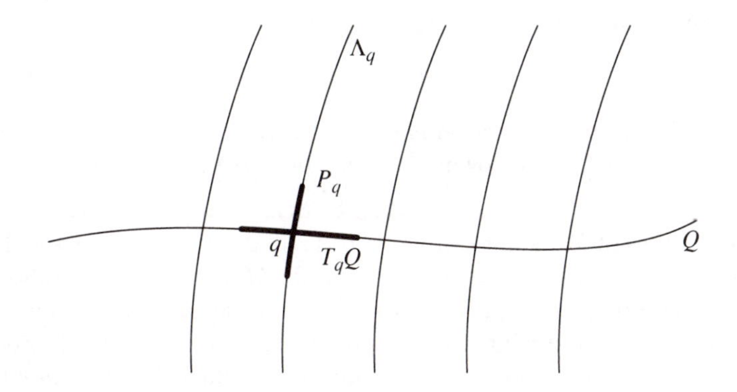

Fig. 4.2. The identification of M with T^*Q.

Proposition (4.7.1). Let P be a real polarization of a symplectic manifold (M, ω) with leaves that are simply-connected and geodesically complete, and let Q be a Lagrangian submanifold of M that intersects each leaf transversally in exactly one point. Then there is a natural identification of M with T^*Q under which ω coincides with the canonical 2-form, P coincides with the vertical foliation, and Q coincides with the zero section of T^*Q.

Proof. Let ω' be the canonical 2-form on T^*Q (we shall revert to our usual notation, in which the canonical 2-form is denoted by ω, when we have shown that it coincides with the symplectic form on M). Let $q \in Q$ and let Λ_q be the leaf of P through q; see Fig. 4.2. Under the topological conditions, the flat affine connection makes Λ_q into an affine space, which becomes a vector space when we take the intersection with Q as origin. Thus we can identify Λ_q with P_q, which is the tangent space to Λ_q at the origin. Now P_q and T_qQ are transverse Lagrangian subspaces of $T_q(T^*Q)$, and therefore ω identifies P_q with T_q^*Q by mapping $X \in P_q$ to the linear form $2\omega(X, .)$ on T_qQ (§1.2). So we can identify T_q^*Q with Λ_q, and, by allowing q to vary, T^*Q with M.

By construction, P coincides with the vertical foliation and Q coincides with the zero section. The connection on the leaves of P coincides with the natural flat connection on the fibres of T^*Q. Also ω coincides with ω' at points of Q (see the proof of Proposition 1.2.5). So it remains to show that ω coincides with ω' elsewhere.

Let $f \in C_P^\infty(M)$ and let X_f and X'_f be the Hamiltonian vector fields generated from f by ω and ω'. Both are tangent to P and $X_f = X'_f$ at points of Q (where ω and ω' coincide). Also both are covariantly constant on the leaves of P. Therefore $X_f = X'_f$ everywhere. Moreover,

$$\mathcal{L}_{X_f}\omega = 0 \quad \text{and} \quad \mathcal{L}_{X'_f}\omega' = 0,$$

so $\omega - \omega'$ is Lie propagated along $X_f = X'_f$. It is possible to reach any point of M from Q by following a trajectory of X_f for some $f \in C^\infty_P(M)$. Therefore $\omega = \omega'$ everywhere on $M = T^*Q$. ∎

If the leaves of P are not complete, but still geodesically convex, then the proposition still holds except that we identify M only with a neighbourhood of the zero section in T^*Q. The existence of a Lagrangian section imposes a strong global constraint on M and P; but the method of proof also establishes a local version of the proposition, which holds without extra conditions on P or M.

Proposition (4.7.2). Let P be a real polarization of a symplectic manifold (M, ω) and let $m \in M$. Then there exists a neighbourhood U of m and a canonical diffeomorphism $\rho : U' \subset T^*Q \to U$, where U' is a neighbourhood of the zero section in the cotangent bundle of some manifold Q, such that ρ^*P is the vertical foliation of U' and $\rho^{-1}(m)$ lies on the zero section in U'.

Proof. Let Q be a Lagrangian submanifold through m that cuts the leaves of P transversally in some neighbourhood of m. Such a Q exists. For example, one can construct it by introducing local canonical coordinates x_a, y^b in a neighbourhood of m and taking the Lagrangian surface $x_a = x_a(m)$. By first making a linear canonical transformation of the coordinates, if necessary, we can ensure that this is transverse to P at m and hence in a neighbourhood of m.

Now choose a neighbourhood U of m so that each leaf of $P|_U$ is geodesically convex and intersects Q in a unique point. Then the same construction as in the proof of proposition (4.7.1) identifies U with a neighbourhood U' of the zero section in T^*Q. If we define $\rho : U' \to U$ to be the map that sends $(p, q) \in U'$ to the corresponding point of U, then it will have the required properties. ∎

Two immediate corollaries are the following.

Proposition (4.7.3). Let P be a real polarization of a symplectic manifold (M, ω). Then it is possible to find a canonical coordinate system p_a, q^b in some neighbourhood of each $m \in M$ such that the leaves of P coincide locally with the surfaces of constant q^a.

Proposition (4.7.4). Let P be a real polarization of a symplectic manifold (M, ω). Then it is possible to find a symplectic potential θ in some neighbourhood of each point such that $\theta|_P = 0$.

These are proved by using ρ to transfer canonical coordinates and the canonical 1-form from T^*Q to M. Canonical coordinates and symplectic potentials with the properties stated in the propositions are said to be *adapted* to P.

Clearly a necessary condition for a symplectic manifold with real polarization to be equivalent to a cotangent bundle is that the polarization should admit a global Lagrangian section: this is because every cotangent bundle has a natural Lagrangian section (the zero section). Proposition (4.7.1) shows that under certain conditions, it is also sufficient. We shall now look at conditions for the existence of such a section.

Proposition (4.7.5). Let P be a reducible real polarization of a symplectic manifold (M, ω) with convex leaves. Then P determines a class $[F] \in H^2(M/P; \mathbb{R})$. If the leaves of P are also complete, then P admits a Lagrangian section if and only if $[F] = 0$.

Proof. First we shall show that it is possible to find some section of P (not necessarily a Lagrangian one). This can be done by using either a general topological argument or the following direct construction. There is no difficulty in finding sections locally. Hence it is possible to find a cover $\{V_i\}$ of M/P and, for each i, a section Q_i of P restricted to $\mathrm{pr}^{-1}(V_i)$ (here $\mathrm{pr} : M \to M/P$ is the projection).

Now if Q_1 and Q_2 are two local sections of P and if $h \in C^\infty(M/P)$ with $0 \le h \le 1$, then we can construct a third section Q_3, which we shall denote by $hQ_2 + (1-h)Q_1$, as follows. Let $q \in M/P$ and let $\Lambda_q = \mathrm{pr}^{-1}(q)$. Then, since Λ_q is geodesically convex, there is a unique geodesic $\gamma : [0,1] \to \Lambda_q$ such that $\gamma(0) = \Lambda_q \cap Q_1$ and $\gamma(1) = \Lambda_q \cap Q_2$. We define Q_3 so that $Q_3 \cap \Lambda_q = \gamma(h(q))$. See Fig. 4.3.

If $\{h_i\}$ is a partition of unity subordinate to $\{V_i\}$, then we can make sense of the formal sum $Q = \sum_i h_i Q_i$ in the same way to define a global section.

Next we define $F \in \Omega^2(M/P)$ by restricting 2ω to Q, and show that its class in $H^2(M/P)$ is independent of Q. Let $\tau : M/P \to Q$ be the map that sends $\Lambda \in M/P$ to $\Lambda \cap Q$. Put

$$F = 2\tau^*\omega \quad \text{and} \quad \omega' = \omega - \tfrac{1}{2}\mathrm{pr}^* F.$$

Then F is a closed 2-form on M/P, ω' is a symplectic structure on M, and Q is Lagrangian with respect to ω'. Hence we can identify (M, ω') as a

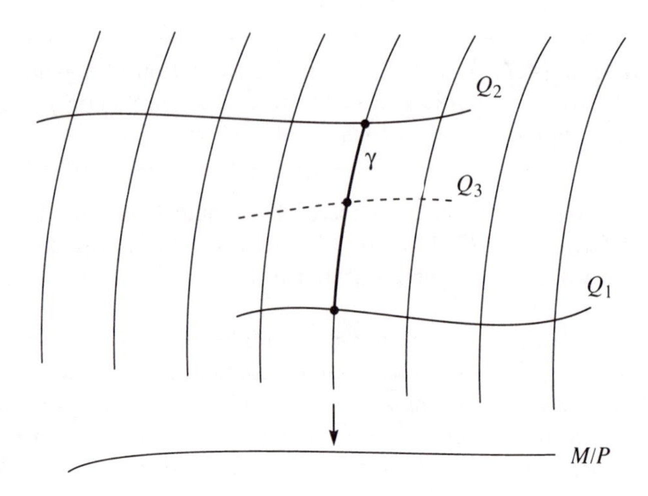

Fig. 4.3. The construction of Q_3.

polarized symplectic manifold with a neighbourhood U of the zero section in T^*Q.

Let Q' be another section of P. Then Q' is the graph in $M = U \subset T^*Q$ of some $\alpha \in \Omega^1(Q)$ and

$$\omega'|_{Q'} = \pi^*(\mathrm{d}\alpha),$$

where $\pi : Q' \to Q$ is the projection along P. It follows that, if $\tau' : M/P \to Q'$ is defined in the same way as τ, then

$$2\tau'^*\omega = \tau'^*\mathrm{pr}^*F + 2\tau'^*\omega' = F + 2\mathrm{d}(\tau^*\alpha)$$

since $\mathrm{pr} \circ \tau'$ is the identity and $\pi \circ \tau' = \tau$. Hence the class $[F]$ of F in $H^2(M/P;\mathbb{R})$ is independent of Q.

If P admits a global Lagrangian section, then we can take Q to be this section to obtain $F = 0$. Conversely, if the leaves of P are complete and if $[F] = 0$, then $U = T^*Q$ and $F = 2\mathrm{d}A$ for some 1-form A. In this case we can choose Q' so that $A = -\tau^*(\alpha)$. Then $\tau'^*(\omega) = 0$ and so Q' is a global Lagrangian section. ∎

It follows that if P is reducible with complete simply-connected leaves and if $H^2(M/P;\mathbb{R}) = 0$, then M is equivalent to a cotangent bundle. It also follows that the second example at the beginning of §4.5 is the most general example of a real reducible polarization with complete simply connected leaves.

These results are illustrated by the Legendre transformation. When L is a regular time-independent Lagrangian, ω_L is a symplectic structure on TQ. The vertical foliation of TQ is a real polarization and θ_L is a potential adapted to the polarization. The surface on which θ_L vanishes is Lagrangian and the corresponding identification of TQ with a neighbourhood of the zero section in T^*Q coincides with the Legendre transformation. The leaves of P are simply connected, but not necessarily complete since the flat affine connection derived from ω_L may not coincide with their vector space structure (as tangent spaces to Q); this happens only when

$$L = \tfrac{1}{2} g_{ab} v^a v^b + \alpha_a v^a - V,$$

that is, when L is a polynomial of degree two in the velocities. When $\alpha = 0$, the zero section of TQ is the Lagrangian section of P on which θ_L vanishes. In the general case $2\mathrm{d}(\alpha_a \mathrm{d}q^a) = F$, where F is the 'electromagnetic' 2-form in the proof of Proposition (4.7.5).

Even when L is regular everywhere on TQ, it is still possible for the leaves of P to be incomplete and for the Legendre transformation to fail to be a diffeomorphism. An example in two dimensions is the physically outlandish Lagrangian $L = \exp(v^1)\cos(v^2)$, for which the Legendre transformation is

$$p_1 + \mathrm{i}p_2 = e^{v^1 - \mathrm{i}v^2},$$

which is many-to-one.

So far, we have only looked at the global geometry of polarizations with simply connected leaves. We shall now consider what happens when the leaves are complete, but not simply connected. For simplicity, we shall suppose that there is a global Lagrangian section Q and that M is a fibre bundle over M/P, which implies that the leaves all have the same topology.

Let $\mathrm{pr} : M \to M/P$ be the projection and let q^a be a local coordinate system on M/P. The Hamiltonian vector fields generated by the functions $q^a \circ \mathrm{pr}$ form a global set of parallel frames on each leaf Λ. It follows that the holonomy group of the flat connection on Λ is trivial and hence that $\Lambda = T^k \times \mathbb{R}^{n-k}$ for some k, where $n = \tfrac{1}{2}\dim M$ (Kobayashi and Nomizu 1963, p. 211).

The construction in the proof of (4.7.1) gives a map $\rho : T^*Q \to M$, which is surjective and canonical (Fig. 4.4). Locally in Q, $\rho^{-1}(Q)$ is a collection of Lagrangian sections, the graphs of integer combinations

$$n_1\alpha_1 + n_2\alpha_2 + \ldots + n_k\alpha_k$$

of a set of closed 1-forms $\alpha_i \in \Omega^1(Q)$. The αs are determined by Q up to order and sign. It follows that M is equivalent to the polarized symplectic manifold $T^*Q/\sim$ in the third example in §4.5. The construction also gives a local canonical coordinate system p_a, q^b in which the leaves of P are the surfaces of constant q and $Q = \{p = 0\}$. The momentum coordinates are multi-valued since passing round a closed loop in a leaf of P adds to p_a an integer combination of $\alpha_{1a}, \alpha_{2a}, \ldots, \alpha_{na}$, where α_{ia} is the ath component of α_i.

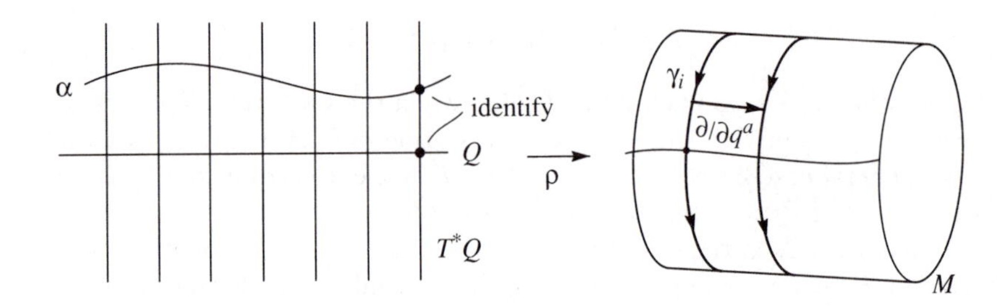

Fig. 4.4. The symplectic manifold $M = T^*Q/\sim$.

The αs are the exterior derivatives of the *action variables*, which are defined as follows. Let θ be a symplectic potential on a neigbourhood in M of one of the leaves—for example, we could take $\theta = -q^a \mathrm{d}p_a$, which is well-defined even though the ps are not single-valued. Each leaf of P is topologically a product of k circles $\gamma_1, \ldots, \gamma_k$ and $n-k$ copies of $\mathbb{R}$. The action variables $j_1, \ldots j_k \in C^\infty(M/P)$ are the functions

$$j_i(\Lambda) = \frac{1}{2\pi} \int_{\gamma_i} \theta.$$

Up to order and sign,[3] the js depend only on Λ, and not on the precise choice of the γs. This is because $\theta|_\Lambda$ is closed. In particular, we can take γ_i to be the path $[0,1] \to M : t \mapsto (p_a - t\alpha_{ia}, q^b)$.

To see that $\mathrm{d}j_i = \alpha_i$, we apply (A.1.17) to the vector field $X = \partial/\partial q^a \in V(M)$. The result is

$$\frac{\partial}{\partial q^a}(j_i) = \int_{\gamma_i} \frac{\partial}{\partial q^a} \lrcorner\, \omega = -\int_{\gamma_i} \mathrm{d}p_a = \int_0^1 \alpha_{ia}\, \mathrm{d}t = \alpha_{ia}.$$

Therefore $\mathrm{d}j_i = \alpha_i$. See Fig. 4.4.

The action variables are unchanged when θ is replaced by $\theta + \mathrm{d}f$; but if M is not simply-connected, it is also possible to replace θ by $\theta + \lambda$, where λ is closed but not exact. The effect is to add a constant to each of the js. (Synge 1960, Arnol'd 1978, Guillemin and Sternberg 1984.)

The js can also be regarded as polarized functions on M. The corresponding Hamiltonian vector fields generate an action of T^k which preserves the polarization.

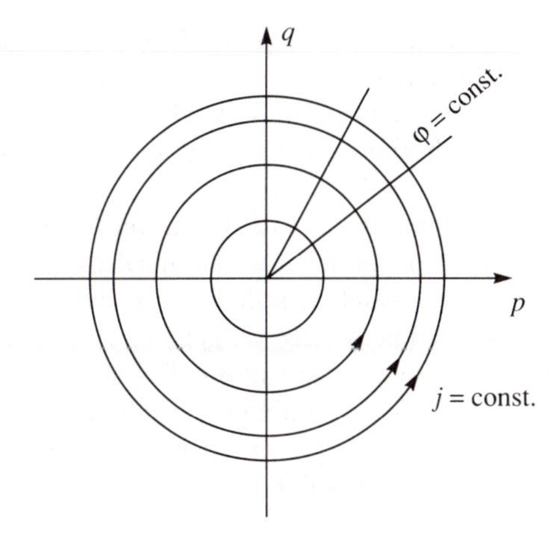

Fig. 4.5. Action-angle coordinates in the plane.

When $k = n$, the leaves are compact and the polarization is a foliation by Lagrangian tori. There are then n action variables, and they form a local coordinate system on Q. We can extend this to a canonical coordinate system on M by introducing the conjugate *angle variables* ϕ^a, defined by

$$X_{j_a}(\phi^b) = \delta_a^b \quad \text{with } \phi^a = 0 \text{ on } Q.$$

The symplectic form[4] is then $\mathrm{d}j_a \wedge \mathrm{d}\phi^a$.

The ϕs are affine linear coordinates on the leaves of P, and run from 0 to 2π round the n circles. They are independent of the choice of θ, but if Q is replaced by another Lagrangian section, they are transformed by $\phi^a \mapsto \phi + \beta^a$, where $\beta^a \mathrm{d}j_a$ is a closed 1-form on M/P.

'Action-angle' coordinates arise in integrable systems: if the Hamiltonian is polarized, then the js are constants of the motion and the dynamical evolution is linear in the ϕs.

Example. Take $M = \mathbb{R}^2 - \{0\}$, $\omega = r\mathrm{d}r \wedge \mathrm{d}\phi$, $\theta = \frac{1}{2}r^2\mathrm{d}\phi$, where r and ϕ are polar coordinates. Let P be the polarization with the circles of constant r as leaves. The action variable is $j = \frac{1}{2}r^2$ (the area of the circle divided by 2π) and the angle variable is ϕ. See Fig. 4.5. ∎

4.8 Polynomials and invariant polarizations

When we come to quantization, we shall be interested in constructing polarizations that behave well under the canonical flow of a Hamiltonian $h \in C^\infty(M)$. In the simplest case, h is *completely integrable*, which means that there is a real polarization P with leaves tangent to X_h. Then X_h

is covariantly constant along P and the flow on the leaves is linear with respect to the natural affine structure. This is a very special situation: in the large, such polarizations exist only for Hamiltonians that are unusually well behaved (although they always exist locally, away from the critical points of h).

The vector field X_h is tangent to P whenever $h \in C_P^\infty(M)$, i.e. whenever h is a function only of the qs in local canonical coordinates adapted to P. The next simplest case, which is still very special, is that P is invariant under the flow, but not necessarily tangent to X_h. The condition for this is that $X_h \lrcorner \, \mathrm{d}f$ should be constant along P whenever f is constant along P. By taking $f = q^a$ in local canonical coordinates adapted to P, this is equivalent to: for each a, $[h, q^a]$ is independent of the ps. In other words, h is a linear function of the ps of the form

$$h = v^a(q)p_a + u(q). \tag{4.8.1}$$

When P is reducible, the canonical flow on M projects onto the flow in M/P of the vector field with components v^a.

In general, since the leaves of a real polarization are affine spaces, it is possible to pick out a special class of *polynomial observables* in $C^\infty(M)$ of the form

$$f = f_k^{ab\ldots c} p_a p_b \ldots p_c + f_{k-1}^{b\ldots c} p_b \ldots p_c + \cdots + f_0, \tag{4.8.2}$$

where p_a, q^b is a canonical coordinate system adapted to P and each $f_i^{ab\ldots c}$ (symmetric in its i upper indices) is independent of the momentum coordinates p_a.

The ps and qs are determined locally by P up to canonical coordinate transformations given by

$$\tilde{q}^a = \tilde{q}^a(q) \quad \text{and} \quad \tilde{p}_a = \frac{\partial q^b}{\partial \tilde{q}^a}\left(p_b - \frac{\partial S}{\partial q^b}\right),$$

where S is a function of the qs (S is the generating function in the p, q system of the Lagrangian submanifold $\tilde{p}_a = 0$). On substituting this into (4.8.2), f becomes a polynomial of degree k in the $\tilde{p}$s, with the coefficients $f_k^{ab\ldots c}$ in the leading term behaving as the components of a kth rank contravariant symmetric tensor in the variables q^a. When P is reducible, this tensor is well defined on M/P; it will be denoted by $\sigma(f)$ (it behaves like the principal symbol of a differential operator). The coefficients in the lower order terms do not transform as tensors.

Example. Let $M = TQ$ and $\omega = \omega_L$, where $L = \frac{1}{2}g_{ab}v^a v^b + \alpha_a v^a$ is the Lagrangian of charged particle motion in a magnetic field (§2.6), and let $h \in C^\infty(TQ)$ be the Hamiltonian. Then $\{p_a, q^b\} = \{g_{ac}v^c + \alpha_a, q^b\}$ is a canonical coordinate system adapted to the vertical polarization and $h = \frac{1}{2}g^{ab}(p_a - \alpha_a)(p_b - \alpha_b)$. The tensor $\sigma(h)$ is the contravariant metric, with components g^{ab}. The linear terms in h are gauge-dependent. ∎

The polynomial observables can also be defined intrinsically by an inductive construction (Kostant 1974b). If U is an open set in M, then we put $S^0_P(U) = C^\infty_P(U)$ and define $S^k_P(U)$ to be the set of smooth functions f on U such that

$$[f, g] \in S_P^{k-1}(U \cap V)$$

for every open set $V \subset M$ and every $g \in C^\infty_P(V)$. Then $f \in S^k_P(U)$ if and only if it is given by (4.8.2) in local canonical coordinates adapted to P.

In particular, $S^0_P(M) = C^\infty_P(M)$ consists of the generators of the Hamiltonian vector fields tangent to P; and $S^1_P(M)$ consists of the generators of Hamiltonian flows that preserve P.

If $g \in S^r_P(M)$ and $h \in S^s_P(M)$, then $[g, h] \in S_P^{r+s-1}(M)$ and $\sigma([g, h])$ has components

$$rg^{a(b\ldots c}\partial_a h^{d\ldots f)} - sh^{a(b\ldots d}\partial_a g^{e\ldots f)}$$

where ∂ is the partial derivative with respect to q, and $g^{a\ldots c}$ and $h^{a\ldots d}$ are the components of $\sigma(g)$ and $\sigma(h)$. This is the *Nijenhuis concomitant* (Nijenhuis 1955) of $\sigma(g)$ and $\sigma(h)$. It is unaltered when ∂ is replaced by an arbitrary torsion-free affine connection on M/P and it reduces to the Lie bracket of vector fields when $r = s = 1$.

4.9 Generating functions of real polarizations

Let $M = T^*Q$, with its natural sympletic structure, and let P be the vertical polarization. Suppose that S is a smooth real function on some open subset of $Q \times \mathbb{R}^n$ such that

$$\det\left(\frac{\partial^2 S}{\partial q^a \partial q'^b}\right) \neq 0, \tag{4.9.1}$$

where the qs are coordinates on Q and the q's are Cartesian coordinates on $\mathbb{R}^n$. For fixed q', S generates a Lagrangian submanifold of T^*Q transverse to P. As the q's vary, we obtain a real polarization P' on a subset of T^*Q. The leaves are the surfaces

$$p_a = \frac{\partial S}{\partial q^a} \quad (q' \text{ constant}).$$

The nondegeneracy condition (4.9.1) ensures that these surfaces do in fact fill out an open set in T^*Q. If we think of the q's as given parameters labelling the leaves of P', then S is determined by P' up to the addition of an arbitrary function of the q's; S is called the local *generating function* of P'. It is clear that any real polarization of T^*Q transverse to P can be obtained locally in this way.

There is another way of looking at S that emphasizes the symmetry between P and P'. Let P and P' be two transverse real polarizations of a symplectic manifold (M, ω) and let θ be a local symplectic potential adapted to P. Let Λ' be a leaf of P'. Since $\omega = \mathrm{d}\theta$ vanishes on restriction to Λ', there exists a local phase function S on Λ' such that $\theta_{\Lambda'} = \mathrm{d}S$. We can make S vary smoothly with Λ'. It then becomes a function on M.

We can introduce local coordinates q^a, q'^b on M such that the leaves of P are the surfaces of constant q and the leaves of P' are the surfaces of constant q'. In these,

$$\theta = \frac{\partial S}{\partial q^a}\mathrm{d}q^a \quad \text{and} \quad \omega = \frac{\partial^2 S}{\partial q'^a \partial q^b}\mathrm{d}q'^a \wedge \mathrm{d}q^b. \tag{4.9.2}$$

Hence if we put

$$p_a = \frac{\partial S}{\partial q^a} \quad \text{and} \quad p'_a = -\frac{\partial S}{\partial q'^a}, \tag{4.9.3}$$

then p_a, q^b is a canonical coordinates system adapted to P and p'_a, q'^b is a canonical coordinate system adapted to P'; S is the generating function of the transformation from p, q to p', q' and $-S$ is the generating function of the reverse transformation; S is determined by P and P' up to the addition of $f + f'$, where $f \in C^\infty_P(M)$ and $f' \in C^\infty_{P'}(M)$. It is closely analogous to the generating function of a Kähler metric.

4.10 The Hamilton-Jacobi condition

We are now in a position to understand the symplectic geometry underlying the Hamilton-Jacobi equation. Let $h \in C^\infty(T^*Q)$ be the Hamiltonian of some time-independent system and let $S \in C^\infty(Q)$. The Hamilton-Jacobi equation is

$$h\left(\frac{\partial S}{\partial q^a}, q^a\right) = \text{constant}, \tag{4.10.1}$$

where on the left-hand side the partial derivatives of S have been substituted for the ps in the coordinate expression for h. This is the condition that h should be constant on the Lagrangian submanifold $\Lambda \subset T^*Q$ generated by S. Now, quite generally, we have

Lemma (4.10.2). Let Λ be a connected Lagrangian submanifold of a symplectic manifold (M, ω) and let $h \in C^\infty(M)$. Then h is constant on Λ if and only if X_h is tangent to Λ.

Proof. Since Λ is connected, h is constant on Λ if and only if $\mathrm{d}h|_\Lambda = 0$. But $\mathrm{d}h = -X_h \lrcorner\, \omega$, so $\mathrm{d}h|_\Lambda = 0$ if and only if $X_h(m) \in T_m\Lambda^\perp = T_m\Lambda$ for every $m \in \Lambda$. That is, if and only if X_h is tangent to Λ. ∎

The Hamilton-Jacobi equation is the condition that the Lagrangian submanifold Λ generated by S should be tangent to X_h. When it holds, the integral curves of X_h in Λ project onto the solution curves of

$$\dot{q}^a = \left.\frac{\partial h}{\partial p_a}\right|_{p=\mathrm{d}S}$$

in Q. Thus a single solution of the Hamilton-Jacobi equation reduces Hamilton's equations from a system of $2n$ first-order differential equations in the ps and qs to a system of n first order equations in the qs.

With an n-parameter family of solutions, we can do better: the dynamical problem becomes trivial. For if $S(q^a, q'^b)$ satisfies (4.10.1) for each fixed q' (in some open subset of $\mathbb{R}^n$), and also the nondegeneracy condition (4.9.1), then S generates a (local) polarization of T^*Q tangent to X_h. The equations of motion become

$$\dot{p}'_a = 0, \qquad \dot{q}'^a = 0,$$

and one obtains explicit solutions by solving these and then transforming back to the original coordinates $\{p_a, q^b\}$, which are related to the p's and q's by (4.9.3).

Definition (4.10.3). A complete integral of the Hamilton-Jacobi equation is an n-parameter family of solutions satisfying (4.9.1).

Example. Let $Q = \mathbb{R}^2$ with polar coordinates $q^1 = r$ and $q^2 = \phi$, and let

$$h = \tfrac{1}{2}(p_1)^2 + \tfrac{1}{2}(p_2/r)^2$$

be the Hamiltonian generating the straight lines in Q ($p_1 = \dot{r}$ and $p_2 = r^2\dot{\phi}$). Then the Hamilton-Jacobi equation is solved by

$$S = \int \sqrt{2E - \frac{a^2}{r^2}}\,\mathrm{d}r + a\phi \tag{4.10.4}$$

where E and a are constants (the energy and the angular momentum about the origin). The solution is double-valued outside the circle C given by $r = a/\sqrt{2E}$, singular on C, and not defined at all inside C. The two congruences of orbits generated by the two branches of S are the two families of tangent half lines to C; these are interchanged as one passes through the caustic singularity on C (Fig. 4.6).

The Langrangian submanifold $\Lambda \subset T^*Q$ generated by S is given by

$$p_2 = a \quad \text{and} \quad 2E = (p_1)^2 + \left(\frac{p_2}{r}\right)^2,$$

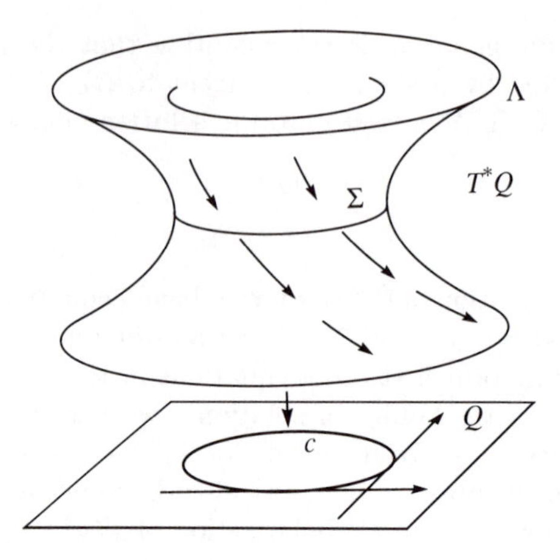

Fig. 4.6. The integral curves of X_h project onto the tangents to C.

and it is everywhere smooth and nonsingular. The projection $\Lambda \to Q$ is well behaved except on the singular set Σ given by $r = a/\sqrt{2E}$, which is mapped onto C. Away from Σ, the projection is two-to-one, mapping $\Lambda - \Sigma$ onto the outside of C in Q. The two branches of S are the generating functions of the two sheets of Λ. ∎

Example. *Time-dependent systems.* Let $L \in C^\infty(Q \times \mathbb{R})$ be a time-dependent Lagrangian. Then, in the notation of §2.4, we can obtain the dynamical evolution from the Hamiltonian $\tilde{h} = h + s$ on $T^*(Q \times \mathbb{R})$, subject to the constraint $\tilde{h} = 0$ (s is the momentum conjugate to t).

Let $\Lambda \subset T^*(Q \times \mathbb{R})$ be a Lagrangian submanifold such that the Hamilton-Jacobi condition $\tilde{h}|_\Lambda = 0$ holds, and let $S(q,t)$ be a local generating function of Λ. Then

$$h\left(q^a, \frac{\partial S}{\partial q^b}\right) + \frac{\partial S}{\partial t} = 0,$$

which is the time-dependent form of the Hamilton-Jacobi equation.

An interesting example is to take Λ to be the submanifold made up of the integral curves of $X_{\tilde{h}}$ in $\tilde{h} = 0$ whose projections into $Q \times \mathbb{R}$ pass through a fixed point (q', t'). Where Λ intersects $T^*_{(q',t')}(Q \times R)$, its tangent space is spanned by $X_{\tilde{h}}$ and by the vertical vectors

$$\frac{\partial}{\partial p_a} - \frac{\partial h}{\partial p_a}\frac{\partial}{\partial s},$$

which are all tangent to $\tilde{h} = 0$. Therefore the tangent space is Lagrangian at these points, and hence everywhere on Λ since $\mathcal{L}_{X_{\tilde{h}}}\tilde{\omega} = 0$, where $\tilde{\omega}$ is the canonical 2-form on $T^*(Q \times \mathbb{R})$.

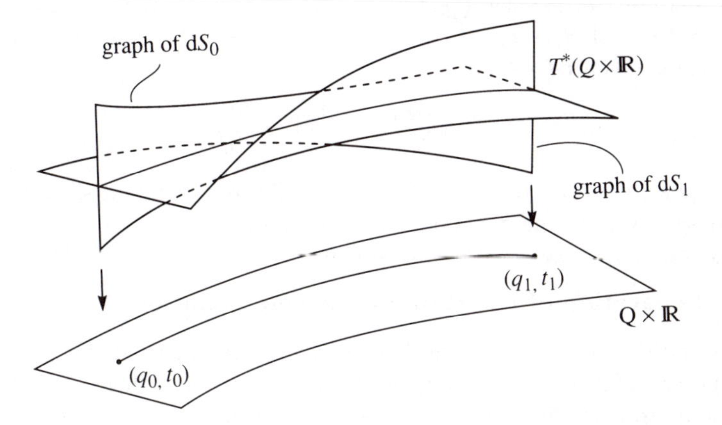

Fig. 4.7. The construction of an orbit from Hamilton's two-point characteristic function.

A local phase function $W \in C^\infty(\Lambda)$ is defined by

$$W(m) = \int \tilde{\theta} = \int (p_a \, \mathrm{d}q^a - h \, \mathrm{d}t)$$

where $\tilde{\theta}$ is the canonical 1-form and the integral is taken from any point in $\Lambda \cap T^*_{(q',t')}(Q \times \mathbb{R})$ along any path in Λ to $m \in \Lambda$. Since $\tilde{\theta}$ vanishes on contraction with vertical vectors, it is irrelevant which point is chosen. We can, in particular, take the path to be the integral curve of $X_{\tilde{h}}$ through m. Then

$$W(m) = \int X_{\tilde{h}} \lrcorner \, (p_a \mathrm{d}q^a - h\mathrm{d}t) \, \mathrm{d}t'' = \int L \, \mathrm{d}t'',$$

where t'' is the parameter along the curve. It follows that Λ is generated by $S \in C^\infty(Q \times \mathbb{R})$, where

$$S(q,t) = \int_{t'}^{t} L \, \mathrm{d}t'' = -\int_{t}^{t'} L \, \mathrm{d}t'',$$

with the integral along the dynamical trajectory joining (q', t') to (q, t). When (q', t') is allowed to vary as well, S becomes Hamilton's two-point characteristic function $S(q, q', t, t')$ (eqn 4.4.2).

With (q', t') fixed, $S(q, q', t, t')$ generates the Lagrangian submanifold made up of orbits through (q', t'); with (q, t) fixed , $-S(q, q', t, t')$ generates the Lagrangian submanifold made up of orbits through (q, t). The intersection of these is the orbit joining (q, t) and (q', t'): see Fig. 4.7.

In analytic terms: let (q_0, t_0), $(q_1, t_1) \in Q \times \mathbb{R}$ and put $S_0(q,t) = S(q, q_0, t, t_0)$ and $S_1(q,t) = -S(q_1, q, t_1, t)$. Then the orbit from (q_0, t_0) to (q_1, t_1) is given by solving

$$\frac{\partial S_0}{\partial q^a} = \frac{\partial S_1}{\partial q^a}, \qquad \frac{\partial S_0}{\partial t} = \frac{\partial S_1}{\partial t}$$

for the qs as functions of t (the second equation is redundant since both S_0 and S_1 satisfy the time-dependent Hamilton-Jacobi equation). ∎

4.11 Separation of the Hamilton-Jacobi equation

In this and the next section, we shall consider some methods for constructing a polarization tangent to a given Hamiltonian vector field on T^*Q. We shall suppose that the Hamiltonian is of the form

$$h = \tfrac{1}{2} g^{ab}(p_a - \alpha_a)(p_b - \alpha_b) + V \tag{4.11.1}$$

where g is a Riemannian metric on Q, α is the potential of an electromagnetic field $F = 2d\alpha$ and V is some scalar potential.

One, rather trivial, possibility is that there exists a system of 'ignorable' coordinates q^a such that h depends on the corresponding mometa p_a, but not on the q^as themselves. Then the surfaces of constant p_a are the leaves of a polarization tangent to X_h. A slight generalization arises when V, g^{ab}, and F_{ab} (but not the components of α) are independent of q, in which case the surfaces

$$p_a + F_{ab} q^b + \alpha_a = \text{constant}$$

are the leaves of a linear polarization tangent to X_h ('linear' in the sense that the leaves are the level surfaces of inhomogeneous linear functions of the momenta).

Without such a high degree of symmetry, one of the few practical ways to solve the Hamilton-Jacobi equation without first integrating Hamilton's equations is to separate the variables. The method is not generally applicable, however, and still requires very special conditions on the Hamiltonian. We shall see that the invariant polarizations of T^*Q that arise in this way are always linear or quadratic: the leaves are the level surfaces of constants of the motion which are at most quadratic in the momenta.

Suppose that the Hamilton-Jacobi equation separates partially in the sense that it has a complete integral of the form

$$S(q, q') = S_1(q^1, q'^1, \ldots, q'^n) + S_2(q^2, \ldots, q^n, q'^1, \ldots, q'^n).$$

The q's are regarded either as parameters labelling the leaves of the polarization generated by S, or as functions on T^*Q, by inverting

$$p_a = \frac{\partial S}{\partial q^a}$$

to express the q's in terms of q and p.

The discussion that follows is local, and makes use of the special relationship between two coordinate systems on T^*Q: the configuration and momentum coordinates p_a, q^b, which are canonical, and the coordinates q^a, q'^b, in which ω is given by (4.9.2).

Let P be the vertical polarization and let P' be the polarization generated by S. In the q, q' system, the leaves of P are the surfaces of constant q and the leaves of P' are the surfaces of constant q'. Because S satisfies the Hamilton-Jacobi equation, h depends only on the q's. Also, because of the special form of S, when p_1 is expressed in terms of the qs and q's, it depends only on $q^1, q'^1, \ldots, q'^n$, while $p_2, \ldots, p_n$ depend only on $q^2, \ldots, q^n, q'^1, \ldots, q'^n$.

In the p, q system, $h = b(p_1)^2 + cp_1 + d$ where b, c, d are functions of the ps and qs independent of p_1. Define a function $w \in C^\infty(T^*Q)$ by

$$w(q'^1, \ldots, q'^n) = \left.\frac{\partial S_1}{\partial q^1}\right|_{q^1=0}.$$

Then $[h, w] = 0$ since w is a function of the q's alone. The choice of $q^1 = 0$ here is not important: any other value would do equally well.

Define b_0, c_0, d_0 by

$$\begin{aligned} b_0 &= b(p_2, \ldots, p_n, 0, q^2, \ldots, q^n), \\ c_0 &= c(p_2, \ldots, p_n, 0, q^2, \ldots, q^n), \\ d_0 &= d(p_2, \ldots, p_n, 0, q^2, \ldots, q^n), \end{aligned}$$

Then b_0, c_0, d_0 are functions of T^*Q. In the p, q system, they are independent of p_1 and q^1 , and are polynomials in $p_2, \ldots, p_n$ of degrees 0,1, and 2 respectively. In the q, q' system, they are independent of q^1.

When $q^1 = 0$,

$$b_0 w^2 + c_0 w + d_0 = h, \tag{4.11.2}$$

by the Hamilton-Jacobi equation and the definition of w. But in the q, q' coordinates, neither side depends on q^1. Therefore (4.11.2) holds everywhere. It can be rewritten

$$w^2 + fw + k = 0, \tag{4.11.3}$$

where $f = c_0/b_0$ and $k = (d_0 - h)/b_0$. In the p, q coordinates, f is linear in the momenta and k is quadratic in the momenta.

By taking the Poisson bracket with h, we obtain $[f, h]w + [k, h] = 0$. Hence by substituting back into (4.11.3), either

$$[k, h]\big([k, h] - f[f, h]\big) = -k[f, h]^2 \quad \text{or} \quad [f, h] = 0 = [k, h].$$

All the terms in the first equation are polynomials in the ps, so, if it holds, $[f, h]$ divides $[k, h]$ as a polynomial in p. In this case, since $[f, h]$ is of degree 2 and $[k, h]$ is of degree 3, w is a linear function of the momenta.

The other possibility is that the second set of equations hold, in which case we get two constants of the motion: $f \in S^1_P(T^*Q)$ and $k \in S^2_P(T^*Q)$, of which k certainly does not vanish identically. It is also true in this case that $[f, q'^a] = [k, q'^a] = 0$, although this is not obvious.[5]

In both cases, therefore, there is a constant of the motion which is independent of h and which is either linear or quadratic in the momenta. In the first case, it is w; in the second, it is k. Note that k, f and w can be written down directly from h without actually solving the Hamilton-Jacobi equation. If all the coordinates separate, then we can apply the argument to each coordinate in turn to conclude that the leaves of P' are spanned by the Hamiltonian vector fields of constants of the motion that are either linear or quadratic in the momenta.

Linear constants arise when one of the coordinates q^a is ignorable (it does not appear in the coordinate expression for g^{ab}, F_{ab}, or V), and so this type of separation is associated with the existence of symmetry.

We shall now look in more detail at the second case, which is the one that gives something beyond the trivial separation arising from the existence of ignorable coordinates. Returning to the separation of q^1, suppose that $[h, f] = [h, k] = 0$ and let k^{ab} and g^{ab} be the components of $2\sigma(k)$ and $2\sigma(h)$. Since the Nijenhuis concomitant of g^{ab} and k^{ab} vanishes,

$$\nabla^{(a}k^{bc)} = 0 \tag{4.11.4}$$

where ∇ is the metric connection of g. That is, k^{ab} must be a Killing tensor (Penrose and Rindler 1986, p. 105). Moreover, by the definition of k,

$$b_0 k^{a1} = -g^{a1}.$$

Therefore the gradient of q^1 is an eigenvector of $\sigma(k)$. The second type of separability can arise only if the level hypersurfaces of the coordinate that separates are orthogonal to an eigenvector of a two-index Killing tensor.

Successive differentiation of the Killing equation leads to algebraic relations between the derivatives of k^{ab} and the curvature tensor of g^{ab}. From these it can be shown that the solution space is finite dimensional (Sommers 1973a,b). In particular it can be shown that in the flat case, the general solution of (4.11.4) is a sum of solutions of the form $k^{ab} = X^a X^b$, where X^a is a Killing vector (i.e. $\mathcal{L}_X g = 0$), although this is not true for a general metric.

Example. In spherical polar coordinates, the Kepler system has Hamiltonian

$$h = \frac{1}{2}\left(p_r^2 + \frac{p_\theta^2}{r^2} + \frac{p_\phi^2}{r^2 \sin^2\theta}\right) - \frac{1}{r}.$$

The corresponding Hamilton-Jacobi equation separates completely.

In particular, the r coordinate separates. Pick a constant r_0 and take $q^1 = r - r_0$, $q^2 = \theta$, $q^3 = \phi$. Then

$$k = \frac{1}{r_0^2}\left(p_\theta^2 + \frac{p_\phi^2}{\sin^2\theta}\right) - \frac{2}{r_0} - 2h$$

is a linear combination of the total energy and the square of the total angular momentum. Note that a different value of r_0 gives a different combination. ∎

Example. In two-dimensional Euclidean space, the Killing vectors are linear combinations (with constant coefficients) of

$$x\frac{\partial}{\partial y} - y\frac{\partial}{\partial x}, \quad \frac{\partial}{\partial x}, \quad \text{and} \quad \frac{\partial}{\partial y},$$

where x and y are Cartesian coordinates. Therefore the general Killing tensor is

$$(k^{ab}) = \begin{pmatrix} Ay^2 - 2By + D & -Axy + Bx - Cy + E \\ -Axy + Bx - Cy + E & Ax^2 + 2Cx + F \end{pmatrix},$$

where A, B, C, D, E, F are constants.

Let h be a quadratic Hamiltonian such that $\sigma(h)$ is the Euclidean metric. Suppose that the Hamilton-Jacobi equation separates in the coordinate system q^1, q^2. Then the curves of constant q^1 must be orthogonal to eigenvectors of k^{ab}, and so must be solution curves of the differential equation

$$\alpha\left(\frac{\mathrm{d}y}{\mathrm{d}x}\right)^2 + \beta\frac{\mathrm{d}y}{\mathrm{d}x} + \gamma = 0,$$

where

$$\begin{aligned} \alpha &= -Axy + Bx - Cy + E \\ \beta &= A(y^2 - x^2) - 2By - 2Cx + D - F \\ \gamma &= Axy - Bx + Cy - E. \end{aligned}$$

It is straightforward to deduce that they must coincide with the level curves of either $\mathrm{Re}(H)$ or $\mathrm{Im}(H)$, where H is a holomorphic function of $z = x + \mathrm{i}y$ such that

$$\frac{\mathrm{d}H}{\mathrm{d}z} = \frac{1}{\sqrt{D - F + 2\mathrm{i}E - 2(B + \mathrm{i}C)z - Az^2}}.$$

It follows that when $A \neq 0$, the solution curves are confocal hyperbolas or ellipses; and that when $A = 0$, they are confocal parabolas.

Apart from 'ignorable' coordinates associated with the Killing vectors, therefore, the only coordinates that can separate in the Hamilton-Jacobi equation when the kinetic energy term in the Hamiltonian is $\frac{1}{2}(p_1)^2+\frac{1}{2}(p_2)^2$ are those whose level curves are a family of confocal conics. Of course this is not sufficient: it is only for very special choices of α and V that separation does actually occur (for example, when $\alpha = 0$ and V is the gravitational potential of two equal fixed masses, the Hamilton-Jacobi equation separates when the level curves of q^1 and q^2 are the confocal ellipses and hyperbolas with foci at the two masses). Eisenhart (1949) discusses these issues in more detail.

Example. Let $Q' \subset Q$ be a hypersurface of constant q^1 and let k', h' be the restrictions of h, k to $T^*Q' \subset T^*Q$. It follows from the definition of k that

$$\frac{\partial k}{\partial q^1}\frac{\partial h}{\partial p_1} = \frac{\partial h}{\partial q^1}\frac{\partial k}{\partial p_1}.$$

Therefore, if $[h, k] = 0$, then $[h', k']' = 0$, where $[\cdot, \cdot]'$ is the Poisson bracket on T^*Q'.

In three-dimensional Euclidean space,

$$(k^{ab}) = \begin{pmatrix} y^2 + z^2 + A & -xy & -xz \\ -xy & z^2 + x^2 + B & -yz \\ -xz & -yz & x^2 + y^2 + C \end{pmatrix}$$

is a Killing tensor for any constant A, B, C. It arises from the separation in the free-particle Hamilton-Jacobi equation of the coordinate q given by

$$\frac{x^2}{q - A} + \frac{y^2}{q - B} + \frac{z^2}{q - C} = 1.$$

There are three roots of this equation, giving three possibilities for q. In each case the level surfaces are a family of confocal quadrics orthogonal to an eigenvector field of the Killing tensor. If Q' is one of these quadrics and $h = \frac{1}{2}(p_x^2 + p_y^2 + p_z^2)$ is the free particle Hamiltonian in Euclidean space, then h' generates the geodesics on Q' and k' is conserved by geodesic motion within the surface: it is the constant that comes from Jacobi's separation of the Hamilton-Jacobi equation for the geodesics on an triaxial ellipsoid (see Arnol'd 1978, pp. 264-6). ∎

4.12 Bihamiltonian systems

Magri's construction gives an invariant polarization in certain completely integrable systems in which the flow in M is canonical with respect to two distinct symplectic structures which satisfy a certain compatibility condition. The first applications were to the Korteweg-de Vries equation and other similar infinite-dimensional systems (Magri 1978). In finite dimensions, the construction is the following (see Brouzet 1989).[6]

Let ω and ω' be symplectic structures on a manifold M and let $[\cdot,\cdot]$ and $[\cdot,\cdot]'$ be the corresponding Poisson brackets. Since both ω and ω' are nondegenerate, there is a unique linear transformation $K : T_m M \to T_m M$ at each $m \in M$ such that

$$\omega(KX, Y) = \omega'(X, Y)$$

for every $Y \in T_m M$. It follows from the antisymmetry of ω and ω' that $\omega(KX, Y) = \omega(X, KY)$ and hence that the nondegenerate bilinear form defined by

$$\omega_i(X, Y) = \omega(K^i X, Y)$$

is also antisymmetric for $i \in \mathbb{Z}$: if $i = 2N$, then $\omega_i(X, Y) = \omega(K^N X, K^N Y)$; if $i = 2N+1$, then $\omega_i(X, Y) = \omega'(K^N X, K^N Y)$. Note that $\omega_0 = \omega$ and $\omega_1 = \omega'$.

Definition (4.12.1). The symplectic structures ω and ω' are *compatible* if $[\cdot,\cdot]'' = [\cdot,\cdot] + [\cdot,\cdot]'$ satisfies the Jacobi identity.

Proposition (4.12.2). If ω and ω' are compatible, then $\mathrm{d}\omega_i = 0$ for every i.

Proof. Suppose that ω and ω' are compatible. Then, for any $f, g, h \in C^\infty(M)$,

$$\sum [f, [g, h]'] = -\sum [f, [g, h]]',$$

where the summation is over the three cyclic permutations of f, g, h. It follows that for each $t \in \mathbb{R}$, $[\cdot,\cdot]_t = t[\cdot,\cdot] + [\cdot,\cdot]'$ satisfies the Jacobi identity. Now $[\cdot,\cdot]_t$ is the Poisson bracket of the skew 2-form ω_t defined by

$$\omega_t(X, Y) = \omega((t + K^{-1})^{-1} X, Y) = \sum_0^\infty (-1)^i t^i \omega_{i+1}(X, Y),$$

where the expansion is valid for small $|t|$. The Jacobi identity on $[\cdot,\cdot]_t$ is equivalent to the closure of ω_t (§1.6); so, by taking the exterior derivative through the expansion of the right-hand side, we obtain $\mathrm{d}\omega_i = 0$ for all $i > 0$. A similar argument applied to $[\cdot,\cdot] + t[\cdot,\cdot]'$ takes care of the negative values of i. ∎

Suppose now that X is a vector field on M which is locally Hamiltonian with respect to both ω and ω'. Then $\mathcal{L}_X\omega = \mathcal{L}_X\omega' = 0$ and so $\mathcal{L}_X K = 0$. Therefore $\mathcal{L}_X\omega_i = 0$ for every i. So if ω and ω' are compatible, then (locally) there exist $h_i \in C^\infty(M)$ such that

$$X \lrcorner\, \omega_i + \mathrm{d}h_i = K^i X \lrcorner\, \omega + \mathrm{d}h_i = 0$$

for $i \in \mathbb{Z}$. The vector fields $K^i X$ are locally Hamiltonian with respect to ω for all i, and are generated by the functions h_i. Moreover the hs are in involution with respect to ω since

$$[h_i, h_j] = 2\omega(K^i X, K^j Y) = 2\omega(K^{i+j}X, X) = \omega_{i+j}(X, X) = 0.$$

The $K^i X$s span an isotropic subspace of $T_m M$ at each $m \in M$. In favourable circumstances, the subspace is actually Lagrangian and is therefore a polarization tangent to X.

Example. Let $X = \partial/\partial p_1$. Then X is Hamiltonian with respect to both

$$\begin{aligned} \omega &= \mathrm{d}p_1 \wedge \mathrm{d}q^1 + \mathrm{d}p_2 \wedge \mathrm{d}q^2 + \cdots + \mathrm{d}p_n \wedge \mathrm{d}q^n \quad \text{and} \\ \omega' &= \mathrm{d}p_1 \wedge \mathrm{d}q^2 + \mathrm{d}p_2 \wedge \mathrm{d}q^3 + \cdots + \mathrm{d}p_{n-1} \wedge \mathrm{d}q^n + \mathrm{d}p_n \wedge \mathrm{d}q^1. \end{aligned}$$

In this case,

$$K\left(\frac{\partial}{\partial p_1}\right) = \frac{\partial}{\partial p_2}, \;\ldots\; , K\left(\frac{\partial}{\partial p_{n-1}}\right) = \frac{\partial}{\partial p_n}, \quad K\left(\frac{\partial}{\partial p_n}\right) = \frac{\partial}{\partial p_1}$$

and the vector fields $K^i X$ span the polarization with leaves of constant q. If, on the other hand,

$$\omega' = \mathrm{d}p_1 \wedge \mathrm{d}q^2 + \mathrm{d}p_2 \wedge \mathrm{d}q^1 + \mathrm{d}p_3 \wedge \mathrm{d}q^3 + \cdots + \mathrm{d}p_n \wedge \mathrm{d}q^n,$$

then $K^i X$ is either $\partial/\partial p_1$ or $\partial/\partial p_2$ as i is odd or even, and we do not get a polarization.

The eigenvalues of K are always constants of the motion. In this example, they are trivial (i.e. they are constant everywhere). ∎

5

COMPLEX POLARIZATIONS

5.1 Polarizations and representations

REAL polarizations play an important part in linking classical and quantum mechanics. We shall see that picking a polarization in a classical phase space amounts to choosing a representation in the underlying quantum theory—'representation' in the sense of a position or momentum space representation. We shall also come across complex representations, which in many ways are easier to handle. An obvious example is the Fock space treatment of the harmonic oscillator, in which the quantum Hilbert space is realized as a space of holomorphic functions of $z = p + \mathrm{i}q$.

Complex representations arise from complex polarizations, which are natural generalizations of real polarizations: one simply relaxes the condition that the Hamiltonian vector fields spanning P should be real. In the extreme case, all the Hamiltonian vector fields are complex and P gives M the structure of a Kähler manifold. In the intermediate cases, P is a mixture of a real polarization and a Kähler structure.

Before making these ideas more precise, we need some basic facts about complex Lagrangian subspaces. We shall make use of the material on real and complex vector spaces outlined in §A.1 and covered in detail in, for example, Kobayashi and Nomizu (1969), Chapter IX.

5.2 Complex structures

Let (V, ω) be an n-dimensional real symplectic vector space. A complex stucture J on V is *compatible* with ω if it is also a linear canonical transformation, in which case,

$$g(X, Y) = 2\omega(X, JY) \qquad X, Y \in V$$

defines a nondegenerate symmetric bilinear form on V; and

$$\langle X, Y\rangle_J = g(X, Y) + 2\mathrm{i}\omega(X, Y)$$

defines a Hermitian inner product on $V_{(J)}$—it is antilinear in the first entry and linear in the second, and it is always nondegenerate. We say that J is *positive* whenever g, or equivalently $\langle\cdot,\cdot\rangle_J$, is positive definite.

By mapping $X \in V$ to $\frac{1}{2}(X - \mathrm{i}JX) \in V_{\mathbb{C}}$, we can identify $V_{(J)}$ as a complex vector space with the Lagrangian subspace $P_J = \{X - \mathrm{i}JX\} \subset V_{\mathbb{C}}$ on which $J = \mathrm{i}$, in such a way that $\langle .,. \rangle_J$ coincides with the inner product $\langle Z, Z'\rangle = 4\mathrm{i}\omega(\overline{Z}, Z')$ on P_J.

Conversely, a complex Lagrangian subspace $P \subset V_{\mathbb{C}}$ such that $P \cap \overline{P} = 0$ determines a unique compatible complex structure J such that $P = P_J$: if $X \in V$, then JX is defined by writing $X = Z + \overline{Z}$, where $Z \in P$, and taking J to be the map $X \mapsto \mathrm{i}Z - \mathrm{i}\overline{Z}$. Since $P \cap \overline{P} = 0$, it follows that $V_{\mathbb{C}} = P \oplus \overline{P}$ and hence that Z is uniquely determined by X.

In discussing complex structures on symplectic spaces, we shall always take 'compatible with ω' as understood.

Example. Let $\{X^a, Y_b\}$ be a symplectic frame in (V, ω) and let (g_{ab}) be a real symmetric nonsingular $n \times n$ matrix. Define J by

$$JX^a = g^{ab}Y_b, \qquad JY_a = -g_{ab}X^b, \tag{5.2.1}$$

$g^{ab}g_{bc} = \delta^a_c$. Then J is a (compatible) complex structure on V, which is positive whenever g is positive definite. ∎

This example shows that complex structures exist on symplectic vector spaces. It is also the generic example since, given V, ω, and J, we can find a basis $\{Y_1, \ldots, Y_n\}$ for $V_{(J)}$ such that

$$\langle Y_a, Y_b \rangle_J = g_{ab}$$

where $(g_{ab}) = \mathrm{diag}(1, \ldots, 1, -1, \ldots, -1)$, with k ones and $n-k$ minus ones, according to the signature of $\langle \cdot, \cdot \rangle_J$. If the X^as are defined by

$$X^a = -g^{ab}JY_b \quad \text{where} \quad g^{ab}g_{bc} = \delta^a_c,$$

then $\{X^a, Y_b\}$ is a real symplectic frame and (5.2.1) holds.

The subgroup of $SP(V, \omega)$ of linear canonical transformations that preserve J is isomorphic to the pseudo-unitary group $U(k, n-k)$, which is the group of complex $n \times n$ matrices $(U^a{}_b)$ such that

$$g_{ab}U^a{}_cU^b{}_d = g_{cd}.$$

The isomorphism is constructed as follows. We use g^{ab} and g_{ab} to raise and lower indices, and we use the symplectic frame X^a, Y_b to identify $SP(V, \omega)$ with $SP(n, \mathbb{R})$, as in §1.3. If ρ is given by (1.3.3), then ρ preserves J whenever

$$C^a{}_b = F^a{}_b, \qquad D^a{}_b = -E^a{}_b, \tag{5.2.2}$$

in which case $U^a{}_b = C^a{}_b + \mathrm{i}E^a{}_b$ is pseudo-unitary. When J is positive, $k = n$, and we can take $g_{ab} = \delta_{ab}$. Then (5.2.2) reduces to $C = F$, $D = -E$, which implies that $C + \mathrm{i}E$ is unitary.

5.3 Complex Lagrangian subspaces

A complex structure on (V, ω) is the same as a Lagrangian subspace $P \subset V_{\mathbb{C}}$ such that $P \cap \overline{P} = 0$. At the other extreme, there are the real Lagrangian

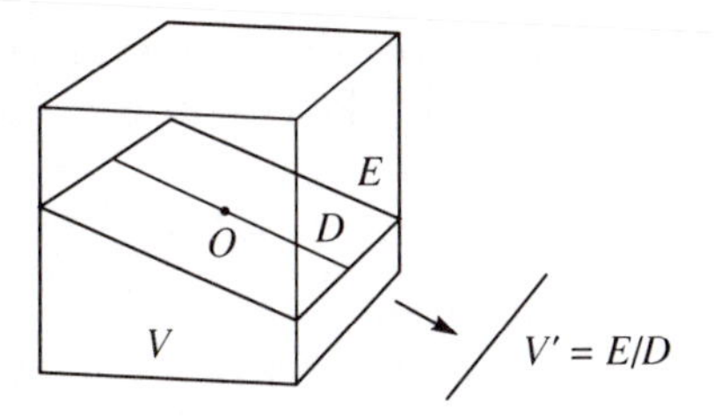

Fig. 5.1. The projection onto V'.

subspaces $P \subset V_{\mathbb{C}}$, for which $P = \overline{P}$: these are complexifications of Lagrangian subspaces of V. We shall also be interested in the intermediate cases, in which the dimension of $P \cap \overline{P}$ is between zero and n.

Let P be a general Lagrangian subspace of $V_{\mathbb{C}}$. Then $D = P \cap \overline{P} \cap V$ is a real isotropic subspace of V. The corresponding coisotropic subspace is $E = D^{\perp} = (P + \overline{P}) \cap V$.

By (1.2.6), $V' = E/D$ is a symplectic vector space—its complexification is

$$V'_{\mathbb{C}} = E_{\mathbb{C}}/D_{\mathbb{C}} = (P + \overline{P})/P \cap \overline{P};$$

and by (1.2.7), $P' = \pi(P)$ is a Lagrangian subspace of $V'_{\mathbb{C}}$, where $\pi : E_{\mathbb{C}} \to V'$ is the projection; see Fig. 5.1. Clearly $P' \cap \overline{P'} = 0$, so P' determines a complex structure J' on V': the reduction from V to V' removes the real directions from P.

Given E and J', we can reconstruct P. Now E is determined up to canonical transformations of V by its dimension; and by a canonical transformation of V', J' can be brought to the standard form in which $\langle \cdot, \cdot \rangle_{J'}$ has matrix $\operatorname{diag}(1, \ldots, 1, -1, \ldots, -1)$, with r ones and s minus ones. Any linear canonical transformation of V' can be induced by a canonical transformation of V that preserves E. Thus P is determined to within a real canonical transformation of V by two integers r and s such that $r + s \leq n$. The dimension of D is $n - r - s$.

Definition (5.3.1). The *type* of P is the pair of integers (r, s). When $r + s = n$, $P \cap \overline{P} = 0$ and P is said to *Kähler*. When $r = n$, P is said to be *positive*; when $s = 0$, P is said to be *nonnegative*; and when $r = s = 0$, P is said to be *real*.

When P is real, it is the complexification of a Lagrangian subspace of V. With some looseness of terminology, we shall refer to $n - r - s$ as the number of real directions in P. Note that J is positive whenever P_J is positive.

Example. Let $V = \mathbb{R}^{2n}$, with linear coordinates $p_1, \ldots, p_n, q^1, \ldots, q^n$, and let $\omega = \mathrm{d}p_a \wedge \mathrm{d}q^a$. In $V_{\mathbb{C}}$, the ps and qs take on complex values. An example of a Lagrangian subspace of type (r, s) is given by taking $q^a = 0$ for $a = 1, \ldots, n - r - s$; $p_a = \mathrm{i}q^a$ for $a = n - r - s + 1, \ldots, n - s$; and $p_a = -\mathrm{i}q^a$ for $a = n - s + 1, \ldots, n$. ∎

5.4 Complex polarizations

Definition (5.4.1). A *complex polarization* of a symplectic manifold (M, ω) is a complex distribution P on M such that

(CP1) for each $m \in M$, $P_m \subset (T_m M)_{\mathbb{C}}$ is Lagrangian;

(CP2) the dimension of $D = P \cap \overline{P} \cap TM$ is constant; and

(CP3) P is integrable.

The terminology is explained in §A.4. The third condition is equivalent to the existence in some neighbourhood of each point of n complex-valued functions $z^1, \ldots, z^n$ such that the Hamiltonian vector fields X_{z^a} span $\overline{P}$. These functions are necessarily in involution.

Definition (5.4.2). A complex polarization P of (M, ω) is *strongly integrable* if $E = D^\perp = (P + \overline{P}) \cap TM$ is integrable. A complex symplectic potential θ is *adapted* to P if $\overline{X} \lrcorner \theta = 0$ for every $X \in V_P(M)$; P is *admissible* if there exists an adapted potential in some neighbourhood of each point. A complex function f is *polarized* if $\overline{X}(f) = 0$ for every $X \in V_P(M)$. The space of polarized functions is denoted by $C_P^\infty(M)$.

Much of the real theory extends to the complex case. In particular, the partial connection

$$\nabla : V_P(M) \times V_P(M) \to V_P(M)$$

is defined in exactly the same way by $(\nabla_X Y) \lrcorner \omega = X \lrcorner \mathrm{d}(Y \lrcorner \omega)$; and the spaces $S_P^k(U) \subset C_{\mathbb{C}}^\infty(U)$, $U \subset M$, of polynomial observables are defined by Kostant's inductive definition, starting with $S_P^0(U) = C_P^\infty(U)$ (§4.8). As in the real case, the Hamiltonian flow of a real function f on M preserves P if and only if $f \in S_P^1(M)$.

Example. *Real polarizations.* The complexification of a real polarization is a complex polarization with $D = E$. ∎

Example. A *Kähler manifold* is a real $2n$-dimensional manifold M with a symplectic structure ω and a complex structure J which is compatible with ω at each point. The metric $g(\cdot,\cdot) = 2\omega(\cdot, J\cdot)$ makes M into a (pseudo) Riemannian manifold.[1]

In local holomorphic coordinates z^a $(a = 1, \ldots, n)$,

$$\omega = \mathrm{i}\omega_{ab}\mathrm{d}z^a \wedge \mathrm{d}\overline{z}^b, \quad \text{where} \quad \omega_{ab} = \overline{\omega}_{ba}.$$

Thus M has two polarizations: the *holomorphic polarization* P spanned by the vectors $\partial/\partial z$, and the *antiholomorphic polarization* $\overline{P}$ spanned by the vectors $\partial/\partial \overline{z}$. Both have $D = 0$, and every complex polarization of a symplectic manifold with this property determines a Kähler structure. The holomorphic functions on M are polarized with respect to the holomorphic polarization.

There exists a real function (a 'Kähler scalar') K in some neighbourhood of each point of M such that

$$\omega = \mathrm{i}\frac{\partial K}{\partial z^a \partial \overline{z}^b}\mathrm{d}z^a \wedge \mathrm{d}\overline{z}^b. \tag{5.4.3}$$

That is, $\omega = \mathrm{i}\partial\overline{\partial}K$ (see, for example, Morrow and Kodaira 1971, or Kobayashi and Nomizu 1969). It follows that $\omega = \mathrm{d}\theta$ where $\theta = -\mathrm{i}\partial K$. Since θ annihilates the vectors $\partial/\partial\overline{z}$, it is a symplectic potential adapted to P; its complex conjugate $\overline{\theta} = \mathrm{i}\overline{\partial}K$ is adapted to $\overline{P}$. Therefore both P and $\overline{P}$ are admissible. Note the similarity between (5.4.3) and the second equation in (4.9.2).

The partial connection along P is the restriction to $V_P(M) \times V_P(M)$ of the Levi-Civita connection of the metric[2] g.

Example. *The Lagrangian Grassmannian.* Let (V, ω) be a $2n$-dimensional real symplectic vector space and let L^+V be the set of all positive complex structures on V compatible with ω. This is called the (positive) *Lagrangian Grassmannian* of V. It has a symplectic form and a Kähler polarization. Both are invariant under $SP(V, \omega)$, which acts transitively on L^+V on the right by $J \mapsto \rho^{-1}J\rho$. We shall look at these structures in detail since they will be used later to construct the metaplectic representation.

The complex structure comes from the map $J \mapsto P_J \subset V_{\mathbb{C}}$, which identifies L^+V with a complex submanifold of the Grassmannian of n-dimensional complex subspaces of $V_{\mathbb{C}}$. Let $\{X_a, Y^b\}$ be a fixed symplectic frame; then for each $J \in L^+V$, there is a unique symmetric $n \times n$ complex matrix z such that the vectors

$$X^a - z^{ab}Y_b \tag{5.4.4}$$

span P_J. Conversely, any symmetric $n \times n$ complex matrix $z = (z^{ab}) = (x^{ab}) + \mathrm{i}(y^{ab})$ such that $y = (y^{ab})$ is positive definite determines an element of L^+V. The independent entries in z are holomorphic coordinates on L^+V. They identify L^+V with a contractible open subset of $\mathbb{C}^{\frac{1}{2}n(n+1)}$, called the Siegel upper half-space. It is a generalization of the 'upper half-plane' of complex numbers with positive imaginary part.

Alternatively, we can represent J by its $2n \times 2n$ matrix in the symplectic frame, which we shall also denote by J. This matrix is given in terms of $z = x + \mathrm{i}y$ by $J = NJ_0N^{-1}$, where

$$N = \begin{pmatrix} 1 & 0 \\ -x & y \end{pmatrix} \quad \text{and} \quad J_0 = \begin{pmatrix} 0 & -1 \\ 1 & 0 \end{pmatrix}$$

(by taking the real and imaginary parts of $J(X^a - z^{ab}Y_b) = \mathrm{i}(X^a - z^{ab}Y_b)$).

A real tangent vector to L^+V at J is a linear map $T : V \to V$ such that $J + tT$ is also a positive complex structure to the first order in t; that is,

$$TJ + JT = 0 \quad \text{and} \quad \omega(JTX, Y) = \omega(JTY, X),$$

for every $X, Y \in V$. We can also think of a tangent vector as a small displacement $z \mapsto z + tw$, where w is a complex symmetric matrix. If we write $w = u + \mathrm{i}v$, and put

$$W = \begin{pmatrix} 0 & 0 \\ -u & v \end{pmatrix}$$

then $T = (WJ_0 - JW)N^{-1}$.

The symplectic structure on L^+V is the 2-form $\Omega(T, T') = \frac{1}{8}\mathrm{tr}(TJT')$. It is clearly skew symmetric, since $TJ = -JT$, and invariant. We shall show both that it is closed and that the complex structure is a polarization by showing that

$$\Omega = \mathrm{i}\partial\bar{\partial}K = \mathrm{i}\frac{\partial^2 K}{\partial z^{ab}\partial \bar{z}^{cd}} \mathrm{d}z^{ab} \wedge \mathrm{d}\bar{z}^{cd} \tag{5.4.5}$$

where $K = -\log\det y$; i.e. that Ω is a Kähler form.[3]

First we note that $TJ = [(WJ_0)J_0 - J(WJ_0)]N^{-1}$; but

$$WJ_0 = \begin{pmatrix} 0 & 0 \\ v & u \end{pmatrix};$$

so the complex structure $w = u + \mathrm{i}v \mapsto \mathrm{i}w = -v + \mathrm{i}u$ in the tangent space to L^+V at J is also given by $T \mapsto TJ$. Therefore, we need only show that $2\Omega(\cdot, \cdot J)$ coincides with the Kähler metric determined by $K = -\log\det(y)$. But

$$\begin{aligned} 2\Omega(T, TJ) = \tfrac{1}{4}\mathrm{tr}(T^2) &= \tfrac{1}{2}\mathrm{tr}\left((WJ_0N^{-1})^2 + (WN^{-1})^2\right) \\ &= \tfrac{1}{2}\mathrm{tr}\left((uy^{-1})^2 + (vy^{-1})^2\right). \end{aligned}$$

On the other hand, since K depends only on the imaginary part of z,

$$
\begin{aligned}
2\frac{\partial^2 K}{\partial z^{ab}\partial \overline{z}^{cd}} w^{ab}\overline{w}^{cd} &= \tfrac{1}{2}\frac{\partial^2 K}{\partial y^{ab}\partial y^{cd}}(u^{ab}u^{cd}+v^{ab}v^{cd}) \\
&= 2\mathrm{tr}\left((uy^{-1})^2+(vy^{-1})^2\right),
\end{aligned}
$$

by using $u^{ab}\partial/\partial y^{ab}(\log\det y) = \mathrm{tr}(uy^{-1})$. ∎

As always, reduction provides a rich source of examples. Let P be a polarization of a symplectic manifold (M,ω). Let C be a coisotropic submanifold of M and let (M',ω') be the reduction of C. We should like to use the algebraic construction in §5.3 to obtain a polarization of M' from P.

We recall from §1.7 that M' is the space of leaves of the characteristic foliation of C. If $m' \in M'$ and if $m \in C$ is a point on the corresponding leaf, then $T_{m'}M' = T_mC/K_m$ where $K = (TC)^{\perp}$. So the algebraic construction certainly gives a Lagrangian subspace $P'_{m'} \subset (T_{m'}M')_{\mathbb{C}}$ by taking the intersection of P_m with the complexified tangent space to C, and then projecting into M'. The difficulty is that a different choice of m will give a different Lagrangian subspace unless the distribution $(P\cap T_{\mathbb{C}}C)+K_{\mathbb{C}}$ on C is itself involutive, which is not always the case. If this distribution is integrable and if, in addition, $\dim P'\cap\overline{P'}$ is constant, then we say that C is *compatible* with P. In this case P' is a well-defined polarization of M'. It is called the *reduction* of P.

The compatibility condition is automatically satisfied if C is given by $f_i = 0$ where $f_1,\ldots,f_k$ are real functions whose canonical flows preserve P. Then P' is a polarization provided only that the dimension of $P'\cap\overline{P'}$ is constant.

In the Kähler case, the Hamiltonian flows preserve P if the vector fields X_{f_i} are the real parts of holomorphic vector fields, or, equivalently, if they are Killing vectors of the metric g.

Example. *The Fubini-Study metric.* Let $M = \mathbb{C}^{n+1}$ with the symplectic form

$$\omega = \mathrm{i}\hbar(\mathrm{d}z^0\wedge\mathrm{d}\overline{z}^0 + \mathrm{d}z^1\wedge\mathrm{d}\overline{z}^1 + \cdots + \mathrm{d}z^n\wedge\mathrm{d}\overline{z}^n).$$

The function $f = z^0\overline{z}^0 + \cdots + z^n\overline{z}^n - 1$ generates the Hamiltonian flow $z \mapsto e^{it/\hbar}z$, which preserves the holomorphic polarization. Take C to be the sphere $f=0$, which is a coisotropic submanifold. Then the reduction M' is the projective space $\mathbb{CP}_{n-1}$ with the Fubini-Study Kähler structure (Kobayashi and Nomizu 1969, p. 160): the symplectic form is

$$\mathrm{i}\hbar\ \frac{\partial^2}{\partial w^a\partial\overline{w}^b}\left(\log(1+w^1\overline{w}^1+\cdots+w^n\overline{w}^n)\right)\mathrm{d}w^a\wedge\mathrm{d}\overline{w}^b$$

in the coordinates $w^a = z^a/z^0$ $(a = 1, 2, \ldots, n)$. This is shown by expressing $\omega|_{\mathbb{C}}$ in terms of the ws. The constant $\hbar$ is included so that the quantization of M' should recover $\mathbb{C}^{n+1}$ (p. 190). ∎

Example. It is not hard to see that the reduction of a positive polarization is always positive, as in the last example. But if, for instance, P is the holomorphic polarization on $\mathbb{C}^2$ and

$$\omega = \mathrm{i}dz^1 \wedge d\bar{z}^2 - \mathrm{i}d\bar{z}^1 \wedge dz^2,$$

then P' is well-defined on the reduction of $C = \{z^2 + \bar{z}^2 = 0\}$, but it is a *real* polarization. To see this, put $z^a = x^a + \mathrm{i}y^a$. Then $x^2 = 0$ on C and

$$\omega|_C = 2dx^1 \wedge dy^2.$$

The characteristic foliation of C is spanned by $\partial/\partial y^1$, and $P \cap T_{\mathbb{C}}C$ is spanned by

$$\frac{\partial}{\partial z^1} = \frac{1}{2}\left(\frac{\partial}{\partial x^1} - \mathrm{i}\frac{\partial}{\partial y^1}\right).$$

In the coordinates x^1, y^2 on M', $\omega' = 2dx^1 \wedge dy^2$ and P' is spanned by $\partial/\partial x^1$. Polarizations can only acquire real directions in this way under reduction if they are not positive. In this example, the Hermitian metric on $\mathbb{C}^2$ has signature $+, -$. ∎

The compatibility condition is satisfied when P is a strongly integrable polarization and $C = \Lambda$, where Λ is a leaf of the corresponding coisotropic distribution E. The characteristic foliation of Λ is $D = E^\perp$, restricted to Λ. If this is reducible, then $M' = \Lambda/D$ is a Kähler manifold with holomorphic polarization P'.

Example. *Product polarizations.* Let M' be a Kähler manifold and let T^*Q be a cotangent bundle. Then the direct sum of the vertical polariztion of T^*Q and the holomorphic polarization of M' is a complex polarization on $T^*Q \times M'$, with as many real directions as the dimension of Q. It is strongly integrable. For each $q \in Q$, $\Lambda = T_q^*Q \times M'$ is a leaf of E with $\Lambda/D = M'$. ∎

This is not the most general example of a strongly integrable polarization (even locally) since the Kähler structure on Λ/D can vary with Λ.

Example. Let $M = \mathbb{R}^{2n} \times \mathbb{C}^{n'}$, with real coordinates p_a, q^b on $\mathbb{R}^{2n}$ $(a, b, \ldots = 1, \ldots, n)$ and complex coordinates z^α on $\mathbb{C}^{n'}$ $(\alpha = 1, \ldots, n')$. Let $K(q, z, \bar{z})$ be a real function of the qs and zs such that

$$\det\left(\frac{\partial^2 K}{\partial z^\alpha \partial \bar{z}^\beta}\right) \neq 0.$$

Then

$$\begin{aligned}\omega &= \mathrm{d}\left(p_a \mathrm{d}q^a - \frac{\mathrm{i}}{2}\frac{\partial K}{\partial z^\alpha}\mathrm{d}z^\alpha + \frac{\mathrm{i}}{2}\frac{\partial K}{\partial \overline{z}^\alpha}\mathrm{d}\overline{z}^\alpha\right) \\ &= \mathrm{d}p_a \wedge \mathrm{d}q^a - \frac{\mathrm{i}}{2}\frac{\partial^2 K}{\partial q^a \partial z^\alpha}\mathrm{d}q^a \wedge \mathrm{d}z^\alpha \\ &\quad + \frac{\mathrm{i}}{2}\frac{\partial^2 K}{\partial q^a \partial \overline{z}^\alpha}\mathrm{d}q^a \wedge \mathrm{d}\overline{z}^\alpha + \mathrm{i}\frac{\partial^2 K}{\partial z^\alpha \partial \overline{z}^\beta}\mathrm{d}z^\alpha \wedge \mathrm{d}\overline{z}^\beta \qquad (5.4.6)\end{aligned}$$

is a symplectic form on M, and the vector fields

$$\frac{\partial}{\partial p_a} \quad \text{and} \quad \frac{\partial}{\partial z^\alpha}$$

span a polarization with n real directions. It is strongly integrable and admissible since

$$\theta = p_a \mathrm{d}q^a - \mathrm{i}\frac{\partial K}{\partial z^\alpha}\mathrm{d}z^\alpha - \frac{\mathrm{i}}{2}\frac{\partial K}{\partial q^a}\mathrm{d}q^a$$

is an adapted symplectic potential.

The leaves of E are the surfaces of constant q and D is spanned by the vector fields $\partial/\partial p$. If Λ is a leaf of E, then the zs project onto holomorphic coordinates on Λ/D and K projects onto Kähler scalar. ∎

Proposition (5.4.7). Let P be a strongly integrable polarization of a symplectic manifold (M, ω). Then in some neighbourhood of each point, there is a coordinate system p_a, q^b, z^α, with p_a, q^b real and z^α complex, in which P is spanned by $\partial/\partial p_a$ and $\partial/\partial z^\alpha$, and ω is given by (5.4.6) for some $K(q, z, \overline{z})$.

Proof. Let E and $D = E^\perp$ be the coisotropic and isotropic foliations associated with P. Since the proposition is local, nothing is lost by assuming that E and D are reducible. Let $q^1, \ldots, q^n$ be independent real functions which are constant along E (n is the number of real directions in P). Since P is integrable, we can find further complex functions $z^1, \ldots, z^{n'}$, where $n + n' = \frac{1}{2}\dim M$, such that $\overline{P}$ is spanned by the Hamiltonian vector fields generated by the qs and zs.

A leaf Λ of E is a surface of constant q. The zs are constant along D and are holomorphic coordinates on Λ/D (Fig. 5.2). Since Λ/D is Kähler, there is a real function $K(z, \overline{z})$ such that

$$\omega|_\Lambda = \mathrm{i}\frac{\partial^2 K}{\partial z^\alpha \partial \overline{z}^\beta}\mathrm{d}z^\alpha \wedge \mathrm{d}\overline{z}^\beta.$$

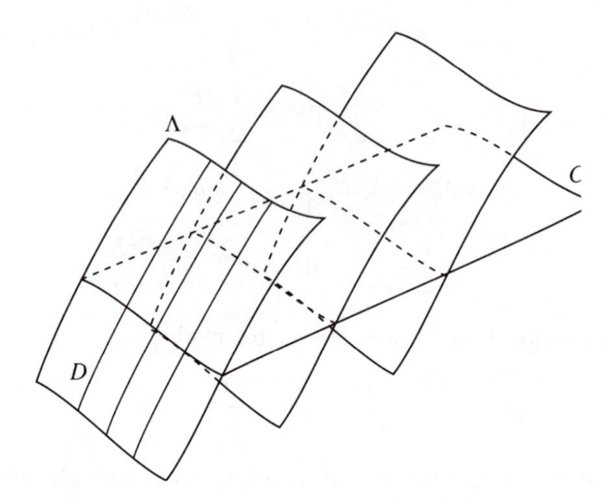

Fig. 5.2. For each leaf Λ of E, $M' = \Lambda/D$ is a Kähler manifold.

As Λ varies, K becomes a function of q as well as z and $\overline{z}$. It is determined by ω up to the addition of $A(q,z) + \overline{A}(q,\overline{z})$, where A is some function of q and z, holomorphic with respect to z.

We define p_a $(a = 1, \ldots, n)$ by picking a section C of D and putting

$$X_{q^a}(p_b) = \delta_{ab} \quad \text{with} \quad p_b = 0 \;\text{ on } C. \tag{5.4.8}$$

This determines the ps in a neighbourhood of C since the Hamiltonian vector fields generated by the qs span D.

Together p_a, q^b, and the real and imaginary parts of z^α make up a local coordinate system on M in which D is spanned by $\partial/\partial p_a$, P is spanned by $\partial/\partial p_a$ and $\partial/\partial z^\alpha$, and ω is of the form

$$\omega = \mathrm{d}p_a \wedge \mathrm{d}q^a + \zeta_{ab} \mathrm{d}q^a \wedge \mathrm{d}q^b + \xi_{a\alpha} dq^a \wedge \mathrm{d}z^\alpha + \overline{\xi}_{a\alpha} \mathrm{d}q^a \wedge \mathrm{d}\overline{z}^\alpha + \mathrm{i}\frac{\partial^2 K}{\partial z^\alpha \partial \overline{z}^\beta} \mathrm{d}z^\alpha \wedge \mathrm{d}\overline{z}^\beta.$$

The closure of ω implies that ζ_{ab} and $\xi_{a\alpha}$ are independent of p and that

$$\zeta_{ab} = \tfrac{1}{2}\left(\frac{\partial \eta_b}{\partial q^a} - \frac{\partial \eta_a}{\partial q^b}\right)$$

for some $\eta_a(q, z, \overline{z})$.

By substituting $p_a - \eta_a$ for p_a, which is equivalent to making a new choice for C, we can take $\zeta_{ab} = 0$. Closure then implies that

$$\frac{\partial \xi_{a\alpha}}{\partial z^\beta} = \frac{\partial \xi_{a\beta}}{\partial z^\alpha}, \qquad \frac{\partial \xi_{a\alpha}}{\partial q^b} = \frac{\partial \xi_{b\alpha}}{\partial q^a},$$

and hence that

$$2\xi_{a\alpha} = -\frac{\partial^2 g}{\partial q^a \partial z^\alpha}$$

for some function $g(q, z, \overline{z})$; again by closure, this must satisfy

$$\frac{\partial^3 (g - \overline{g})}{\partial q^a \partial z^\alpha \partial \overline{z}^\beta} = 2\mathrm{i}\frac{\partial^3 K}{\partial q^a \partial z^\alpha \partial \overline{z}^\beta}.$$

Hence $K - \mathrm{Im}(g)$ is of the form $-A(q, z) - \overline{A}(q, \overline{z}) + B(z, \overline{z})$. However, we are free to replace K by $K + A + \overline{A}$ and g by $g + B$, so without loss of generality, we can assume that $K = \mathrm{Im}(g)$. We can also make a further change in the origin of the p coordinates by substituting

$$p_a - \frac{\partial}{\partial q^a}\left(\frac{g + \overline{g}}{4}\right) \quad \text{for} \quad p_a.$$

Then ω is given by (5.4.6). ∎

Corollary (5.4.9). Every strongly integrable polarization is admissible.

When ω is in the form (5.4.6), the vector fields $\partial/\partial q^a$ are not Hamiltonian unless

$$\frac{\partial^3 K}{\partial q^a \partial z^\alpha \partial \overline{z}^\beta} = 0,$$

in which case it is possible to reduce ω to the form

$$\omega = \mathrm{d}p_a \wedge \mathrm{d}q^a + \mathrm{i}\frac{\partial^2 K}{\partial z^\alpha \partial \overline{z}^\beta}\mathrm{d}z^\alpha \wedge \mathrm{d}\overline{z}^\beta, \tag{5.4.10}$$

where $K = K(z, \overline{z})$, by changing p_a and K. This is the special case in which, locally, M is the product of a cotangent bundle and a Kähler manifold.

The decomposition into a product is related to the existence of the submanifold $C = \{p = 0\}$ with the properties

(1) C is a coisotropic section of D; and

(2) C is compatible with P.

If we have such a section, then we can immediately reduce ω to the form (5.4.10) by taking C to be the section of D in the proof of the proposition. The leaves of D are the cotangent spaces in the cotangent bundle, with the intersection with C picking out the origin; the base space of the cotangent bundle is the space of leaves of E; and the Kähler manifold is the reduction of C.

Under certain conditions, the decomposition is global. For example, a straightforward extension of the proof of Proposition (4.7.1) establishes the following.

Proposition (5.4.11). Let P be a strongly integrable polarization of (M, ω) and let E and D be the corresponding coisotropic and isotropic foliations. Let C be a reducible coisotropic submanifold of M which is compatible with P and which intersects each leaf of D transversally in exactly one point. Let M' be the reduction of C. Suppose that the leaves of D are geodesically convex and that E is reducible. Then M' is a Kähler manifold and, as a symplectic manifold with polarization, M can be identified with an open subset of $T^*Q \times M'$, where $Q = M/E$.

It is always possible to find a coisotropic section of D locally, but the existence of a global coistropic section involves the vanishing of certain topological obstructions. On the other hand, the second condition, compatibility with P, is local.

Aside. There is a strong analogy between the topological conditions on D, E, and M under which D admits a global coisotropic section and the conditions under which a vector bundle admits a flat connection. Ignoring awkward details (which are given in Woodhouse 1982), the basis of the analogy is the following. Suppose that D and E are reducible. Put $Q = M/E$. Then each $q \in Q$ determines a leaf Λ_q of E. By taking the quotients $M'_q = \Lambda_q/D$, we obtain a bundle of symplectic manifolds over Q. A 'connection' on this bundle is an object that determines parallel propagation of points of the fibres M'_q along curves in Q. Locally, the bundle is a product of (a subset of) Q and a symplectic manifold M', and a 'connection' assigns a Hamiltonian vector field on M' to each $X \in T_qQ$. By taking generators, we can represent it by a 1-form A on Q which takes values in $C^\infty(M')$. The 'curvature' is the 2-form on Q with values in $C^\infty(M')$ defined by

$$\Omega(X, Y) = 2\mathrm{d}A(X, Y) + [A(X), A(Y)]$$

where the square bracket is the Poisson bracket on M'.

Any section C of D determines such a 'connection': by taking the intersections with the leaves of E, we identify C with the total space of the bundle of symplectic manifolds over Q. The connection is defined so that a curve in C is 'parallel' if it is orthogonal to E with respect to the symplectic structure; and the curvature vanishes if and only if C is coisotropic.

Like many similar suggestive analogies, this one becomes sharper on quantization: under certain conditions, mixed real and complex polarizations become Hermitian vector bundles and sections of D become connections. We shall look at some relativistic examples, where the M is the phase space of a particle with internal degrees of freedom and the sections of the Hermitian vector bundle are spinor-valued wave functions. ∎

Example. *Induced symplectic actions.* Mixed polarizations can arise in the construction of representations of semi-direct products. The following is a 'classical' version of Wigner's method of finding representations by inducing from a representation of a 'little group' (Wigner 1939, Mackey 1968). Quantization of this example recovers Wigner and Mackey's construction. (See also: Sternberg 1977, Guillemin and Sternberg 1978, Weinstein 1978.)

Let G be a Lie group which acts transitively on the right on a manifold Q. Suppose that we are given a G-invariant subspace $V \subset C^\infty(Q)$, which we think of as an abelian group under addition. The semi-direct product of G and V is the group $H = G \times V$, with the group law

$$(g, f) \cdot (g', f') = (gg', f \circ (g')^{-1} + f'),$$

where $g, g' \in G$ and $f, f' \in V$.

Pick a base point $q_0 \in Q$ and let G_0 be the stabilizer of q_0. Then Q is the space of right cosets $G_0 g$, $g \in G$. Let (M', ω') be a symplectic manifold on which G_0 has a Hamiltonian action on the right, with moment $\mu : M' \to \mathcal{G}_0^*$, where $\mathcal{G}_0$ is the Lie algebra of G_0. Let P' be an invariant polarization of M'. An example to bear in mind is when G is the identity component of the Lorentz group, Q is a sheet of the mass hyperboloid $p_a p^a = m^2$ in the space of four-momenta, q_0 is $m(1, 0, 0, 0)$, G_0 (the 'little group') is $SO(3)$, and $M' = \mathbb{CP}_1$. The elements of V are the translations in space-time, which are identified with the functions $T^a p_a$ on Q, and H is the Poincaré group.

We shall construct from Q and M' a symplectic manifold with polarization on which the whole of H acts. We begin by taking the product symplectic manifold $\overline{T^*G} \times M'$. This has a polarization which is the product of the vertical polarization of $\overline{T^*G}$ and P'. Here $\overline{T^*G}$ denotes the cotangent bundle of G with the reversed symplectic structure $-\omega$, where ω is the canonical 2-form. For $A \in \mathcal{G}_0$, define $h_A \in C^\infty(T^*G)$ by

$$h_A(g, p) = -R_A(g) \lrcorner\, p,$$

where $(g, p) \in T^*G$, and R_A is the right-invariant vector field on G generated by A (recall that $[R_A, R_B] = -R_{[A,B]}$; it is because of the minus sign here that we have reversed the symplectic structure on T^*G). Denote by μ the corresponding moment. Then $\mu + \mu'$ is a moment for a canonical action of G_0 on $\overline{T^*G} \times M'$. Let M be the Marsden-Weinstein reduction of the constraint $\mu + \mu' = 0$. The polarization is compatible with the constraint, and descends to a polarization P of M. If P' is Kähler, then $M/E = Q$ and the Kähler manifolds Λ/D are copies of M'.

The group G acts on itself by right translation, and hence on $\overline{T^*G}$. A function $f \in V \subset T^*Q$ pulls back by the projection $G \to Q$ to a function on G, which we shall also denote by f. We get an action of V on T^*G by mapping (g, p) to $(g, p + \mathrm{d}f)$. By combining these with the trivial action on

M', we obtain an action of H on $\overline{T^*G} \times M'$ which preserves the constraint, and descends to a canonical action on M which preserves P.

In the relativistic example, in which H is the Poincaré group, M is the phase space of a spinning massive particle (§6.6). ∎

So far, we have considered in detail only strongly integrable polarizations. We shall now look at a way of constructing polarizations which do not satisfy this condition.

Example. *CR structures.* Let Q be a real hypersurface in $\mathbb{C}^n$, given by $f = 0$ where $f = f(z, \overline{z})$ is a real function such that[4]

$$\Omega = i\partial\overline{\partial}f|_Q$$

has the maximum possible rank (which is $2n-2$). Then Q is an embedded CR manifold (Chern and Moser 1975).

Let H be the complex distribution on Q spanned by the $(1,0)$-vectors tangent to Q (H has complex dimension $n-1$); H is integrable, but $H+\overline{H}$ is not. Put

$$M = \{(p, q) \in T^*Q \mid p|_{H+\overline{H}} = 0\}$$

and let ω be the restriction to M of the canonical 2-form on T^*Q. Then (M, ω) is a symplectic manifold and the vectors that project into H under $M \to Q$ span a polarization P containing one real direction. It is not strongly intregrable since $H + \overline{H}$ is not integrable.[5]

An alternative way to construct M and P is to add another complex variable z^0 to the coordinates $z^1, \ldots, z^n$ on $\mathbb{C}^n$ and put $\Omega = \mathrm{i}\partial\overline{\partial}(z^0 f + \overline{z}^0 f)$. Let $C = \{f = 0\} \subset \mathbb{C}^{n+1}$. Then, provided that $\mathrm{d}f \neq 0$ on C, Ω is a symplectic structure on a neighbourhood of C. The symplectic manifold (M, ω) is the reduction of C, and P is the reduction of the holomorphic polarization. The reduction map sends $(z^0, \ldots, z^n) \in C$ to $(p, q) \in M$, where $q = (z^1, \ldots, z^n)$ and $p = \frac{1}{2}\mathrm{i}(z^0 + \overline{z}^0)(\overline{\partial} f - \partial f)$. This construction makes it easy to see that although P is not strongly integrable, it is admissible: $-\mathrm{i}(z^0 + \overline{z}^0)\partial f|_C$ is the pull-back of a symplectic potential on M adapted to P. ∎

5.5 Invariant polarizations

In order to make precise the analogy between coadjoint orbits and irreducible representations of a Lie group, we shall need to construct invariant polarizations on the orbits. There is sometimes a natural way to do this by using the structure of the Lie algebra.

Let G be a real connected Lie group with Lie algebra $\mathcal{G}$. The complexification $\mathcal{G}_{\mathbb{C}}$ is a complex Lie algebra, which we shall identify with the space of complex tangent vectors to G at the identity. Each $A \in \mathcal{G}_{\mathbb{C}}$ determines left and right invariant vector fields $L_A, R_A \in V_{\mathbb{C}}(G)$ by translation.

Let $f \in \mathcal{G}^*$ and let (M, ω) be the reduction of (G, ω_f). That is, $M = G/(G_f)_0$, where G_f is the stabilizer of f under the coadjoint action, and the zero subscript denotes the identity component.[6] Suppose that $\mathcal{B}$ is a subalgebra of $\mathcal{G}_{\mathbb{C}}$ such that

(IP1) $f([A, B]) = 0 \quad \forall\, A, B \in \mathcal{B}$; and

(IP2) $2 \dim_{\mathbb{C}} \mathcal{B} = \dim G + \dim \mathcal{G}_f$.

Let P' be the complex distribution on G spanned by the vector fields R_A, $A \in \overline{\mathcal{B}}$, and let K be the characteristic distribution of ω_f, which is spanned by the vector fields R_A, $A \in \mathcal{G}_f$. The first condition on $\mathcal{B}$ implies that $\omega_f(R_A, R_B) = 0$ for every $A, B \in \overline{\mathcal{B}}$ and hence that P' is isotropic with respect to ω_f. The second condition is that P' has the maximum possible dimension for which such isotropy is possible (since $\dim K = \dim \mathcal{G}_f$). It follows that $\mathcal{B}$ contains the complexification of $\mathcal{G}_f$: otherwise $P' + K_{\mathbb{C}}$ would be isotropic with respect to ω_f and of higher dimension than P'.

The projection of P' into M is the distribution P given by

$$P_m = \mathrm{pr}_* P'_g,$$

where $\mathrm{pr} : G \to M = G/K$ is the reduction map and g is any element of $\mathrm{pr}^{-1}(m)$. Since $P' \supset K$, the choice of g is immaterial. Clearly P is isotropic and of dimension

$$\tfrac{1}{2} \dim M = \tfrac{1}{2}(\dim G - \dim G_f).$$

It is also integrable because it is analytic and involutive ($\mathcal{B}$ is a subalgebra); and invariant under the action of G on M because P' is invariant under right translation. Consequently the dimension of $P \cap \overline{P} \cap TM$ is constant, and so P is an invariant polarization of M (Kostant 1970b, Auslander and Kostant 1971, Ozeki and Wakimoto 1972).

If $\mathcal{B}$ is invariant under the adjoint action of G_f on $\mathcal{G}$, then P descends further to an invariant polarization on the coadjoint orbit of f, which is the symplectic manifold G/G_f.

Compact Lie groups.

Let G be a compact connected Lie group and let $f \in \mathcal{G}^*$. Then G contains a maximal torus T, which is unique up to conjugation $T \mapsto gTg^{-1}$ (see Adams 1969). 'Maximal' here is in the sense that T is not a proper subset of any other connected abelian subgroup of G. The dimension of T is called the *rank* of the group. We can use the freedom in the choice of maximal torus to ensure that $G_f \supset T$.

For example, the rotation group $SO(3)$ has rank one. The rotations about the z axis form a maximal torus (circle). The unitary group $U(n)$

has rank n: the diagonal subgroup $T = \{\mathrm{diag}(t_1, \ldots, t_n)\}$, where $t_j \in \mathbb{C}$, $|t_j| = 1$, is a maximal torus.

Much is known about the way in which the representations and orbits of G break up under the action of T (a very clear summary is given in Pressley and Segal 1986, Chapter 2). The starting point of this theory is the decomposition $\mathcal{G}_\mathbb{C}$ into simultaneous eigenspaces of the generators of T, which act on $\mathcal{G}_\mathbb{C}$ by the adjoint representation. The corresponding eigenvalues are imaginary, and determine a collection of linear forms on $\mathcal{T}$, the Lie algebra of T.

Definition (5.5.1). The *roots* of $\mathcal{G}$ are the nonzero linear forms $\alpha \in \mathcal{T}_\mathbb{C}^*$ such that

$$\mathcal{G}_\alpha = \{A \in \mathcal{G}_\mathbb{C} \mid [Z, A] = \alpha(Z)A \quad \forall Z \in \mathcal{T}\}$$

is nontrivial (has dimension greater than zero). The set of roots is denoted by Δ.

The principal facts about roots are as follows. If α is a root, then so is $-\alpha = \overline{\alpha}$, and $\mathcal{G}_{-\alpha} = \overline{\mathcal{G}}_\alpha$. The eigenspaces $\mathcal{G}_\alpha$ corresponding to the roots are one-dimensional, and $\mathcal{G}_\mathbb{C}$ is a direct sum

$$\mathcal{G}_\mathbb{C} = \mathcal{T}_\mathbb{C} \oplus \mathcal{G}_\alpha \oplus \mathcal{G}_\beta \oplus \ldots,$$

where $\alpha, \beta, \ldots \in \Delta$. If $A \in \mathcal{G}_\alpha$ and $B \in \mathcal{G}_\beta$, then

$$\begin{aligned} [A, B] &\in \mathcal{G}_{\alpha+\beta} \quad \text{whenever} \quad \alpha + \beta \in \Delta; \\ [A, B] &\in \mathcal{T}_\mathbb{C} \quad \text{whenever} \quad \alpha = -\beta. \end{aligned} \tag{5.5.2}$$

Otherwise $[A, B] = 0$.

If A is a nonzero element of $\mathcal{G}_\alpha$, then $Z_\alpha = \frac{1}{2}\mathrm{i}[\overline{A}, A]$ is also nonzero, and lies in $\mathcal{T}$. By rescaling A we can arrange that $\alpha(Z_\alpha) = \mathrm{i}$. If we then put $A = X_\alpha + \mathrm{i}Y_\alpha$, where X_α and Y_α are real, we have

$$[X_\alpha, Y_\alpha] = -Z_\alpha, \quad [Y_\alpha, Z_\alpha] = -X_\alpha, \quad [Z_\alpha, X_\alpha] = -Y_\alpha.$$

So $X_\alpha, Y_\alpha, Z_\alpha$ span a subalgebra $\mathcal{H}_\alpha \subset \mathcal{G}$ which is isomorphic to the Lie algebra of $SU(2)$. The Zs are completely determined by the roots, while the Xs and Ys are determined up to plane rotations

$$(X, Y) \mapsto (X \cos\phi + Y \sin\phi, -X \sin\phi + Y \cos\phi).$$

Note that $Z_{-\alpha} = -Z_\alpha$.

Since $\mathcal{T} \subset \mathcal{G}_f$, we have $f([Z, \cdot]) = 0$ for every $Z \in \mathcal{T}$. It follows that, if $A \in \mathcal{G}_\alpha$, then

$$0 = f([Z, A]) = \alpha(Z)f(A) \quad \forall\, Z \in \mathcal{T},$$

and hence that $f(A) = 0$. Therefore f vanishes on $\mathcal{G}_\alpha \oplus \mathcal{G}_\beta \oplus \cdots$. If $f(Z_\alpha) \neq 0$ for every root α, then $\mathcal{G}_f = \mathcal{T}$. In this case f is said to be *regular*. In general, $\mathcal{G}_f$ is the sum of $\mathcal{T}$ and the $\mathcal{H}_\alpha$s for which $f(Z_\alpha) = 0$.

Whether or not f is regular, we can define an invariant polarization on $M = G/(G_f)_0$ as follows. Let Δ_f^+ and Δ_f^- be the subsets of Δ of roots such that $f(Z_\alpha) > 0$ and $f(Z_\alpha) < 0$, respectively. Put

$$\mathcal{B} = \mathcal{T}_\mathbb{C} \oplus \bigoplus_{\alpha \notin \Delta_f^+} \mathcal{G}_\alpha .$$

Since α and $-\alpha$ lie in Δ_f^+ only if $f(Z_\alpha) = 0$, this satisfies (IP1) as a consequence of (5.5.2); it also satisfies (IP2). In the regular case, $\mathcal{G}_\alpha \subset \overline{\mathcal{B}}$ only if $\alpha \in \Delta_f^+$ (in which case, α is said to be *positive*).

Let $m_0 = \mathrm{pr}(e)$. Then the tangent space to M at m_0 is $\mathcal{G}/\mathcal{G}_f$, which we can identify with the subspace of $\mathcal{G}$ spanned by the Xs and Ys of the roots $\alpha \in \Delta_f^+$. The polarization at m_0 is the sum over Δ_f^+ of $\mathcal{G}_\alpha$. If $\alpha \in \Delta_f^+$, then

$$\mathrm{i} f([X_\alpha - \mathrm{i}Y_\alpha, X_\alpha + \mathrm{i}Y_\alpha]) = 2f(Z_\alpha) > 0.$$

So P is always a positive polarization, whatever the choice of f.

The Lie algebra $\mathcal{B} \subset \mathcal{G}_\mathbb{C}$ generates a subgroup B of $G_\mathbb{C}$ (the complexification of G). Such a subgroup is called a *Borel* subgroup when f is regular, or a *parabolic* subgroup otherwise. As a complex manifold, M is the same as the homogeneous space of right cosets of B in $G_\mathbb{C}$.

The maximal torus has a Hamiltonian action on M. The image of the moment in $\mathcal{T}^*$ is a convex polytope (§3.4, p. 48). One vertex is f, which defines an element of $\mathcal{T}^*$ by restriction, and the others make up the orbit of f under the action on $\mathcal{T}$ of the *Weyl group*, which is the group of of linear transformations of $\mathcal{T}$ induced by adjoint transformations $\mathcal{G} \to \mathcal{G} : A \mapsto gA$ such that $g\mathcal{T} = \mathcal{T}$ (Kostant 1973). There is an analogy between the polytope in $\mathcal{T}^*$ and the weight diagram of the representation, which we shall make precise by quantization.

One can construct other nonpositive polarizations by replacing $\mathcal{B}$ by $g\mathcal{B}$, where g is any $g \in G$ such that $g\mathcal{T} = \mathcal{T}$ (Bott 1957).

Example. *The rotation group and* $SU(2)$. The fundamental example is $SU(2)$. In the notation of §3.5 (p. 54), take $f = (0, 0, 1)$ and take T to be the diagonal subgroup, which is one-dimensional. Then $\mathcal{T}$ is spanned by Z. There are two roots: the positive root α, which is given by $\alpha(Z) = \mathrm{i}$, and $-\alpha$. In this case $X_\alpha = X$ and $Y_\alpha = Y$.

The elements of G are labelled by $(z^0, z^1) \in \mathbb{C}^2$, where $z^0\overline{z}^0 + z^1\overline{z}^1 = 1$. The points of M are equivalence classes of elements of G, where g and g' are regarded as equivalent whenever $g' = tg$ for some diagonal $t \in SU(2)$, that is, whenever $(z'^0, z'^1) = \mathrm{e}^{\mathrm{i}\phi}(z^0, z^1)$ for some real ϕ. Thus M is the projective space $\mathbb{CP}_1$, with z^0 and z^1 as homogeneous coordinates. The symplectic structure is the reduction of

$$2\mathrm{i}(\mathrm{d}z^0 \wedge \mathrm{d}\bar{z}^0 + \mathrm{d}z^1 \wedge \mathrm{d}\bar{z}^1),$$

and the polarization is the holomorphic one. The complexified group is $SL(2, \mathbb{C})$ and the Borel subgroup is the subgroup of lower triangular matrices.

Example. *Unitary groups.* In higher dimensions, the canonical example is $U(n)$, with T as the diagonal subgroup, which has Lie algebra

$$\{Z = \mathrm{diag}(z_1, \ldots, z_n)\}$$

where the zs are real. The roots are the linear maps $\alpha_{jk}(Z) = \mathrm{i}(z_j - z_k)$ $(j \neq k)$. If $f(Z) = f_1 z_1 + \cdots + f_n z_n$, where $f_1 > f_2 > \cdots > f_n$, then f is regular and $\alpha_{jk} \in \Delta_f^+$ for $k > j$. The corresponding Borel subgroup of $G_{\mathbb{C}} = GL(n, \mathbb{C})$ is the group of lower triangular matrices and M is a flag manifold. At the other extreme, if $f = \mathrm{diag}(1, 0, \ldots, 0)$, then M is $\mathbb{CP}_{n-1}$ and B is the parabolic subgroup of complex matrices with zeros in the entries 12, 13, ..., $1n$.

6

ELEMENTARY RELATIVISTIC SYSTEMS

6.1 Spin

IT is often said that spin is a purely quantum concept, with no analogue in classical mechanics; and this is certainly true in the sense that the notion of 'internal degress of freedom' is not easy to reconcile with classical particle physics. Nevertheless, it is still possible to construct classical phase spaces for particles with spin from coadjoint orbits of the Poincaré group. By quantizing these, we can recover the usual wave equations for spinning particles. Geometric methods are essential here because the internal degrees of freedom cannot be separated into configuration and momentum variables. Although this is a formal procedure rather than an attempt to understand internal structure in terms of classical physics, it is useful because it enables us to consider the relationship between the classical and quantum theories of particles with spin within the same general framework as scalar systems. It also leads to a clearer and more direct picture of the connection between the quantum theory of spin and the geometry of space-time than is allowed by the traditional methods of group theory.

In this chapter, we shall look at the structure of the classical phase spaces of massive and massless particles, using two-component spinors. The next three sections are a brief outline of spinor algebra and calculus which should be just adequate for the applications. A full account is given by Penrose and Rindler (1984) (we shall follow their conventions, except that we shall not need to use abstract indices); see also Pirani (1965).

6.2 Two-component spinors

Spinor algebra is based on the following observation. Let X be a real four-vector in Minkowski space with components

$$(X^0, X^1, X^2, X^3) = (t, x, y, z)$$

in some inertial frame, and let $(X^{AA'})$ be the matrix

$$\left(X^{AA'}\right) = \frac{1}{\sqrt{2}}\begin{pmatrix} t+z & x+\mathrm{i}y \\ x-\mathrm{i}y & t-z \end{pmatrix}. \tag{6.2.1}$$

Then

$$\det\left(X^{AA'}\right) = \tfrac{1}{2}\left(t^2 - x^2 - y^2 - z^2\right) = \tfrac{1}{2}g_{ab}X^aX^b,$$

where g is the Minkowski metric. The superscripts $A, B, \ldots, A', B', \ldots$ are *spinor indices.* They take the values 0 and 1, with the range and summation conventions. The significance of the primes ($'$) will be explained shortly.

Equation (6.2.1) determines a real linear correspondence between four-vectors and 2×2 Hermitian matrices. Because of the relationship between the determinant and the metric, the correspondence behaves in a natural way under Lorentz tranformations. Let $(L^A{}_B) \in SL(2, \mathbb{C})$ and let $(\overline{L}^{A'}{}_{B'})$ be the complex conjugate matrix—the matrix obtained by replacing the entries in $(L^A{}_B)$ by their complex conjugates, without transposition. Then

$$\tilde{X}^{AA'} = L^A{}_B \overline{L}^{A'}{}_{B'} X^{BB'}$$

determines a second set of four-vector components $\tilde{X}^a$ with $g_{ab}\tilde{X}^a\tilde{X}^b = g_{ab}X^aX^b$. In other words,

$$X^{AA'} \mapsto L^A{}_B \overline{L}^{A'}{}_{B'} X^{BB'}, \quad \text{equivalently,} \quad X \mapsto LX\overline{L}^t, \tag{6.2.2}$$

defines a Lorentz transformation $X^a \mapsto \tilde{X}^a$. It is not hard to see that every proper orthochronous Lorentz transformation can be represented in this way and that the construction gives a two-to-one local isomorphism from $SL(2, \mathbb{C})$, as a real Lie group, to the identity component of the Lorentz group, with the identity Lorentz transformation represented by both the identity and minus the identity in $SL(2, \mathbb{C})$.

Four-vectors can be represented by objects with two spinor indices which transform as in (6.2.2) under change of inertial coordinates. A *spinor* α is an object with one spinor index and two complex components which transform by

$$\alpha^A \mapsto L^A{}_B \alpha^B.$$

The spinors form a two dimensional complex vector space $\mathcal{S}$, which carries a symplectic structure ϵ, defined by

$$\epsilon(\alpha, \beta) = \epsilon_{AB}\alpha^A\beta^B = \alpha^0\beta^1 - \alpha^1\beta^0 \quad \text{where} \quad (\epsilon_{AB}) = \begin{pmatrix} 0 & 1 \\ -1 & 0 \end{pmatrix}.$$

This is invariant under $SL(2, \mathbb{C})$ because

$$\epsilon_{AB} L^A{}_C L^B{}_D = \det L \, \epsilon_{CD} = \epsilon_{CD}.$$

There are three other complex symplectic spaces associated with $\mathcal{S}$: the dual space $\mathcal{S}^*$, the complex conjugate space $\overline{\mathcal{S}}$, and the dual conjugate space $\overline{\mathcal{S}}^*$. The elements of $\mathcal{S}^*$ have components with one lower spinor index which transform by

$$\beta_A \mapsto M_A{}^B \beta_B$$

where $L^B{}_A M_B{}^C = \delta^C_A$. That is, $M = (M_A{}^B) = (L^t)^{-1}$. Thus if $\alpha \in \mathcal{S}$, then $\beta_A\alpha^A = M_A{}^B\beta_B L^A{}_C\alpha^C$, so the pairing $\beta_A\alpha^A$ is invariant.

Elements of $\overline{\mathcal{S}}$ transform by the complex conjugate of the matrix $(L^A{}_B)$, and it is conventional to label their components by primed spinor indices. Thus if $\gamma \in \overline{\mathcal{S}}$, then γ has components $\gamma^{A'}$, with the transformation rule

$$\gamma^{A'} \mapsto \overline{L}^{A'}{}_{B'}\gamma^{B'} \quad \text{where} \quad \overline{L}^{A'}{}_{B'} = \overline{L^A{}_B}$$

(that is, $\overline{L}^0{}_1 = \overline{L^0{}_1}$, and so on). Similarly, $\xi \in \overline{\mathcal{S}}^*$ has components $\xi_{A'}$, which transform by

$$\xi_{A'} \mapsto \overline{M}_{A'}^{\ B'}\xi_{\ B'} \quad \text{where} \quad \overline{M}_{A'}^{\ B'} = \overline{M_A^{\ B}}.$$

The use of primed indices is a bookkeeping device that prevents the misuse of the summation convention. Summation over an upper unprimed index and a lower unprimed index is an invariant operation, as is summation over an upper primed index and a lower primed index, but not summation over two indices of different types or over two indices in the same position. Thus $\xi_{A'}\gamma^{A'}$ is an invariant, but $\xi_0\alpha^0 + \xi_1\alpha^1$ and $\alpha^0\alpha^0 + \alpha^1\alpha^1$ are not (for $\xi \in \overline{\mathcal{S}}^*$, $\alpha \in \mathcal{S}$). Primed and unprimed indices are distinct: in the expression $\xi_{A'}\alpha^A$, there is no summation over A' and A.

The symplectic forms on $\mathcal{S}^*$, $\overline{\mathcal{S}}$ and $\overline{\mathcal{S}}^*$ are ϵ^{AB}, $\epsilon_{A'B'}$, and $\epsilon^{A'B'}$, where

$$\left(\epsilon^{AB}\right) = \left(\epsilon_{A'B'}\right) = \left(\epsilon^{A'B'}\right) = \begin{pmatrix} 0 & 1 \\ -1 & 0 \end{pmatrix}.$$

All are invariant.

There are antilinear conjugation maps $\mathcal{S} \to \overline{\mathcal{S}}$ and $\mathcal{S}^* \to \overline{\mathcal{S}}^*$, given by

$$\alpha^A \mapsto \overline{\alpha}^{A'} = \overline{\alpha^A} \quad \text{and} \quad \beta_A \mapsto \overline{\beta}_{A'} = \overline{\beta_A},$$

where the components $\overline{\alpha}^{A'}$ and $\overline{\beta}_{A'}$ are the complex conjugates of the components α^A and β_A. The replacement of an unprimed by a primed index is necessary to keep track of the different transformation rules. There are also conjugations going in the opposite directions, which convert primed indices to unprimed indices.

We can use the ϵs to raise and lower spinor indices, and so to define linear correspondences between $\mathcal{S}$ and $\mathcal{S}^*$ and between $\overline{\mathcal{S}}$ and $\overline{\mathcal{S}}^*$, with the conventions

$$\alpha_B = \alpha^A\epsilon_{AB}, \quad \beta^A = \epsilon^{AB}\beta_B, \quad \gamma_{B'} = \gamma^{A'}\epsilon_{A'B'}, \quad \xi^{A'} = \epsilon^{A'B'}\xi_{B'}.$$

Some care is needed because of the asymmetry of the ϵ. Although raising followed by lowering brings one back to ones's starting point, note that

$$\alpha_A\beta^A = -\alpha^A\beta_A \quad \text{and} \quad \gamma_{A'}\xi^{A'} = -\gamma^{A'}\xi_{A'}.$$

Note also that $\alpha_A\alpha^A = 0$.

6.3 Spinors and tensors

Eqn (6.2.1) identifies the space V of four-vectors with $\mathcal{S}\otimes\overline{\mathcal{S}}$. Similarly, the dual space V^* is identified with the $\mathcal{S}^*\otimes\overline{\mathcal{S}}^*$ by

$$(Y_{AA'}) = \frac{1}{\sqrt{2}}\begin{pmatrix} Y_0+Y_3 & Y_1-\mathrm{i}Y_2 \\ Y_1+\mathrm{i}Y_2 & Y_0-Y_3 \end{pmatrix}.$$

This is consistent with the conventions for raising and lowering indices since if $Y_b = X^a g_{ab}$, then $Y_{BB'} = X^{AA'}\epsilon_{AB}\epsilon_{A'B'}$.

The correspondence extends to space-time tensors. Any space-time tensor with q upper indices and p lower indices determines an element of

$$(\mathcal{S}\otimes\overline{\mathcal{S}})^q\otimes(\mathcal{S}^*\otimes\overline{\mathcal{S}}^*)^p = \mathcal{S}^q\otimes\overline{\mathcal{S}}^q\otimes\mathcal{S}^{*p}\otimes\overline{\mathcal{S}}^{*p}$$

where $\mathcal{S}^q = \mathcal{S}\otimes\cdots\otimes\mathcal{S}$, and so on. Conversely, we can construct a tensor from any $\alpha\in\mathcal{S}^q\otimes\overline{\mathcal{S}}^q\otimes\mathcal{S}^{*p}\otimes\overline{\mathcal{S}}^{*p}$. Such a 'multi-index' spinor has q upper unprimed indices, q upper primed indices, p lower unprimed indices, and p lower primed indices ('multi-index' will usually be dropped). We shall denote the relationship between the spinor and tensor components by

$$\alpha^{AB\dots CA'B'\dots C'}{}_{DE\dots FD'E'\dots F'} \sim \alpha^{ab\dots c}{}_{de\dots f},$$

where each lower case tensor index corresponds to two upper case spinor indices, one unprimed, the other primed, according to the rule

$$a\leftrightarrow AA',\quad b\leftrightarrow BB',\dots.$$

For example, (6.2.1) now reads $X^{AA'}\sim X^a$. In general the tensor equivalent of α is complex. However, the conjugations $\mathcal{S}\to\overline{\mathcal{S}}$ and $\mathcal{S}^*\to\overline{\mathcal{S}}^*$ extend to an antilinear map

$$\mathcal{S}^q\otimes\overline{\mathcal{S}}^q\otimes\mathcal{S}^{*p}\otimes\overline{\mathcal{S}}^{*p}\to\mathcal{S}^q\otimes\overline{\mathcal{S}}^q\otimes\mathcal{S}^{*p}\otimes\overline{\mathcal{S}}^{*p}$$

under which

$$\alpha^{A\dots CA'\dots C'}{}_{D\dots FD'\dots F'}\mapsto\overline{\alpha}^{A'\dots C'A\dots C}{}_{D'\dots F'D\dots F} = \overline{\alpha^{A\dots CA'\dots C'}{}_{D\dots FD'\dots F'}}.$$

Real tensors correspond to self-conjugate spinors. Note that conjugation interchanges primed and unprimed indices. For example, in the case of a four-vector, the reality condition is $\overline{X}^{A'A} = X^{AA'}$, which is the condition that $(X^{AA'})$ should be a Hermitian matrix and *not* the condition that it should have real entries.

A number of tensor calculations become much simpler in spinor form. One reason is that, because $\mathcal{S}$ is two-dimensional, any skew-symmetric two-index spinor must be proportional to ϵ and any skew-symmetric three-index spinor must vanish. Therefore, if $\phi^{AB} = -\phi^{BA}$, then

$$\phi^{AB} = \tfrac{1}{2}\phi_C{}^C\epsilon^{AB}$$

since $\epsilon_A{}^A = 2$; and if $\phi^{ABC} = \phi^{[ABC]}$, then $\phi^{[ABC]} = 0$. It follows that for any three-index spinor ψ^{ABC},

$$\psi_A{}^{A}{}^{B} + \psi^{BA}{}_A + \psi_A{}^{BA} = 0,$$

a very useful identity that has no straightforward tensor equivalent.

I shall end this section by summarizing some properties of spinors that we shall need. Detailed derivations can be found in Penrose and Rindler (1984).

Covariant derivative. The spinor equivalent of the covariant derivative ∇_a is the operator

$$(\nabla_{AA'}) = \frac{1}{\sqrt{2}} \begin{pmatrix} \partial_t + \partial_z & \partial_x - \mathrm{i}\partial_y \\ \partial_x + \mathrm{i}\partial_y & \partial_t - \partial_z \end{pmatrix},$$

where the ∂s are partial derivatives with respect to inertial coordinates.

Null vectors. A four-vector X is null if and only if $\det(X^{AA'}) = 0$. That is, if and only if $X^{AA'} = \alpha^A \beta^{A'}$ for some $\alpha \in \mathcal{S}$ and $\beta \in \overline{\mathcal{S}}$, with $\beta = \overline{\alpha}$ whenever X is real and future pointing.

The alternating tensor. The alternating tensor is defined by $e_{abcd} = e_{[abcd]}$, with $e_{0123} = 1$. Its spinor equivalent is

$$\mathrm{i}\epsilon_{AB}\epsilon_{CD}\epsilon_{A'C'}\epsilon_{B'D'} - \mathrm{i}\epsilon_{AC}\epsilon_{BD}\epsilon_{A'B'}\epsilon_{C'D'}.$$

Bivectors. The complex bivectors are the two-index contravariant tensors F^{ab}. They have spinor equivalents of the form

$$F^{ab} \sim \phi^{AB}\epsilon^{A'B'} + \psi^{A'B'}\epsilon^{AB},$$

where ϕ^{AB} and ψ^{AB} are symmetric, with $\psi^{A'B'} = \overline{\phi}^{A'B'}$ whenever F is real. The *dual* bivector $F^{*ab} = \frac{1}{2}e^{abcd}F_{cd}$ is the tensor equivalent of

$$-\mathrm{i}\phi^{AB}\epsilon^{A'B'} + \mathrm{i}\psi^{A'B'}\epsilon^{AB};$$

F is *self-dual* ($F^* = \mathrm{i}F$) whenever $\phi = 0$, and *anti-self-dual* ($F^* = -\mathrm{i}F$) whenever $\psi = 0$.

Symmetric spinors. Let $\phi_{AB\ldots C}$ be a totally symmetric spinor with p indices, so that $\phi_{AB\ldots C} = \phi_{(AB\ldots C)}$. Put

$$\phi(x, y) = \phi_{(AB\ldots C)}\zeta^A\zeta^B \ldots \zeta^C \quad \text{where} \quad (\zeta^A) = \begin{pmatrix} x \\ y \end{pmatrix}.$$

Then $\phi(x, y)$ is a homogeneous polynomial of degree p in the complex variables x, y. Consequently it can be written as a product of p linear factors in the form

$$\phi(x, y) = (\alpha_0 x + \alpha_1 y)(\beta_0 x + \beta_1 y) \ldots (\gamma_0 x + \gamma_1 y).$$

It follows that $\phi_{AB\ldots C} = \alpha_{(A}\beta_B \cdots \gamma_{C)}$, or, in other words, that any totally symmetric spinor can be written as a symmetric product of one-index spinors. The factors are unique up to proportionality and permutation.

6.4 The Lorentz group

The proper orthochronous Lorentz transformations act on $\mathcal{S}$ by the corresponding $SL(2, \mathbb{C})$ matrices. There is a sign ambiguity in the choice of $(L^A{}_B)$ assigned to a given Lorentz matrix, which is significant in contexts where the phases of the spinors are important. At the classical level, however, we shall be interested mainly in the action on the Riemann sphere $\mathbb{P}\mathcal{S}$, where the sign is immaterial.

The full Lorentz group has four components, and is generated by the proper orthochronous transformations, which preserve the space-time orientation and the direction of time, and which make up the identity component, together with space reflections and time reversals. To understand the action of time-reversing and improper transformations, we shall consider the geometric interpretation of a spinor $\alpha \in \mathcal{S}$ in terms of a 'null flag' (Penrose 1968a, Penrose and Rindler 1984). From α, we can construct a bivector

$$N^{ab} \sim \alpha^A \alpha^B \epsilon^{A'B'} + \overline{\alpha}^{A'} \overline{\alpha}^{B'} \epsilon^{AB}. \tag{6.4.1}$$

This is real and null in the sense that $N^{ab}N_{ab} = N^{ab}N^*_{ab} = 0$. It is therefore of the form $N^{ab} = 2n^{[a}m^{b]}$, where n is a future-pointing null four-vector and m is spacelike and orthogonal to n. If we normalize m by imposing $m^a m_a = -2$, then n is uniquely determined by α, and m is determined up to the addition of a multiple of n. In fact $n^a \sim \alpha^A \overline{\alpha}^{A'}$.

Conversely, any four-vectors n and m that satisfy these conditions determine a null bivector of the form (6.4.1), and hence, up to sign, a spinor $\alpha \in \mathcal{S}$. We can therefore picture a spinor in terms of Penrose's null flag (Fig. 6.1). The 'flag pole' is the null vector n; and the 'flag' is the half-plane spanned by n and the positive multiples of m. The flag can be rotated about n (one dimension in suppressed in the diagram), an operation that corresponds to mutiplication of α by a complex number of unit modulus. However the flag plane rotates through *twice* the argument of the complex factor, and returns to its original configuration when $\alpha \mapsto -\alpha$. The picture does not distinguish between a spinor and its negative.

The null bivector N can be written[1]

$$(N^{ab}) = \begin{pmatrix} 0 & -u_1 & -u_2 & -u_3 \\ u_1 & 0 & -v_3 & v_2 \\ u_2 & v_3 & 0 & -v_1 \\ u_3 & -v_2 & v_1 & 0 \end{pmatrix},$$

where $\mathbf{u} \cdot \mathbf{u} = \mathbf{v} \cdot \mathbf{v}$ and $\mathbf{u} \cdot \mathbf{v} = 0$. If we express n in terms of its spatial and temporal parts as $n = (n^0, \mathbf{n})$ and use the freedom in m to bring it into the

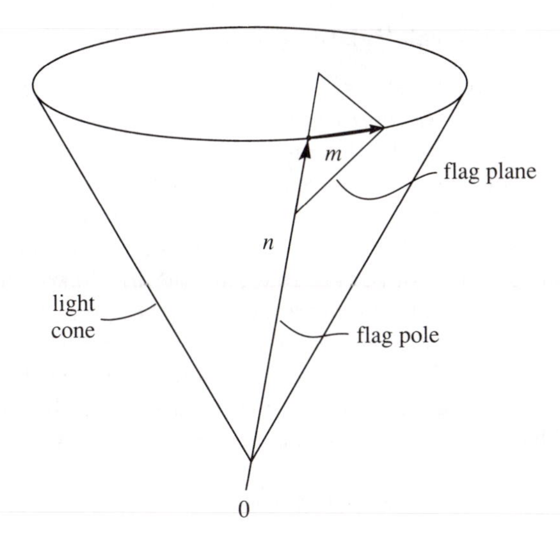

Fig. 6.1. A null bivector.

form $(0, \mathbf{m})$, then

$$\mathbf{u} = -n^0\mathbf{m} \quad \text{and} \quad \mathbf{v} = \mathbf{m} \wedge \mathbf{n}.$$

In terms of the spinor α, $\mathbf{u}$ and $\mathbf{v}$ are given by

$$\mathbf{u} - \mathrm{i}\mathbf{v} = \left(\alpha^1\alpha^1 - \alpha^0\alpha^0,\ \mathrm{i}(\alpha^1\alpha^1 + \alpha^0\alpha^0),\ 2\alpha^0\alpha^1\right).$$

All Lorentz transformations (not just those in the identity component) map null bivectors to null bivectors. Consequently there is a geometrically determined action of the full Lorentz group on the null flags, and hence, to within a sign ambiguity, on the two component spinors. The time reversal $(t, x, y, z) \mapsto (-t, x, y, z)$ acts on N by $\mathbf{u} \mapsto -\mathbf{u}$, $\mathbf{v} \mapsto \mathbf{v}$, and hence on $\mathcal{S}$ by the *antilinear* transformation $\alpha^A \mapsto \tilde{\alpha}^A$ where

$$\text{either } \tilde{\alpha}^0 = -\overline{\alpha}^1, \ \ \tilde{\alpha}^1 = \overline{\alpha}^0 \ \text{ or } \ \tilde{\alpha}^0 = \overline{\alpha}^1, \ \ \tilde{\alpha}^1 = -\overline{\alpha}^0.$$

That is, $\tilde{\alpha}^A = \pm t^{AA'}\overline{\alpha}_{A'}$, where $(t^a) = (\sqrt{2}, 0, 0, 0)$. The spatial reflection $(t, x, y, z) \mapsto (t, -x, -y, -z)$ induces the same transformation of N and hence of α. Note that the transformations are antilinear since they reverse the sense of rotations about the null flag pole.

Any isometry ρ of Minkowski space determines a Lorentz transformation $\rho_* : V \to V$, and hence (up to sign) a map $\mathcal{S} \to \mathcal{S}$, which will also be denoted by ρ_*. It is linear or antilinear according to whether or not ρ preserves the overall orientation.

6.5 Elementary relativistic systems

The elementary relativistic systems—the phase spaces of the classical elementary particles—are represented by orbits in the dual of the Lie algebra of the Poincaré group. In this and the next two sections we shall look at the geometry of the orbits as a prelude to the derivation of the various relativistic wave equations in Chapter 9. The ideas are taken from Bacry (1967), Penrose (1968b), Souriau (1970), and Arens (1971).

We shall denote Minkowski space by $(\mathbb{M}, g)$, where the metric g has signature $+---$. The Poincaré group is the group G of isometries $\rho : \mathbb{M} \to \mathbb{M}$, with the group law defined so that G acts on $\mathbb{M}$ on the right. It is generated by the Lorentz transformations (including the improper and time-reversing transformations) and the translations. Its Lie algebra is the space $\mathcal{G}$ of Killing vector fields. That is vector fields $X \in V(\mathbb{M})$ such that

$$\mathcal{L}_X g_{ab} = \nabla_a X_b - \nabla_b X_a = 0.$$

Such vector fields are of the form $X^a = x^b L_b{}^a + T^a$, where the xs are inertial coordinates, and L and T are constants, with $L_{ab} = -L_{ba}$; L generates infinitesimal Lorentz transformations and T generates translations. We can therefore represent $\mathcal{G}$ as the set of pairs (L_{ab}, T^c). The representation depends, however, on the choice of coordinates. Under Lorentz transformations of the xs, L_{ab} and T^c behave as tensors; but if we move the origin, by making the coordinate transformation $x^a \mapsto \tilde{x}^a = x^a - K^a$, where K is a constant four-vector, then we must replace (L_{ab}, T^c) by

$$(\tilde{L}_{ab}, \tilde{T}^c) = (L_{ab}, T^c + K^d L_d{}^c).$$

The dual space $\mathcal{G}^*$ is the set of pairs (M^{ab}, p_c), with $M^{ab} = -M^{ba}$, with the pairing with $\mathcal{G}$ given by

$$\tfrac{1}{2}\mathrm{tr}(ML) - p(T) = -\tfrac{1}{2}M^{ab}L_{ab} - p_a T^a. \tag{6.5.1}$$

Again M and p behave as tensors under Lorentz transformations, but under translations, we must adopt the transformation rule

$$(M^{ab}, p_c) \mapsto (\tilde{M}^{ab}, \tilde{p}_c) = (M^{ab} + p^a K^b - K^a p^b,\ p_c), \tag{6.5.2}$$

in order to preserve the pairing; that is in order to have

$$\tfrac{1}{2}\tilde{M}^{ab}\tilde{L}_{ab} + \tilde{p}_a \tilde{T}^a = \tfrac{1}{2}M^{ab}L_{ab} + p_a T^a.$$

Setting aside some not entirely straightforward issues concerning parity and time reversal, p and M are the four-momentum and the total angular momentum about the origin. Eqn (6.5.2) is the usual transformation rule for angular momentum under change of origin.

We can describe $\mathcal{G}^*$ in an intrinsic way as the space of tensor fields on $\mathbb{M}$ of the form[2]

$$f^{ab} = M^{ab} + p^a x^b - x^a p^b. \qquad (6.5.3)$$

That is, the value of f at $x \in \mathbb{M}$ is the total angular momentum about x. We recover M and p from f by $M = f(0)$ and $p_a = \frac{1}{3}\nabla_b f_a{}^b$. Note that $\nabla_b f_a{}^b$ is constant, so the expression for p can be evaluated at any point of $\mathbb{M}$. The pairing between $\mathcal{G}^*$ and $\mathcal{G}$ is

$$f(X) = -\tfrac{1}{2} f^{ab}\nabla_a X_b - \tfrac{1}{3} X_a \nabla_b f^{ab}.$$

Again, the right-hand side is constant, and so can be evaluated anywhere.

By thinking of the elements of $\mathcal{G}$ as vector fields and the elements of $\mathcal{G}^*$ as tensor fields, we avoid having to specify an origin. We are also able to express the adjoint action of $\rho \in G$ on $\mathcal{G}$ and the coadjoint action on $\mathcal{G}^*$ very simply as

$$\rho X = (\rho_*^{-1})(X), \qquad f\rho = \rho_*(f).$$

In particular, the coadjoint orbits are sets of tensor fields on $\mathbb{M}$ of the form $\{\rho_* f \mid \rho \in G_0\}$, where G_0 is the identity component of G; that is, the subgroup generated by the translations and the proper orthochronous Lorentz transformations.

Two invariants are used in the classification of the orbits. The first is the rest mass m, given by $m^2 = p_a p^a$. The second is constructed from the Pauli-Lubanski vector

$$S^a = \tfrac{1}{2} e^{abcd} p_b M_{cd}$$

(note that S is independent of the choice of origin). When $m^2 > 0$, S is orthogonal to p, and the second invariant is the spin s, which is defined by $m^2 s^2 = -S^a S_a$. The orbits on which $m^2 < 0$ correspond to inadmissible particles with imaginary rest mass. Those with $m = 0$ and $S^a S_a \neq 0$ are also unphysical: they are the classical analogues of the continuous spin representations of the Poincaré group and they correspond to elementary particles of a type not observed in nature (for example, see Bogolubov, Logonov, and Todorov 1975). The remaining orbits are those for which $m = 0$ and $S^a = sp^a$ for some s. The second invariant in this case is the scalar s, which is now called the *helicity*. The use of the same symbol for spin and helicity should not cause confusion since the meaning will be clear from the context. Note, however, that spin must be nonnegative, while helicity can be of either sign, or zero.

6.6 Massive particles

When the rest mass is positive, there are two orbits for each m and s. They are distinguished by the direction of p, which can be either future or past pointing.

Massive scalar particles

In the scalar case, $s = 0$ and the Pauli-Lubanski vector vanishes. Consequently $p^{[a}f^{bc]} = 0$ everywhere in $\mathbb{M}$, and therefore f must be of the form $f^{ab} = 2p^{[a}w^{b]}$ for some four-vector field w. By comparing with (6.5.3), we see that there must exist constants q^a such that

$$w^a - x^a + q^a$$

is everywhere proportional to p^a. There is therefore a unique timelike geodesic on which $f = 0$. It is called the centre-of-mass worldline, and it is given by $x^a = q^a + \lambda p^a$, where λ is a parameter.

If we know the centre-of-mass worldline and the value of $m > 0$, then we can recover p and f up to sign: p is the tangent four-vector, with the normalization $p_a p^a = m^2$, and

$$f^{ab} = 2p^{[a}x^{b]} - 2p^{[a}q^{b]},$$

where the qs are the coordinates of any point on the worldline. Since any timelike geodesic can be transformed into any other by an element of G_0, there are just two zero-spin orbits for each $m > 0$. They are denoted by M^+_{0m} and M^-_{0m}; p is future-pointing on M^+_{0m}, and past-pointing on M^-_{0m}.

We can parametrize M^+_{0m} and M^-_{0m} by the qs and the covariant components of p, subject to the equivalence relation

$$(p_a, q^b) \sim (p_a, q^b + \lambda p^b), \tag{6.6.1}$$

for any $\lambda \in \mathbb{R}$. The two orbits are six-dimensional: they are the quotients of the two components of the hypersurface

$$C_m = \{p_a p^a = m^2\}$$

in $T^*\mathbb{M} = \{(p, q)\}$ by $\sim$. The coadjoint action of $\rho \in G$ is given by

$$q \mapsto \rho(q), \quad p \mapsto \rho_*(p). \tag{6.6.2}$$

We shall derive the symplectic structures by using the method described at the end of §3.5 (p. 55). The action of G on $\mathbb{M}$ lifts to $T^*\mathbb{M}$, where it preserves the constraint $p_a p^a = m^2$. Therefore each $X \in \mathcal{G}$ lifts to a vector field

$$X' = X^a \frac{\partial}{\partial q^a} - p_b \nabla_a X^b \frac{\partial}{\partial p_a}$$

on $T^*\mathbb{M}$, which is tangent to C_m and which generates the action of the corresponding one-parameter subgroup on $T^*\mathbb{M}$. Since the action of G on $T^*\mathbb{M}$ coincides with (6.6.2), X' projects from C_m onto the Hamiltonian vector field generated by X on M^+_{0m} and M^-_{0m}. By using (6.5.1), and by evaluating $f(X)$ at an event q on the centre-of-mass worldline, at which $f^{ab} = 0$ and $\frac{1}{3}\nabla_b f_a{}^b = p_a$,

$$f(X) = -p_a X^a(q) = X' \lrcorner\, \theta',$$

where $\theta' = -p_a \mathrm{d}q^a|_{C_m}$. It follows that M^+_{0m} and M^-_{0m} are the two components of the reduction of the constraint hypersurface $p_a p^a = m^2$ in $(T^*\mathsf{M}, -\omega)$, where $\omega = \mathrm{d}p_a \wedge \mathrm{d}q^a$ is the canonical 2-form.

The origin of the minus sign can be understood by looking at the nonrelativistic limit, in which the covariant vector p has spatial part $-m\mathbf{v}$, where $\mathbf{v}$ is the velocity. In this limit, $-\omega$ coincides with the symplectic form $\tilde{w}$ on the cotangent bundle of the space of events in §1.8, provided that we take Q to be Euclidean space, and identify $m\mathbf{v}$ with the nonrelativistic momentum, and p_0 with $h = -s$.

The vertical polarization of $T^*\mathsf{M}$ does not survive reduction, but both M^+_{0m} and M^-_{0m} have invariant real polarization with leaves $p_a = $ constant. The spaces of leaves are the future and past components of the 'mass shell' $p_a p^a = m^2$ in the four-dimensional space of momentum vectors, each of which is diffeomorphic to $\mathbb{R}^3$. The leaves are simply-connected and complete, so the orbits can be identified as symplectic manifolds with the cotangent bundles of the mass shells, but not in a way that respects the action of G.

Massive particles with spin

When m and s are both positive, the equation $f^{ab} = 0$ no longer has real solutions: if a particle has spin, then it has nonzero angular momentum about every event. But $f^{ab}p_b$ vanishes on the timelike geodesic

$$x^a = m^{-2} M^{ab} p_b + \lambda p^a.$$

We define this to be the centre-of-mass worldline, and interpret $f^{ab}p_b = 0$ as the condition that the orbital part of the angular momentum should vanish.

The tensor field f^{ab} is constant along the centre of mass worldline, where it is equal to

$$m^{-2} e^{abcd} p_c S_d.$$

If we fix p and an event q at the centre-of-mass, then the only remaining freedom in f is the direction of S, which can be any four-vector such that $p_a S^a = 0$ and $S_a S^a = -m^2 s^2$, that is, any vector of length ms in the three-space of the centre-of-mass rest frame. Since any such vector can be transformed into any other by a rotation that fixes p and q, there are two orbits for each set of positive values of m and s. They will be denoted by M^+_{sm} and M^-_{sm}, according to whether p is future or past pointing. Both are eight-dimensional and their points are labelled by p, q, and S, subject to the same equivalence as before (6.6.1). They have the structure of sphere bundles over M^+_{0m} and M^-_{0m}. The spheres are the additional 'internal' degrees of freedom represented by the direction of S, the same freedom as a spinning particle at rest (§3.5).

The symplectic forms are derived by introducing a spinor parametrization of the orbits and by using the same method as in the scalar case . The spinor equivalent of f^{ab} on the centre-of-mass worldline can be written

$$f^{ab} \sim \mathrm{i}s z^{(A}w^{B)}\epsilon^{A'B'} - \mathrm{i}s\overline{z}^{(A'}\overline{w}^{B')}\epsilon^{AB},$$

where z^A and w^A are constant. The spinors z and w are determined by f^{ab} up to $z \mapsto \lambda z$, $w \mapsto \lambda^{-1}w$. By using this freedom and the conditions $f^{ab}p_b = 0$ and $S^aS_a = -m^2s^2$, we can impose

$$w^A = \pm\frac{\sqrt{2}}{m}p^{AA'}\overline{z}_{A'} \quad \text{and} \quad z_Aw^A = 1,$$

where the sign is + or − according to whether p is future or past pointing. Thus z is determined by f up to phase; and f is determined by p, q and z.

The orbits are parametrized by p_a, the coordinates q^a of an event at the centre of mass, and z^A, subject to the identification

$$(p_a, q^b, z^C) \sim (p_a, q^b + rp^b, \mathrm{e}^{\mathrm{i}\phi}z^C)$$

where r and ϕ are real. The coadjoint action of $\rho \in G$ is given by $q \mapsto \rho(q)$, $p \mapsto \rho_*(p)$, $z \mapsto \rho_*(z)$.

This time we have a projection from the two connected components of

$$C_{sm} = \{p_ap^a = m^2,\ \sqrt{2}p_{AA'}z^A\overline{z}^{A'} = \pm m\} \subset \mathsf{T}^*\mathsf{M} \times \mathcal{S}$$

onto M^+_{sm} and M^-_{sm} ($\mathcal{S}$ is the space of two component spinors).

Again the action of G lifts to $T^*\mathsf{M} \times \mathcal{S}$, where it preserves C_{sm}, and projects onto the coadjoint action on the orbits. Each $X \in \mathcal{G}$ lifts to a vector field

$$X' = X^a\frac{\partial}{\partial q^a} - p_b\nabla_aX^b\frac{\partial}{\partial p_a} + z^A\Lambda_A^{\ B}\frac{\partial}{\partial z^B} + \overline{z}^{A'}\overline{\Lambda}_{A'}^{\ B'}\frac{\partial}{\partial \overline{z}^{B'}}$$

on $T^*\mathsf{M} \times \mathcal{S}$, where

$$\Lambda_A^{\ B} = \tfrac{1}{2}\nabla_{AB'}X^{BB'}.$$

For each $X \in \mathcal{G}$, X' is tangent to C_{sm} and projects onto the Hamiltonian vector field generated by X on M^+_{sm} and M^-_{sm}. Moreover, $f(X) = X' \lrcorner\, \theta'$, where θ' is the restriction to C_{sm} of

$$\pm\frac{\sqrt{2}\mathrm{i}s}{m}p_{AA'}(z^A\mathrm{d}\overline{z}^{A'} - \overline{z}^{A'}\mathrm{d}z^A) - p_a\mathrm{d}q^a.$$

Therefore, by another application of the method of §3.5 (p. 55), M^+_{sm} and M^-_{sm} are the two components of the reduction of $(C_{sm}, \mathrm{d}\theta')$. The plus sign is taken when p is future pointing. Consequently, the orbits are obtained by treating C_{sm} as a first class constraint in $T^*\mathbb{M} \times \mathcal{S}$, and reducing with respect to the symplectic structure

$$\omega' = \pm \frac{\sqrt{2}\,\mathrm{i}s}{m}(p_{AA'}\mathrm{d}z^A \wedge \mathrm{d}\overline{z}^{A'} + \overline{z}^{A'}\mathrm{d}z^A \wedge \mathrm{d}p_{AA'}) + \mathrm{cc} - \mathrm{d}p_a \wedge \mathrm{d}q^a, \quad (6.6.3)$$

where 'cc' stands for 'complex conjugate of the preceding terms'.

The restriction of the symplectic structure to the sphere in M^+_{sm} on which $q = 0$, $p = (m, 0, 0, 0)$ is the reduction of the constraint $z^0\overline{z}^0 + z^1\overline{z}^1 = 1$ in $\mathbb{C}^2$, with respect to the symplectic form

$$2\mathrm{i}s(\mathrm{d}z^0 \wedge \mathrm{d}\overline{z}^0 + \mathrm{d}z^1 \wedge \mathrm{d}\overline{z}^1),$$

which is the same as the phase space of a particle of spin s at rest (p. 54).

The orbits M^+_{sm} and M^-_{sm} have invariant nonnegative polarizations, which are the reductions of the polariztion of $T^*\mathbb{M} \times \mathcal{S}$ spanned by the vector fields $\partial/\partial q^a$ and $\partial/\partial z^A$. They combine the constant momentum polarization for a scalar particle with the holomorphic polarization for a spinning particle at rest (p. 105).

6.7 P, C, and T

To understand relativistic quantization, we must consider not only the infinitesimal generators of the Poincaré group, but also the discrete symmetries, parity and time reversal. We must consider how the non-identity components of G should act on the coadjoint orbits.

In the scalar case, there is an obvious choice, namely the coadjoint action itself, under which $f \mapsto \rho_*(f)$, $q \mapsto \rho(q)$, $p \mapsto \rho_*(p)$. This is canonical for all $\rho \in G$, and interchanges M^+_{0m} and M^-_{0m} whenever ρ reverses time. Unfortunately, it does not lead to the correct action of G at the quantum level.

The manifolds M^+_{0m} and M^-_{0m} are the classical phase spaces of a particle of rest mass m and its associated antiparticle; and so, after quantization, they should go over to the corresponding single-particle wave-function spaces. However, both the particle and the antiparticle must be physically realizable in the sense that they must both have positive energy. At the quantum level, the energy operator relative to some inertial observer is the generator of a one-parameter group of time translations; and this will be positive only if the corresponding classical family of canonical transformations is generated by a strictly positive function on both M^+_{0m} and M^-_{0m}.

At this point, we run into an annoying consequence of standard conventions. In order to obtain the energy operator

$$E = \mathrm{i}\hbar \frac{\partial}{\partial t}$$

at the quantum level, it is necessary to identify the classical energy E_u measured by an observer with four-velocity u with the generator of the one-parameter group of translations into the *past*:

$$x^a \mapsto x^a - tu^a .$$

This same awkwardness occurs in nonrelativistic mechanics. In the notation of §1.8, the generator of $\partial/\partial t$ on extended phase space is s, which is equal to $-h$ on the constraint manifold $h + s = 0$. We should therefore identify the nonrelativistic energy with $-s$, which generates translations into the past.

The energy E_u on the relativistic orbits is the generator of the Hamiltonian vector field

$$X = -u^a \frac{\partial}{\partial q^a} .$$

On M^+_{0m}, the energy is the positive function $E_u = u^a p_a$. On M^-_{0m}, however, it is negative. To obtain a positive generator when p is past pointing, we must reverse the symplectic structure on M^-_{0m} and take the classical phase space of the particle and its antiparticle to be

$$M_{0m} = M^+_{0m} \cup \overline{M}^-_{0m},$$

where, if (M, ω) is a symplectic manifold, $\overline{M}$ is a shorthand notation for the symplectic manifold $(M, -\omega)$. The symplectic form on M_{0m} is the reduction of

$$-\mathrm{d}p_a \wedge \mathrm{d}q^a \ \text{ for } \ p_0 > 0, \ \text{ and } \ \mathrm{d}p_a \wedge \mathrm{d}q^a \ \text{ for } \ p_0 < 0,$$

and the energy is $E_u = |u^a p_a|$.

There are still a number of ways of making G act on M_{0m} since it is always possible to combine time reversals or parity transformations with the *charge conjugation* symmetry $C : f \mapsto -f$, a canonical transformation of M_{0m} which interchanges particle and antiparticle. It does not arise from an isometry of $\mathbb{M}$. In coordinates,

$$C : (p_a, q^b) \mapsto (-p_a, q^b).$$

Whatever the choice, M_{0m} is an elementary system for the enlarged group $G \times \mathbb{Z}_2$ generated by G and C. The important issue is not the labelling of the elements of $G \times \mathbb{Z}_2$, but the identification of the symmetries that survive quantization.

The choice implied by the standard conventions of relativistic quantum field theory is to make G act on M_{0m} by

$$\rho : f \mapsto \xi \rho_*(f), \tag{6.7.1}$$

where $\xi = 1$ or $\xi = -1$ according to whether or not ρ reverses the direction of time. Then every $\rho \in G$ maps M^+_{0m} to M^+_{0m} and M^-_{0m} to M^-_{0m}, with transformations that reverse time acting anticanonically (they become antiunitary at the quantum level). Space reflections that preserve the direction of time still act canonically. Only C interchanges particle and antiparticle.

One can understand the sense in which C 'conjugates charge' by introducing an external electromagnetic field F_{ab} and replacing the canonical 2-form on $T^*\mathbb{M}$ by

$$\omega_F = \mathrm{d}p_a \wedge \mathrm{d}q^a + \frac{e}{2} F_{ab} \mathrm{d}q^a \wedge \mathrm{d}q^b .$$

One then obtains a new two-component symplectic manifold, the phase space of a charged particle and its antiparticle, by imposing the constraint $p_a p^a = m^2$ and reducing the symplectic form $-\omega_F$ for $p_0 > 0$, or ω_F for $p_0 < 0$. That is by factoring out the flow of

$$v^a \frac{\partial}{\partial q^a} + eF_{ab} v^b \frac{\partial}{\partial p_a}$$

on the constraint hypersurface, where $mv^a = \pm p^a$, with the sign chosen to make v^a future pointing. The integral curves of this vector field project into $\mathbb{M}$ as the solution curves of

$$m\ddot{x}^a = eF^{ab}\dot{x}_b \ \ (p_0 > 0), \quad \text{or} \quad m\ddot{x}^a = -eF^{ab}\dot{x}_b \ \ (p_0 < 0),$$

which are orbits under the Lorentz force law. Thus including F in this way is equivalent to giving the particle charge e and its antiparticle charge $-e$. When $e \neq 0$, $C : f \mapsto -f$ is a symmetry of the reduced phase space only if it is accompanied by a charge reversal $e \mapsto -e$.

Much the same considerations apply when $s \neq 0$. The system then consists of a massive particle with spin, and its associated antiparticle, with the classical phase space

$$M_{sm} = M^+_{sm} \cup \overline{M}^-_{sm} .$$

The action of a general isometry is given by (6.7.1). Again it is canonical or anticanonical according to whether or not ρ reverses the arrow of time; and again M^+_{sm} and $\overline{M}^-_{sm}$ separately form elementary systems for G.

In terms of the parametrization p, q, z, the identity component of G acts by

$$p \mapsto \rho_*(p), \quad q \mapsto \rho(q), \quad z \mapsto \rho_*(z)$$

(the sign ambiguity in $\rho_*(z)$ disappears on reduction). When ρ reverses the temporal but not the spatial orientation,

$$p \mapsto -\rho_*(p), \quad q \mapsto \rho(q), \quad z \mapsto -\rho_*(z).$$

For example, the time reversal,

$$T : (x^0, x^1, x^2, x^3) \mapsto (-x^0, x^1, x^2, x^3)$$

generates the the anticanonical transformation

$$T \begin{cases} (p_0, p_1, p_2, p_3) \mapsto (p_0, -p_1, -p_2, -p_3) \\ (q^0, q^1, q^2, q^3) \mapsto (-q^0, q^1, q^2, q^3) \\ (z^0, z^1) \mapsto (-\overline{z}^1, \overline{z}^0). \end{cases}$$

It is slightly harder to derive the effect of parity

$$P : (x^0, x^1, x^2, x^3) \mapsto (x^0, -x^1, -x^2, -x^3),$$

but the result is the canonical transformation

$$P \begin{cases} (p_0, p_1, p_2, p_3) \mapsto (p_0, -p_1, -p_2, -p_3) \\ (q^0, q^1, q^2, q^3) \mapsto (q^0, -q^1, -q^2, -q^3) \\ (z^0, z^1) \mapsto (-\overline{w}^1, \overline{w}^0). \end{cases}$$

where $w^A = m^{-1}\sqrt{2}p^{AA'}\overline{z}_{A'}$. Together with the elements of G_0, these account for the entire action of the Poincaré group.

The remaining discrete symmetry, the charge conjugation $C : f \mapsto -f$, is

$$C \begin{cases} p \mapsto -p \\ q \mapsto q \\ z \mapsto -w \end{cases}.$$

Again this can be understood as a charge reversal by including an external electromagnetic field in the symplectic structure. However the details are rather complicated because of the need to add to the rest mass a correction term representing the rest energy of a dipole in an external electromagnetic field (Souriau 1970, Künzle 1972).

Before looking at massless particles, it is worth pausing briefly to consider what we have done so far in the light of the general theory of §3.6. The two components M^+_{sm} and $\overline{M}^-_{sm}$ of M_{sm} are not essentially different. In fact charge conjugation is a natural canonical diffeomorphism from M^+_{sm} to $\overline{M}^-_{sm}$. Moreover the moments determined by the actions of G_0 map both

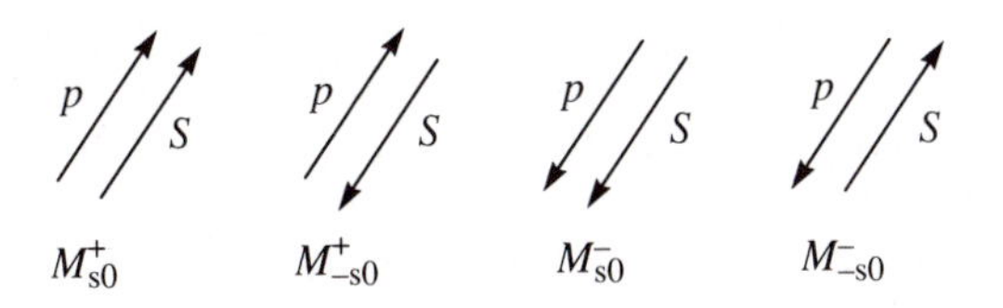

Fig. 6.2. The four components of a massless coadjoint orbit of the Poincaré group.

M^+_{sm} and $\overline{M}^-_{sm}$ onto M^+_{sm}, as an orbit in $\mathcal{G}^*$. This is an immediate consequence of the requirement that energy should be positive. Thus we could equally well describe M_{sm} as the union of two copies of M^+_{sm}. This would have the advantage of emphasizing that there is no fundamental distinction between particles and antiparticles: taking the electron as fundamental and calling the positron its antiparticle is simply a convention. The disadvantage is that it does not fit in quite as well with some standard conventions of quantum field theory. Our description, on the other hand, has the displeasing feature that the physical four-momentum is p on M^+_{sm} and $-p$ on $\overline{M}^-_{sm}$. There does not seem to be any really satisfactory way out. Some inelegance in the formalism is inevitable once it is realized that the natural way of fitting M^+_{sm} and M^-_{sm} together to form an elementary symplectic manifold on which all of G acts canonically is physically incorrect.

There are, of course, particles that are the same as their antiparticles. For these, we have to factor out the action of C on M_{sm} by identifying f with $C(f) = -f$. Then the phase space is a single copy of M^+_{sm} and the distinction between the two points of view disappears.

6.8 Massless particles

When $m = 0$ and the Pauli-Lubanski vector is proportional to p, there is one orbit of G in $\mathcal{G}^*$ for each positive value of the helicity. Each has four connected components: M^+_{s0}, M^+_{-s0}, M^-_{s0}, and M^-_{-s0}, where the superscript indicates the time orientation of p and the first subscript is the helicity (Fig. 6.2).

For a fixed value of s, $\rho \in G$ interchanges M^+_{s0} and M^+_{-s0} whenever it reverses the spatial orientation, but preserves the arrow of time; it interchanges M^+_{s0} and M^-_{-s0} whenever ρ preserves spatial orientation, but reverses the arrow of time; and interchanges M^+_{s0} and M^-_{s0} whenever ρ reverses both space and time orientations.

One might expect, therefore, the corresponding quantum theory to describe a system with four different states of particle and antiparticle. However, nature does not work like that. For example, in the two most familiar systems, the electromagnetic field and the electron neutrino field, there are just two: the left-handed circular polarization of the photon (which has $s < 0$) and right-handed polarization ($s > 0$); and the left-handed neutrino, and its associated right-handed antineutrino.

The classical phase space M_{s0} that leads to the correct quantum theory is obtained from the union

$$M_{s0}^{+} \cup M_{-s0}^{+} \cup \overline{M}_{s0}^{-} \cup \overline{M}_{-s0}^{-}$$

by factoring out the canonical transformation $f \mapsto -f$, which interchanges $+$ and $-$ and reverses the sign of s. Alternatively, and more simply, we can regard M_{s0} as the union of M_{s0}^{+} and M_{-s0}^{+}, with the action of G given by $\rho : f \mapsto \xi\rho_*(f)$, where, as before, $\xi = 1$ or -1 according to whether or not ρ preserves the arrow of time. Either way, M_{s0} is a two-component symplectic manifold on which G acts transitively, with time-reversing isometries generating anticanonical transformations.

When $s > 0$, M_{s0}^{+} is interpreted as the phase space of a right-hand postive helicity particle, and M_{-s0}^{+} as the phase space of a left-handed negative helicity particle. Superficially, this is very like the massive case. The differences are that, first, the two components of the classical phase space are not canonically diffeomorphic in any natural way; the transformation $f \mapsto -f$ which in the massive case gave rise to charge conjugation now reduces to the identity. This reflects the fact that, even in the absence of interactions, the helicity or 'handedness' allows a geometric distinction between particle and antiparticle. Second, transformations that reverse the orientation of space, but preserve the orientation of time, interchange the two components: a neutrino observed through a mirror looks like an antineutrino, and not simply another state of the same particle.

The geometry of the two components of $M_{s0} = M_{s0}^{+} \cup M_{-s0}^{+}$ (for positive or negative s) can be unravelled by the same techniques as in the massive case. The centre of mass is no longer a single worldline, but an entire null hyperplane tangent to p, since $f^{ab}p_b = 0$ whenever

$$M^{ab}p_b + p^a x^b p_b = 0$$

(this equation has solutions because S^a is proportional to p^a; and if $x^a = q^a$ is a solution, then so is $x^a = q^a + k^a$ whenever $p_a k^a = 0$). Thus a classical massless elementary particle with nonvanishing helicity can only be localized to a single wave front and does not have an invariant centre-of-mass worldline.

Following Penrose and Rindler (1986), we begin by writing $p_{AA'} = \overline{\pi}_A \pi_{A'}$. Then the condition that S^a should be parallel to p^a forces f^{ab} to be of the form

$$f^{ab} \sim \mathrm{i}z^{(A}\overline{\pi}^{B)}\epsilon^{A'B'} - \mathrm{i}\overline{z}^{(A'}\pi^{B')}\epsilon^{AB},$$

where z is a spinor field such that

$$z^A\overline{\pi}_A + \overline{z}^{A'}\pi_{A'} = 2s.$$

Eqn (6.5.3) forces $z(x)$ to be of the form

$$z^A = \omega^A - \mathrm{i}x^{AA'}\pi_{A'} \tag{6.8.1}$$

where ω^A is constant (ω^A is the value of z^A at the origin). The centre-of-mass hyperplane is given by $z^A\overline{\pi}_A = s$; or in other words, by $x^a = q^a$, where

$$q^a p_a = \frac{\omega^A\overline{\pi}_A - \overline{\omega}^{A'}\pi_{A'}}{2\mathrm{i}}.$$

If we know ω^A and $\pi_{A'}$, then we can determine p_a, f^{ab} and the centre-of-mass hyperplane. Conversely, f^{ab} determines ω^A and $\pi_{A'}$ up to phase. Therefore M_{s0}^+ can be parametrized by the four components of the two spinors ω and π, subject to the equivalence

$$(\omega^A, \pi_{A'}) \sim (\mathrm{e}^{\mathrm{i}\phi}\omega^A, \mathrm{e}^{\mathrm{i}\phi}\pi_{A'})$$

for real ϕ. So if $\mathbb{T}$ (for 'twistor space') denotes the four-dimensional complex vector space on which ω^A and $\pi_{A'}$ are independent linear coordinates, then factoring out the phase gives a projection from

$$C_{s0} = \{w^A\overline{\pi}_{A'} + \overline{\omega}^{A'}\pi_{A'} = 2s\} \tag{6.8.2}$$

onto M_{0s}^+.

The construction extends to the case $s = m = 0$ by taking M_{00} to be the union of two copies of M_{00}^+, the orbit of G_0 on which $s = 0$ and p is future pointing null; or, equivalently, $M_{00} = M_{00}^+ \cup \overline{M}_{00}^-$. The parametrization is the same, with $s = 0$. The only substantial difference is that one of the null geodesics in the centre-of-mass hyperplane is singled out by the vanishing of f^{ab}, so the particle is localized on a single null geodesic worldline.

Proper orthochronous Lorentz transformations act on M_{s0}^+ by the corresponding $SL(2,\mathbb{C})$ transformations of ω^A and $\pi_{A'}$. From (6.8.1), translation through T^a acts by $\omega^A \mapsto \omega^A + \mathrm{i}T^{AA'}\pi_{A'}$, leaving $\pi_{A'}$ unchanged. At the infinitesimal level, therefore, the flow of $X \in \mathcal{G}$ is generated by the projection from C_{s0} of

$$X' = \omega^A\Lambda_A^{\ B}\frac{\partial}{\partial\omega^B} - \pi_{A'}\overline{\Lambda}^{A'}_{\ B'}\frac{\partial}{\partial\pi_{B'}} + \mathrm{i}T^{AA'}\pi_{A'}\frac{\partial}{\partial\omega^A} + \mathrm{cc},$$

where, $\Lambda_A^{\ B} = \frac{1}{2}\nabla_{AB'}X^{BB'}$, $T^{AA'} = X^{AA'}(0)$ and 'cc' denotes the complex of the preceding terms. Now X' is tangent to C_{s0} and

$$f(X) = -\mathrm{i}\omega^A\overline{\pi}^B\Lambda_{AB} + \mathrm{i}\overline{\omega}^{A'}\pi^{B'}\overline{\Lambda}_{A'B'} - \pi_{A'}\overline{\pi}_{A'}T^{AA'} = X' \lrcorner\, \theta',$$

where θ' is the restriction to C_{s0} of

$$\mathrm{i}\overline{\pi}_A \mathrm{d}\omega^A - \mathrm{i}\pi_{A'}\mathrm{d}\overline{\omega}^{A'} - \mathrm{i}\omega^A \mathrm{d}\overline{\pi}_A + \mathrm{i}\overline{\omega}^{A'}\mathrm{d}\pi_{A'}.$$

Therefore, $\mathrm{d}\theta'$ is the pull-back of the symplectic structure on M^+_{s0}. It follows that, for any s, the orbit M^+_{s0} is the reduction of C_{s0}, as a constraint in $(\mathbb{T}, \Omega)$, where

$$\Omega = -\mathrm{i}\mathrm{d}\omega^A \wedge \mathrm{d}\overline{\pi}_A + \mathrm{i}\mathrm{d}\overline{\omega}^{A'} \wedge \mathrm{d}\pi_{A'}$$

(Crampin and Pirani 1971, Penrose and MacCallum 1972, Tod 1977).

The discrete symmetries P and T act on ω^A and $\pi_{A'}$ by

$$P : (\omega^A, \pi_{A'}) \mapsto (t^{AA'}\overline{\omega}_{A'}, -t_{AA'}\overline{\pi}^A)$$
$$T : (\omega^A, \pi_{A'}) \mapsto (t^{AA'}\overline{\omega}_{A'}, t_{AA'}\overline{\pi}^A)$$

where $(t^a) = (\sqrt{2}, 0, 0, 0)$. The first is canonical, and the second is anti-canonical. It is customary, however, to label the first transformation as CP rather than P. This is because it interchanges particles and antiparticles. In interacting systems it is not a symmetry unless it is accompanied by a CP transformation of the interacting massive particles (and even then it is not always a symmetry).

Another major difference between this and the massive case is that there are two different invariant polarizations of M_{0s}. One (P_1) is the real polarization spanned by the projections of

$$\frac{\partial}{\partial\omega^A} \quad \text{and} \quad \frac{\partial}{\partial\overline{\omega}^{A'}}.$$

Here the space of leaves can be identified with the two halves of the light cone in momentum space and, as we shall see, quantization leads to the conventional relativistic quantum theory based on the Fourier analysis of massless fields. The second is the complex 'twistor polarization' (P_2) spanned by the projections of

$$\frac{\partial}{\partial\pi_{A'}} \quad \text{and} \quad \frac{\partial}{\partial\omega^A}.$$

When $s \neq 0$, P_2 is Kähler, but nonpositive; when $s = 0$, it has one real direction (this is a further example of the way in which nonpositive polarizations can acquire real directions on reduction). Here quantization leads to the twistor description of massless fields (Penrose 1968b).

Only the first polarization is invariant under the full Poincaré group. However, unlike P_1, P_2 is invariant under the identity component of the group of conformal isometries of Minkowski space. This group is isomorphic to $SU(2,2)$, which acts on $\mathbb{T}$ by complex linear transformations that preserve Ω (Penrose and Rindler 1986, Carey and Hannabuss 1978).

6.9 Classical particles in curved space-time

A gravitational field acts on a particle with spin not only through its mass but also through a force that couples the spin with the space-time curvature. The corresponding equations of motion, the 'Papapetrou equations' (Mathisson 1937, Papapetrou 1951), can be derived by constructing a generally covariant phase space for a massive particle with spin (Künzle 1972, Souriau 1974a).

In the flat case,

$$\theta' = \frac{\sqrt{2}\,\mathrm{i}s}{m} p_{AA'}\left(z^A \mathrm{d}\bar{z}^{A'} - \bar{z}^{A'}\mathrm{d}\bar{z}^{A}\right) - p_a \mathrm{d}q^a$$

is a potential for a symplectic structure on $T^*\mathbb{M}\times\mathcal{S}$, and M^+_{sm} is constructed by fixing the values of the functions

$$f_1 = p_{AA'}z^A\bar{z}^{A'}, \quad \text{and} \quad f_2 = -\frac{1}{2m}p_a p^a$$

and factoring out their canonical flows. The flow of f_1 changes the phase of z^A, leaving p and q fixed. The flow of f_2 is the dynamical evolution: q moves along the centre-of-mass geodesic with four-velocity $m^{-1}p^a$, and z^A is transported by parallel propagation.

When $(\mathbb{M}, g)$ is replaced by a curved space-time (Q, g), we can do the same constructuction on the bundle $T^*Q \oplus \mathcal{S}$, where $\mathcal{S} \to Q$ is the bundle of two component spinors.[3] The constraint functions remain well-defined, and θ' can be replaced by

$$\theta' = \frac{\sqrt{2}\,\mathrm{i}s}{m} p_{AA'}\left(z^A \mathrm{D}\bar{z}^{A'} - \bar{z}^{A'}\mathrm{D}z^{A}\right) - p_a \mathrm{d}q^a,$$

where D is the generally covariant exterior derivative.

Aside. This needs some explanation. First the operator D. Let $\pi : V \to Q$ be a vector bundle, with a connection given in local coordinates by

$$\nabla v^\alpha = \mathrm{d}v^\alpha + \Gamma^\alpha{}_\beta v^\beta = \left(\frac{\partial v^\alpha}{\partial q^a} + \Gamma^\alpha{}_{\beta a}v^\beta\right)\mathrm{d}q^a$$

where the vs are linear coordinates on the fibres and the Γs are the connection 1-forms. Then π^*V is a bundle over the total space of V with connection $\mathrm{d} + \pi^*\Gamma$. The operator D acts on differential forms on V with values in π^*V by $\mathrm{D}\xi = \mathrm{d}\xi + (\pi^*\Gamma)\wedge\xi$. That is,

$$\mathrm{D}\xi^\alpha = \mathrm{d}\xi^\alpha + \Gamma^\alpha{}_\beta \wedge \xi^\beta,$$

where we have dropped the distinction between $\pi^*\Gamma^\alpha{}_\beta$, which is a 1-form on V, and $\Gamma^\alpha{}_\beta$, which is a 1-form on Q. For example, $\mathrm{D}v^\alpha$ is the 1-form

$dv^\alpha + \Gamma^\alpha{}_\beta v^\beta$. Unlike the ordinary exterior derivative, $\mathrm{D}^2 \neq 0$, but $\mathrm{D}^2\xi^\alpha = \frac{1}{2}F^\alpha{}_\beta \wedge \xi^\beta$, where F is the curvature 2-form.

We can also define D in the obvious way for forms on V that take values in the various tensor products of π^*V and π^*V^*. For an ordinary real-valued form on V, D and d coincide. If V is the tangent bundle of space-time and ∇ is the metric connection, then $v^a \mathrm{D} v_a = \frac{1}{2}\mathrm{d}(v^a v_a)$.

Second, the 'bundle of two component spinors'. This is constructed by associating a two-dimensional complex symplectic space $\mathcal{S}_q$ with the space of four-vectors at each q in exactly the same way as in Minkowski space. There is the same correspondence between tensors and spinors with equal numbers of primed and unprimed indices, and it is given by the same formulas as in flat space provided that the tensor components are calculated in an orthonormal frame at each event. We shall follow the usual practice, however, of taking the tensor components in a coordinate basis and the spinor components in some symplectic frame. Then the relationship between, for example, the covector components p_a and their spinor equivalent $p_{AA'}$ is more complicated, but need never be written out explicitly.

Each $\alpha \in \mathcal{S}_q$ determines a null bivector N at q; and conversely, each null bivector determines, up to sign, a spinor α. Parallel transport of spinors is defined by parallel transport of the corresponding null bivectors. Hence the covariant derivative

$$\nabla_c \alpha^A = \frac{\partial \alpha^A}{\partial q^c} + \Gamma^A{}_{Bc}\alpha^B$$

is well defined. We shall not need to know the Γs explicitly—they depend both on the choice of coordinates on Q and on the choice of basis in the individual spaces $\mathcal{S}_q$. The curvature form 2-form $F^A{}_{Bcd}$, which is defined by

$$2\nabla_{[c}\nabla_{d]}\alpha^A = F^A{}_{Bcd}\alpha^B,$$

is related to the space-time curvature by

$$\alpha^A \alpha^B F_{ABcd} + \overline{\alpha}^{A'}\overline{\alpha}^{B'}\overline{F}_{A'B'cd} = -\tfrac{1}{2}R_{abcd}N^{ab}.$$

This can be deduced by comparing the holonomy of α and N around small loops in Q. The covariant derivative extends in an obvious way to multi-index spinors, so as to preserve the conjugations. The symplectic forms ϵ are covariantly constant, so the order of covariant differentiation and of raising, lowering, and contracting can be interchanged.

The only additional difficulty that is not present in flat space-time is that it may not be possible to make a consistent choice for the sign of α when N is carried around a loop in Q: spinors can be defined only if certain global toplogical conditions are satisfied (Penrose and Rindler 1984). ∎

By operating on θ' with D and by using

$$\mathrm{D}^2 z^A = \tfrac{1}{2} z^B F^A{}_{Bcd} \mathrm{d}q^c \wedge \mathrm{d}q^d,$$

we deduce that

$$\begin{aligned}\mathrm{d}\theta' = {} & \frac{\sqrt{2}\,\mathrm{i}s}{m}\left(z^A \mathrm{D}p_{AA'} \wedge \mathrm{D}\overline{z}^{A'} + p_{AA'}\mathrm{D}z^A \wedge \mathrm{D}\overline{z}^{A'}\right) \\ & -\frac{\mathrm{i}s}{2} w^A z^B F_{ABcd}\mathrm{d}q^c \wedge \mathrm{d}q^d + \mathrm{cc} - \mathrm{d}p_a \wedge \mathrm{d}q^a.\end{aligned}$$

where $w^A = \sqrt{2}\, m^{-1} p^{AA'} \overline{z}_{A'}$. The Hamiltonian vector field of f_1 has integral curves of the form

$$t \mapsto \left(p_a(t), q^b(t), z^A(t)\right).$$

If we let $v^a = \dot{q}^a$, and use a dot also to denote the covariant derivative $v^a \nabla_a$, then $v^a - m^{-1} p^a$ is the vector equivalent of

$$\frac{\sqrt{2}\,\mathrm{i}s}{m}\left(\dot{\overline{\alpha}}^{A'} \alpha^A - \overline{\alpha}^{A'} \dot{\alpha}^A\right)$$

and

$$\dot{p}_a = \mathrm{i}s F_{CDab} w^C z^D v^b - \mathrm{i}s \overline{F}_{C'D'ab} \overline{w}^{C'} \overline{z}^{D'} v^b, \quad p_{AA'} \dot{z}^A + \overline{z}^{A'} \dot{p}_{AA'} = 0.$$

The spinor z which labels the internal degrees of freedom is no longer parallel-propagated along the worldline $q^a(t)$ and the direction of spin rotates at a rate determined by the curvature. The change in z produces, in turn, a discrepancy between mv and p. The motion is generally not geodetic.

The equation of motion is only an approximation (Dixon 1974), and is certainly not valid when the radius of curvature exceeds s/m, for there is then no guarantee that v is timelike. For an elementary particle, however, s is of the same order as $\hbar$; and when the radius of curvature exceeds $\hbar/m$, the gravitational tidal forces are strong enough to separate particle-antiparticle pairs, and so a single-particle theory must in any case break down.

The phase space of a single massive spinning particle is the reduction of the constraints on f_1 and f_2 (in the case $s = 0$, it is simply the reduction of the constraint $p_a p^a = m^2$, p future-pointing, in the cotangent bundle). There is, however, no generally covariant analogue of the invariant polarization of M^+_{sm}, which is one of the many obstacles to understanding quantum theory in curved space-time. In the massless case, there is not even a good covariant generalization of the single-particle phase spaces.

7

CLASSICAL FIELDS

7.1 Infinite-dimensional phase spaces

AT a formal level, much of the symplectic geometry that underlies the Hamiltonian formulation of classical mechanics extends to field theories, and can be used as the starting point for quantization. In this chapter, we shall look at relativistic field theories, and at the connection between the Lagrangian density and the symplectic structure on the space of solutions.[1] To prepare the ground for quantization, we shall look at how the splitting of the fields into their positive and negative frequency parts can be interpreted as a complex polarization in the classical phase space.

The aim is to gain a clearer understanding of the heuristic arguments that are used to construct quantum field theories. We shall not attempt to develop any of the theory in a rigorous way: where this can be done, it is not easy; and where it has not yet been done, it is not clear that the way forward towards a mathematically sound and self-consistent theory is to make rigorous the process of quantization—although the correspondence principle and the classical limit will have to be incorporated in whatever theory eventually emerges.

7.2 The space of solutions

The basic object that we shall consider is a collection of fields on Q, which is either flat or curved space-time. This will be represented by $\phi^\alpha(x^a)$ where the xs are space-time coordinates and α is an index labelling the various spinor and tensor components of the individual fields. In the case of spinor and complex tensor fields, the components and their complex conjugates are assigned separate values of α (e.g. for a complex scalar field, $\phi^1 = \phi$ and $\phi^2 = \overline{\phi}$). In geometric terms, ϕ is a section of a vector bundle $E \to Q$.

The dynamical behaviour is determined by the Lagrangian density, which in the simplest case is an invariant function of the form

$$L = L(\phi^\alpha, \nabla_b \phi^\beta, x^a),$$

where ∇ is the connection of the space-time metric. Here, and in the shorthand version $L = L(\phi, \nabla\phi, x)$, it is understood that ϕ and $\nabla\phi$ are evaluated at $x \in Q$. Thus L is a map that assigns a real number to each event at which the values of the fields and their first covariant derivatives have been specified. More precisely, it is a function on the first jet bundle $J^1(E)$ (Trautman 1972).

There are two ways of deriving the dynamical theory from L. One can form the *action integral*

$$I = \int_Q L\,\epsilon,$$

where ϵ is the space-time volume element, and then obtain the field equations by requiring that δI should vanish for all variations of the ϕs that vanish outside a compact subset of Q: this is the starting point for the explicitly covariant Lagrangian formalism. Alternatively, one can violate covariance by singling out the time coordinate x^0 and by treating

$$\int_{x^0=t} L\,\mathrm{d}\sigma$$

as a time-dependent Lagrangian, as in classical mechanics ($\mathrm{d}\sigma$ is the volume form on the hypersurface $x^0 = t$). The values of the fields at each point in space are the configuration coordinates and their time derivatives are the velocity coordinates. At least formally, one can then go over to momentum phase space and the canonical formalism.

The second method is the traditional starting point for canonical quantization. However the details are messy and, particularly in curved space-time, a number of artificial difficulties arise that seem to be of no physical relevance. For example, before talking about 'the value of a field at a point in space' it is necessary first to decide which events on different constant time hypersurfaces happen at the same place and how field values at distinct but spatially coincident events are to be compared. This can be done, but the details are not transparent.

In classical mechanics, the symplectic manifold naturally associated with a time-dependent Lagrangian is the space of motions—the classical analogue of the quantum mechanical phase space in the Heisenberg picture. The corresponding object in field theory is the infinite-dimensional symplectic manifold of classical solutions of the field equations. The trouble with the canonical formalism is that it gives too much structure on this manifold: it gives not only the symplectic form, but also a particular canonical coordinate system (the field values at each point in space and their conjugate momenta). The violation of covariance is not inherent in the canonical approach; it is simply an apparent consequence of singling out this particular coordinate system.

In the geometric approach, the coordinates are irrelevant: all that we need is the symplectic structure. And this can be found directly from the Lagrangian without violating covariance.

The 2-form on the space of solutions emerges from a construction very similar to that in §2.3. First, let us consider the variational problem in more detail. Fix ϕ^α and let $\phi^\alpha + tX^\alpha$ be a variation, where t is some small parameter and X is a section of E vanishing outside of a compact set D_X. Then ϕ^α is a solution of the field equations if and only if

$$\frac{\mathrm{d}}{\mathrm{d}t}\left[\int L(\phi + tX, \nabla\phi + t\nabla X, x)\,\epsilon\right]_{t=0} = 0$$

for every choice of X. That is, if and only if

$$\begin{aligned} 0 &= \int_{D_X}\left(\frac{\partial L}{\partial\phi^\alpha}X^\alpha + \frac{\partial L}{\partial\phi^\alpha_a}\nabla_a X^\alpha\right)\epsilon \\ &= \int_{D_X}\left[\frac{\partial L}{\partial\phi^\alpha} - \nabla_a\left(\frac{\partial L}{\partial\phi^\alpha_a}\right)\right]X^\alpha\,\epsilon, \end{aligned}$$

where $\phi^\alpha_a = \nabla_a\phi^\alpha$. We have here integrated by parts,[2] discarding a boundary integral which vanishes since $X = 0$ on ∂D_X. Thus the field equations are

$$\frac{\partial L}{\partial\phi^\alpha} - \nabla_a\left(\frac{\partial L}{\partial\phi^\alpha_a}\right) = 0. \tag{7.2.1}$$

Let $\mathcal{M}$ be the infinite-dimensional manifold of solutions of (7.2.1). A tangent vector to $\mathcal{M}$ at ϕ is a solution X of (7.2.1), linearized about ϕ. That is,

$$\frac{\partial^2 L}{\partial\phi^\beta\partial\phi^\alpha}X^\beta + \frac{\partial^2 L}{\partial\phi^\beta_b\partial\phi^\alpha}\nabla_b X^\beta = \nabla_a\left(\frac{\partial^2 L}{\partial\phi^\beta\partial\phi^\alpha_a}X^\beta + \frac{\partial^2 L}{\partial\phi^\beta_b\partial\phi^\alpha_a}\nabla_b X^\beta\right).$$

We shall not consider the technical issues raised by the definition of $\mathcal{M}$, beyond requiring that it should only contain solutions that fall off sufficiently rapidly at spatial infinity to justify the integrations that follow.

Let $\Sigma \subset Q$ be a spacelike hypersurface, which is either a hyperplane, in the case of Minkowski space, or a general Cauchy surface, in the case of a curved space-time; and let θ be the 1-form on $\mathcal{M}$ defined by

$$X \lrcorner\, \theta = \int_\Sigma X^\alpha \frac{\partial L}{\partial\phi^\alpha_a}\,\mathrm{d}\sigma_a$$

($\mathrm{d}\sigma_a = n_a\mathrm{d}\sigma$, where $\mathrm{d}\sigma$ is the volume element on Σ and n^a is the unit normal). Of course, θ depends on the choice of Σ. However, if θ' is defined in the same way, with Σ replaced by Σ', then

$$X \lrcorner\, \theta' - X \lrcorner\, \theta = X \lrcorner\, \mathrm{d}\left(\int_D L(\phi, \nabla\phi, x)\,\epsilon\right),$$

where D is the region of space-time[3] between Σ and Σ'. The integral is interpreted as a function on $\mathcal{M}$, with ϕ varying and D fixed; the expression on the right-hand side is the derivative of this function along X. It follows that the difference between θ and θ' is exact and hence that the 2-form $\omega = \mathrm{d}\theta$ is independent of Σ.

Proposition (7.2.2). The 2-form ω on the space of solutions is

$$\omega(X,Y) = \tfrac{1}{2}\int_\Sigma \omega^a \,\mathrm{d}\sigma_a$$

where X and Y are solutions of the linearized equations and

$$\omega^a = \frac{\partial^2 L}{\partial\phi^\beta \partial\phi^\alpha_a}(X^\beta Y^\alpha - Y^\beta X^\alpha) + \frac{\partial^2 L}{\partial\phi^\beta_b \partial\phi^\alpha_a}(Y^\alpha \nabla_b X^\beta - X^\alpha \nabla_b Y^\beta).$$

Proof. Let X and Y be vector fields on $\mathcal{M}$; in other words, they are maps that assign solutions $X(\phi)$ and $Y(\phi)$ of the linearized equations to each $\phi \in \mathcal{M}$. Their Lie bracket $[X,Y]$ is the linearized solution $\partial_X Y^\alpha - \partial_Y X^\alpha$, where

$$\partial_X Y^\alpha = \frac{\mathrm{d}}{\mathrm{d}t}\Big(Y^\alpha(\phi + tX)\Big)_{t=0};$$

and the derivative of $Y \lrcorner\, \theta$ along X is

$$\int_\Sigma \left[Y^\alpha(\phi)\frac{\mathrm{d}}{\mathrm{d}t}\left(\frac{\partial L}{\partial\phi^\alpha_a}\right)_{\phi+tX} + \left(\frac{\partial L}{\partial\phi^\alpha_a}\right)_\phi \frac{\mathrm{d}}{\mathrm{d}t}\Big(Y^\alpha(\phi+tX)\Big)\right]_{t=0} \mathrm{d}\sigma_a$$

$$= \int_\Sigma \left(Y^\alpha X^\beta \frac{\partial^2 L}{\partial\phi^\alpha_a \partial\phi^\beta} + Y^\alpha \nabla_b X^\beta \frac{\partial^2 L}{\partial\phi^\alpha_a \partial\phi^\beta_b} + \frac{\partial L}{\partial\phi^\alpha_a}\partial_X Y^\alpha\right) \mathrm{d}\sigma^a.$$

Therefore,

$$2\omega(X,Y) = X(Y \lrcorner\, \theta) - Y(X \lrcorner\, \theta) - [X,Y] \lrcorner\, \theta = \int_\Sigma \omega^a \,\mathrm{d}\sigma_a. \qquad \blacksquare$$

One can also check directly that $\nabla_a \omega^a = 0$ as a consequence of the linearized field equation, and hence that $\omega(X,Y)$ is independent of Σ provided that the linearized fields fall off fast enough at spatial infinity.

Thus ω is a natural closed 2-form on $\mathcal{M}$. But it is not necessarily nondegenerate: this is closely related to the question of whether or not the Cauchy problem for (7.2.1) is well posed. When it is degenerate, we must reduce $\mathcal{M}$ by factoring out the characteristic distribution, which generally involves the removal of gauge freedom.

Aside. There are inequivalent notions of nondegeneracy in infinite dimensions (Chernoff and Marsden 1974). Weak nondegeneracy—the condition we are interested in—requires only that $\omega(X,Y)$ should vanish for all Y only when $X = 0$. Strong nondegeneracy is a subtler condition involving the topologies of $\mathcal{M}$ and $T_\phi\mathcal{M}$ in a nontrivial way: it requires that

$$T_\phi\mathcal{M} \to T^*_\phi\mathcal{M} : X \mapsto X \lrcorner\, \omega$$

should be an isomorphism. It is used in the infinite-dimensional version of Darboux's theorem (Weinstein 1971). ∎

The following examples indicate how the symplectic theory works in practice, and how it relates to the canonical formalism.

Example. *The real scalar field.* Here ϕ is a single real function on Q and

$$L = \tfrac{1}{2}(\nabla_a\phi\nabla^a\phi - \mu^2\phi^2). \tag{7.2.3}$$

The field equation is the Klein-Gordon equation[4]

$$\Box\phi + \mu^2\phi = 0, \tag{7.2.4}$$

which is already linear. Thus $\mathcal{M}$ is a vector space and ω is the bilinear form

$$\omega(\phi, \phi') = \tfrac{1}{2}\int_\Sigma (\phi'\nabla_a\phi - \phi\nabla_a\phi')\,\mathrm{d}\sigma^a.$$

The inverse of ω is an element of $\mathcal{M}\wedge\mathcal{M}$. It is the skew-symmetric function $\Delta(x,y)$ of two points $x, y \in Q$ defined by

$$2\omega\big(\phi(\cdot), \Delta(\cdot, y)\big) = \phi(y) \tag{7.2.5}$$

where $\phi \in \mathcal{M}$, and the symplectic form is evaluated by treating ϕ and Δ as functions of x, with y held fixed. In flat space-time, Δ is the *commutator distribution*

$$\Delta(x,y) = \left(\frac{1}{2\pi}\right)^3 \int \sin\left(k_a x^a - k_a y^a\right) \frac{\mathrm{d}k_1\mathrm{d}k_2\mathrm{d}k_3}{k_0},$$

where $k_a k^a = \mu^2$ and $k_0 > 0$ (for example, see Bjorken and Drell 1964).

Example. *The complex scalar field.* Here ϕ is a complex function on Q and

$$L = \tfrac{1}{2}(\nabla_a\phi\nabla^a\overline{\phi} - \mu^2\phi\overline{\phi}).$$

The symplectic form is

$$\omega(\phi, \phi') = \tfrac{1}{4}\int_\Sigma (\phi'\nabla_a\overline{\phi} + \overline{\phi}'\nabla_a\phi - \phi\nabla_a\overline{\phi}' - \overline{\phi}\nabla_a\phi')\,\mathrm{d}\sigma^a. \tag{7.2.6}$$

Example. *The electromagnetic field.* The ϕs are the components of the electromagnetic potential Φ and the Lagrangian density is

$$L_M = -\tfrac{1}{4}F_{ab}F^{ab} = \tfrac{1}{2}(E^2 - B^2),$$

where $F_{ab} = 2\nabla_{[a}\Phi_{b]}$ is the electromagnetic field tensor. This generates the linear field equation $\nabla_a F^{ab} = 0$ and the bilinear form

$$\omega(\Phi, \Phi') = \tfrac{1}{2}\int_\Sigma (\Phi^b F'_{ab} - \Phi'^b F_{ab})\,\mathrm{d}\sigma^a,$$

which is degenerate: if $\Phi'_a = \nabla_a f$ for some function f, then $\omega(\Phi, \Phi') = 0$ for every Φ satisfying the field equations. To obtain a nondegenerate form, we must remove the gauge freedom by identifying Φ with $\Phi + \nabla f$. The result is a symplectic vector space, each point of which corresponds uniquely to an electromagnetic field F satisfying the source-free Maxwell equations $\nabla_a F^{ab} = 0$ and $\nabla_{[a} F_{bc]} = 0$. The symplectic form is

$$\omega(F, F') = \tfrac{1}{2} \int_\Sigma (\Phi^b F'_{ab} - \Phi'^b F_{ab})\, \mathrm{d}\sigma^a, \tag{7.2.7}$$

where Φ and Φ' are now arbitrary potentials for F and F'. It is easy to see that $\omega(F, F')$ is invariant under *independent* gauge transformations of Φ and Φ'.

The comparision with the canonical treatment is instructive. Take Q to be flat, with inertial coordinates t, x, y, z. Then the configuration space variables are the values of the scalar and vector potentials, ϕ and $\mathbf{A}$, at each point of space. Since $\Phi = (\phi, \mathbf{A})$, the (time-independent) Lagrangian is

$$\mathrm{Lgr} = \tfrac{1}{2} \int (\dot{\mathbf{A}} \cdot \dot{\mathbf{A}} + 2\dot{\mathbf{A}} \cdot \nabla\phi + \nabla\phi \cdot \nabla\phi - \mathrm{curl}\mathbf{A} \cdot \mathrm{curl}\mathbf{A})\, \mathrm{d}\sigma,$$

where $\dot{\mathbf{A}}$ and $\dot{\phi}$ denote the time derivatives and $\mathrm{d}\sigma = \mathrm{d}x\mathrm{d}y\mathrm{d}z$.

To obtain the canonical formalism, we must treat the values of $\phi, \mathbf{A}, \dot{\phi}, \dot{\mathbf{A}}$ at each point (x, y, z) as independent coordinates on the velocity phase space V. Since V is a vector space, a tangent vector at $X \in V$ is the same as an element of V.

The 1-form θ_{Lgr} generated by Lgr is given at $X' \in V$ by

$$X \lrcorner\, \theta_{\mathrm{Lgr}}(X') = \int (\mathbf{A} \cdot \dot{\mathbf{A}}' + \mathbf{A} \cdot \nabla\phi')\, \mathrm{d}\sigma,$$

where $X = (\phi, \mathbf{A}, \dot{\phi}, \dot{\mathbf{A}})$ and $X' = (\phi', \mathbf{A}', \dot{\phi}', \dot{\mathbf{A}}')$ (§2.1). Its exterior derivative is the 2-form

$$\omega_{\mathrm{Lgr}}(X, X') = \tfrac{1}{2} \int (\dot{\mathbf{A}} \cdot \mathbf{A}' - \mathbf{A} \cdot \dot{\mathbf{A}}' + \mathbf{A}' \cdot \nabla\phi - \mathbf{A} \cdot \nabla\phi')\, \mathrm{d}\sigma,$$

which, with a different interpretation, is the same as (7.2.7). The Hamiltonian is

$$h = \int (\dot{\mathbf{A}} \cdot \dot{\mathbf{A}} - \nabla\phi \cdot \nabla\phi + \mathrm{curl}\mathbf{A} \cdot \mathrm{curl}\mathbf{A})\, \mathrm{d}\sigma.$$

A formal application of the finite-dimensional theory gives the equations of motion

$$\ddot{\mathbf{A}} + \nabla\dot{\phi} + \nabla(\mathrm{div}\mathbf{A}) - \nabla^2\mathbf{A} = 0, \qquad \mathrm{div}\dot{\mathbf{A}} + \nabla^2\phi = 0.$$

The Lagrangian is irregular and the second equation is a constraint—it picks out a submanifold $C \subset V$. The first equation determines $\ddot{\mathbf{A}}$, leaving $\dot{\phi}$ arbitrary.

The reduction of C involves the identification of

$$(\phi, \mathbf{A}, \dot{\phi}, \dot{\mathbf{A}}) \quad \text{with} \quad (\phi - f, \mathbf{A} + \nabla g, \dot{\phi} + k, \dot{\mathbf{A}} + \nabla f)$$

where f, g, k are any functions of x, y, z. This leaves h well defined, as the general theory of §2.5 predicts, so the dynamical evolution in the reduced phase space is also well defined.

The reduced phase space is, of course, the same as the space of solutions of Maxwell's equations, with the symplectic form (7.2.7). The correspondence is derived by choosing a value of t, and mapping a solution F of Maxwell's equations with potential $\Phi = (\phi, \mathbf{A})$ to the point of C determined by the values of $\phi, \mathbf{A}, \dot{\phi}, \dot{\mathbf{A}}$ at t.

Example. *The Yang-Mills field.* The nonabelian theory is very similar, except that the field equations are nonlinear. We start with a 4-potential Φ which takes values in the Lie algebra of some group G, and which is to be thought of as the local representative of a connection on a vector bundle over Q with structure group G. For simplicity, we shall take $G = SL(2, \mathbb{C})$. Then the Lagrangian density is $L = -\frac{1}{4}\mathrm{tr}\,(F_{ab}F^{ab})$. Here

$$F_{ab} = 2\nabla_{[a}\Phi_{b]} + 2g\Phi_{[a}\Phi_{b]}$$

is the curvature, which is a 2-form with values in sl$(2, \mathbb{C})$, and g is a coupling constant, which we shall set to unity. In the general case, the trace is replaced by an invariant inner product, and the second term in the definition of the curvature is replaced by the Lie algebra bracket.

The covariant derivative of a scalar or tensor field T with values in the Lie algebra (the trace-free matrices) is a combination of the space-time covariant derivative and the connection Φ. It is defined by $DT = \nabla T + \Phi T - T\Phi$, so that, for example,

$$D_a F_{bc} = \nabla_a F_{bc} + \Phi_a F_{bc} - F_{bc}\Phi_a.$$

Note that $\mathrm{tr}\,\big(TDT' + (DT)T'\big) = \nabla \mathrm{tr}\,(TT')$.

If X is an sl$(2, \mathbb{C})$-valued 1-form, then the curvature of $\Phi' = \Phi + tX$ is $F_{ab} + 2tD_{[a}X_{b]} + O(t^2)$ and

$$\begin{aligned} -\tfrac{1}{4}\mathrm{tr}\,(F'_{ab}F'^{ab}) &= L - t\,\mathrm{tr}\,(F^{ab}D_a X_b) + O(t^2) \\ &= L - t\nabla_a \mathrm{tr}\,(X_b F^{ab}) + t\,\mathrm{tr}\,(X_b D_a F^{ab}) + O(t^2). \end{aligned}$$

Hence the field equations are $D_a F^{ab} = 0$. From its definition, the curvature also satisfies the Bianchi identity $D_{[a}F_{bc]} = 0$.

The 2-form on the space of solutions is

$$\omega(X,Y) = \tfrac{1}{2}\int_\Sigma \mathrm{tr}\,(X_b D^{[a}Y^{b]} - Y_b D^{[a}X^{b]})\,\mathrm{d}\sigma_a,$$

where X and Y are solutions of the linearized equations. Note that ω depends on Φ through D.

When $X_a = D_a f$, where f takes values in the Lie algebra, $D_{[a}X_{b]} = \frac{1}{2}(F_{ab}f - fF_{ab})$ and

$$\omega(X,Y) = -\int \nabla_b \mathrm{tr}\,(fD^{[a}Y^{b]})\,\mathrm{d}\sigma_a = 0,$$

under appropriate boundary conditions. Here we have used the linearized field equation

$$D_b(D^{[a}Y^{b]}) = \tfrac{1}{2}(F^{ab}Y_b - Y_bF^{ab}).$$

Conversely, if $\omega(X,Y) = 0$ for every Y, then $X = Df$ for some f (García 1980, Moncrief 1980).

The vector fields on $\mathcal{M}$ (the space of solutions) of the form $X = Df$ are the generators of gauge transformations $\Phi \mapsto g^{-1}\Phi g + g^{-1}\mathrm{d}g$, where $g : Q \to SL(2,\mathbb{C})$. Again reduction removes gauge freedom. ∎

By writing the action as an integral of $L(\phi, \nabla\phi, x)$, we are emphasizing the role of the background metric. This simplifies calculations when the field equations are hyperbolic with null geodesic characteristics, but can be inconvenient in other contexts (in the case of Einstein's equations, there is no background space-time). It can also hide symmetries of the equations, such,as conformal invariance. We can instead absorb ϵ into L and write the action as the integral of a volume form $\Lambda(\phi, \nabla\phi, x)$ on Q which depends on the value of ϕ and its first derivative at each point. If the Q is n-dimensional and Λ has components $\Lambda_{ab\ldots c}$, then $X \lrcorner\, \theta$ is the integral over a Cauchy surface of the $(n-1)$-form with components

$$nX^\alpha \frac{\partial \Lambda_{ab\ldots c}}{\partial \phi^\alpha_a}.$$

If we replace Λ by $\Lambda + \mathrm{d}\lambda$, where $\lambda(\phi, x)$ is an $(n-1)$-form on Q that depends on x and the value of ϕ at x, then the field equations and the 2-form ω on $\mathcal{M}$ are unchanged, but θ is replaced by $\theta + (2n-1)\mathrm{d}f$, where f is the integral of λ over Σ.

Example. *The conformally invariant wave equation.* The conformally invariant wave equation in four dimensions is generated by

$$\Lambda = \tfrac{1}{2}(g^{ab}\nabla_a\phi\nabla_b\phi + \tfrac{1}{6}R\phi^2)\epsilon,$$

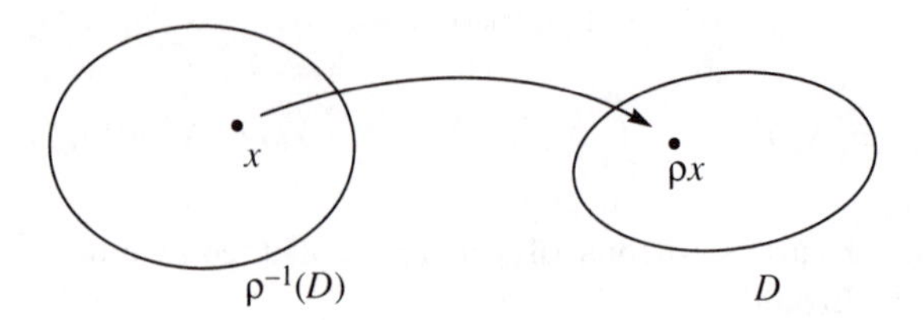

Fig. 7.1. Pulling back an integral.

where R is the scalar curvature and ϵ is the metric volume element. If we replace g_{ab} by $\hat{g}_{ab} = \Omega^2 g_{ab}$ and ϕ by $\hat{\phi} = \Omega^{-1}\phi$, then

$$\Lambda = \hat{\Lambda} + \tfrac{1}{2}\mathrm{d}(\phi^2 v \lrcorner \epsilon)$$

where $\hat{\Lambda}$ is given by replacing g and ϕ by $\hat{g}$ and $\hat{\phi}$ in the expression for Λ, and $v^a = \nabla^a \log \Omega$. Therefore the field equation and the symplectic structure on the space of solutions are conformally invariant, but the symplectic potential θ is not.

7.3 Noether's theorem

As in classical mechanics, Noether's theorem associates conservation laws with symmetries of the Lagrangian.

We shall consider first the symmetries of the background space-time. Let v be a Killing vector on Q, generating the flow $\rho_t : Q \to Q$, where t is the parameter along the integral curves of v. If the fields are tensors or spinors, then the ρ_ts act by the 'push forward' map on the ϕs and their covariant derivatives. We shall denote the push-forward by R_t, so that, for example, for a scalar field $R_t(\phi) = \phi \circ \rho_t^{-1}$. Since ρ_t preserves the metric connection, $R_t(\nabla\phi^\alpha) = \nabla R_t \phi^\alpha$.

Definition (7.3.1). The Killing vector v is an (infinitesimal) symmetry of the Lagrangian density if $L(\phi, \nabla\phi, x) = L(R_t\phi, \nabla R_t\phi, \rho_t x)$ for every t.

Now for any diffeomorphism $\rho : Q \to Q$,

$$\int_D f\,\epsilon = \int_{\rho^{-1}(D)} (f \circ \rho)\,\rho^* \epsilon, \tag{7.3.2}$$

where $D \subset Q$ and $f \in C^\infty(Q)$ (Fig. 7.1). If $f(x) = L(\phi, \nabla\phi, x)$, then $(f \circ \rho)(x) = L(\phi, \nabla\phi, \rho x)$, because of our convention that in the expression

$L(\phi, \nabla\phi, x)$, the fields and their derivatives are evaluated at x. So if v is a symmetry of the Lagrangian, then

$$\begin{aligned} \int_D L(R_t\phi, R_t\nabla\phi, x)\,\epsilon &= \int_{\rho_t^{-1}(D)} L(R_t\phi, \nabla R_t\phi, \rho_t x)\,\epsilon \\ &= \int_{\rho_t^{-1}(D)} L(\phi, \nabla\phi, x)\,\epsilon. \end{aligned}$$

since $\rho_t^*\epsilon = \epsilon$. It follows that if ϕ is a solution of the variational problem, then so is $R_t\phi$. Therefore R_t maps solutions of the field equations to solutions, and so induces a flow on $\mathcal{M}$. It is generated by the vector field

$$X^\alpha = \frac{\mathrm{d}}{\mathrm{d}t}\Big(R_t\phi^\alpha\Big)_{t=0} = -\mathcal{L}_v\phi^\alpha,$$

which assigns to each $\phi \in \mathcal{M}$ a solution of the linearized field equations about ϕ.

Let D be a compact region of space-time bounded by a hypersurface ∂D. Then for any $\phi \in \mathcal{M}$,

$$\begin{aligned} \frac{\mathrm{d}}{\mathrm{d}t}\left(\int_D L(R_t\phi, R_t\nabla\phi, x)\,\epsilon\right)_{t=0} &= \int_D \left(X^\alpha \frac{\partial L}{\partial\phi^\alpha} + \nabla_a X^\alpha \frac{\partial L}{\partial\phi^\alpha_a}\right)\epsilon \\ &= \int_{\partial D} X^\alpha \frac{\partial L}{\partial\phi^\alpha_a}\,\mathrm{d}\sigma_a, \end{aligned}$$

by integrating by parts and using the fact that ϕ is a solution of the field equations. On the other hand,

$$\begin{aligned} \frac{\mathrm{d}}{\mathrm{d}t}\left(\int_D L(R_t\phi, R_t\nabla\phi, x)\,\epsilon\right)_{t=0} &= \frac{\mathrm{d}}{\mathrm{d}t}\int_{\rho_t^{-1}(D)} L(\phi, \nabla\phi, x)\,\epsilon \\ &= -\int_{\partial D} Lv^a\,\mathrm{d}\sigma_a \end{aligned} \qquad (7.3.3)$$

(Fig. 7.2). It follows that

$$\int_{\partial D}\left(X^\alpha \frac{\partial L}{\partial\phi^\alpha_a} + v^a L\right)\mathrm{d}\sigma_a = 0.$$

Since this holds for any D, we have the following restricted form of Noether's theorem.

Proposition (7.3.4). If v is a Killing vector and a symmetry of the Lagrangian density $L(\phi, \nabla\phi, x)$, and if ϕ is a solution of the field equations, then

$$\nabla_a\left(X^\alpha \frac{\partial L}{\partial\phi^\alpha_a} + v^a L\right) = 0,$$

where $X^\alpha = -\mathcal{L}_v\phi^\alpha$.

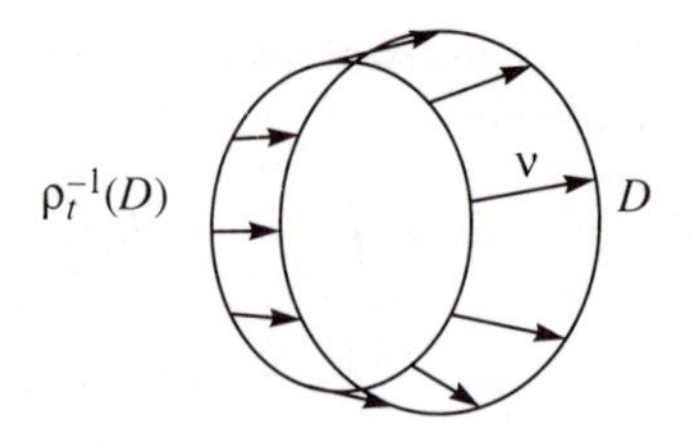

Fig. 7.2. The proof of Noether's theorem.

An infinitesimal symmetry gives a map from $\mathcal{M}$ to the divergence-free vector fields on Q. By integrating over a Cauchy surface Σ, we obtain a function

$$h(\phi) = \int_\Sigma \left(X^\alpha \frac{\partial L}{\partial \phi_a^\alpha} + v^a L \right) \mathrm{d}\sigma_a \tag{7.3.5}$$

on $\mathcal{M}$. Because the integrand is divergence-free, h does not depend on Σ.

Example. A time-independent Lagrangian $L(q, \dot{q})$ in classical mechanics provides an example. The 'space-time' is the time axis with metric $\mathrm{d}t^2$, and the configuration variables q^a are the fields. The space of solutions is the space of motions (no integration is involved in evaluating the symplectic form because a Cauchy surface is a single point). The Killing vector $-\partial/\partial t$ generates the flow $R_t(q^a(\cdot)) = q^a(\cdot + t)$ on the space of motions. The corresponding function h is the Hamiltonian.

Example. *Scalar fields.* All the generators of the Poincaré group are symmetries of the Lagrangian (7.2.3) in Minkowski space. The vector field X on $\mathcal{M}$ assigns to $\phi \in \mathcal{M}$ the solution $-v^a \nabla_a \phi$ of the Klein-Gordon equation. The divergence-free vector on Minkowski space is $-T^{ab} v_b$, where T^{ab} is the energy-momentum tensor

$$T^{ab} = \nabla^a \phi \nabla^b \phi - \tfrac{1}{2} g^{ab} (\nabla_c \phi \nabla^c \phi - \mu^2 \phi^2). \qquad \blacksquare$$

When ω is symplectic, the vector field X is Hamiltonian and its generator is h (otherwise, the projection of X into the reduced phase space is Hamiltonian, and h is the pull-back of its generator). There are a number of calculations that will prove this directly, but the connection between h and X can be made obvious by looking at the construction of ω in a more abstract setting, in which we shall also see that Noether's theorem can be applied more widely.

Let $\mathcal{F}$ denote the manifold of all fields; that is, $\mathcal{F}$ is the space of all appropriately behaved sections of the vector bundle $E \to Q$, not just those that obey the field equations. The data for the variational problem are the action, which assigns a function $S_D : \mathcal{F} \to \mathbb{R}$ to each oriented region D in Q, and the boundary conditions—the information about what is to be

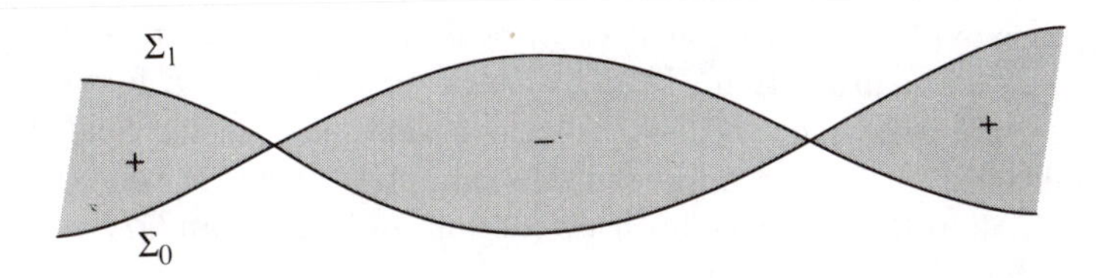

Fig. 7.3. The region D_{01} between Σ_0 and Σ_1.

kept fixed on the boundary of D when the action is varied. The boundary conditions assign a foliation P_Σ of $\mathcal{F}$ to each hypersurface $\Sigma \subset Q$. Two fields lie on the same leaf of P_Σ if they have the same boundary data on Σ. For example, for a scalar field, the action is

$$S_D = \tfrac{1}{2} \int_D (\nabla_a \phi \nabla^a \phi - \mu^2 \phi)\, \epsilon,$$

and the boundary condition is that the values of ϕ on ∂D should be kept fixed when the action is varied. Two fields ϕ and ϕ' lie on the same leaf of P_Σ whenever $\phi = \phi'$ at every point of Σ.

The action determines a function $S_{01} : \mathcal{F} \to \mathbb{R}$ for each pair of Cauchy surfaces Σ_0 and Σ_1, by taking D to be the region D_{01} between Σ_0 and Σ_1 (with positive orientation where Σ_0 is to the past of Σ_1, and negative orientation elsewhere; see Fig. 7.3). The space of solutions is characterized by the condition: $\phi \in \mathcal{M}$ whenever $Z \lrcorner \, \mathrm{d}S_{01} = 0$ for every pair of Cauchy surfaces Σ_0 and Σ_1, and for every vector $Z \in T_\phi \mathcal{F}$ which is tangent to both $P_0 = P_{\Sigma_0}$ and $P_1 = P_{\Sigma_1}$.

When Σ_0 and Σ_1 are disjoint, $T\mathcal{F} = P_0 + P_1$. This allows us to decompose the gradient of the action at $\phi \in \mathcal{M}$ by putting

$$\mathrm{d}S_{01}(\phi) = \theta_1 - \theta_0 \tag{7.3.6}$$

where $Z \lrcorner \, \theta_i = 0$ whenever Z is tangent to P_i. Because the gradient vanishes in directions that are tangent to both foliations, the decomposition is well defined. The restriction of θ_0 to $\mathcal{M}$ is a 1-form on $\mathcal{M}$; its exterior derivative is the 2-form ω.

Apart from giving explicit formulas for the θs, the 'integration by parts' derivation of ω serves one other crucial purpose: it establishes the 'locality property' that θ_0 does not depend on the choice of Σ_1. Because θ_0 is expressed in terms of the values of ϕ and its derivatives on Σ_0, it is unchanged when Σ_1 is replaced by some other Cauchy surface. This, together with the fact that $\theta_0 - \theta_1$ is exact, implies that ω is a well-defined 2-form on $\mathcal{M}$, and is independent of the choice of Cauchy surfaces.

Suppose now that v is a vector field on Q and that $\rho_t : Q \to Q$ is the flow generated by v. Let v' be a vector field on E such that $\pi_* v' = v$, where $\pi : E \to Q$ is the projection. Then by dragging sections of E forwards along v', we obtain a flow $R_t : \mathcal{F} \to \mathcal{F}$. Its generator is a vector field X on $\mathcal{F}$. We shall say that v is an infinitesimal symmetry whenever there is a lift v' of v such that R_t preserves both the action and the boundary conditions. That is, for every t, D, and Σ,

$$S_{D_t}(R_t\phi) = S_D(\phi) \quad \text{and} \quad P_{\Sigma_t} = R_{t*}P_\Sigma,$$

where $D_t = \rho_t(D)$ and $\Sigma_t = \rho_t(\Sigma)$.

Choose a Cauchy surface Σ and let v be an infinitesimal symmetry. Then

$$R_t(\mathcal{M}) = \mathcal{M}, \quad \text{and} \quad R_t^*\theta_t = \theta_0, \tag{7.3.7}$$

where $\theta_t \in \Omega^1(\mathcal{M})$ is the 1-form constructed from Σ_t. Hence $R_t^*\omega = \omega$, since $\omega = \mathrm{d}\theta_0 = \mathrm{d}\theta_t$ and $R_t^*\omega = R_t^*\mathrm{d}\theta_t = \mathrm{d}\theta_0 = \omega$. So R_t maps solutions to solutions and preserves the 2-form ω. Also, by the second equation in (7.3.7),

$$0 = \mathcal{L}_X\theta_t = X \lrcorner \mathrm{d}\theta_t + \mathrm{d}(X \lrcorner \theta_t) + \partial_t\theta_t \tag{7.3.8}$$

(the notation is explained in §A.1). Because $\mathrm{d}\theta_t = \omega$ for all t, the 1-forms $\partial_t\theta_t$ are closed for each value of t. In fact, if $D_{tt'}$ denotes the region between Σ_t and $\Sigma_{t'}$, then

$$\theta_{t'} - \theta_t = \mathrm{d}S_{D_{tt'}}.$$

Hence $\partial_t\theta_t = \mathrm{d}f_t$ where f_t is the time-dependent function on $\mathcal{M}$ defined by

$$f_t = \frac{\mathrm{d}}{\mathrm{d}t'}\Big(S_{D_{tt'}}\Big)_{t'=t} = \int_{\Sigma_t} Lv^a \,\mathrm{d}\sigma_a.$$

Put $h = X \lrcorner \theta_0 + f_0$. Then by taking $t = 0$ in (7.3.8), $X \lrcorner \omega + \mathrm{d}h = 0$, so h is the Hamiltonian generating X—either on $\mathcal{M}$ itself when ω is nondegenerate or on the reduction of $\mathcal{M}$ in general.[5] When v is a Killing vector and v' is defined by Lie dragging, h coincides with (7.3.5).

Example. *Scalar fields.* In the example of the Klein Gordon field in Minkowski space, the integral of $-T^{ab}v_b$ over a spacelike hyperplane generates the flow of the Killing vector v on the space of solutions. In particular,

$$\int_{x^0=0} T^{00}\,\mathrm{d}x^1\mathrm{d}x^2\mathrm{d}x^3$$

generates the flow $\phi(x^0, x^1, x^2, x^3) \mapsto \phi(x^0 - t, x^1, x^2, x^3)$.

There is a shift here from the usual viewpoint of the canonical formalism, where one regards the energy (the integral of T^{00}) as the generator of the evolution of the individual solutions of the field equations, rather than as a flow on the space of solutions. The state of the field at time t is described by its Cauchy data (the values of ϕ and $\nabla_0\phi$ on $x^0 = t$), and ω is regarded as a symplectic form on the space of Cauchy data. As t increases, the state changes. In less simple systems, this traditional interpretation has the drawback that it is necessary first to pick out the constant time hypersurfaces and then to choose a way of identifying which Cauchy data on different hypersurfaces represent the same state. ■

Our construction of ω and h is of great generality: the background geometry of the space-time is not important and it is not necessary that $\mathcal{F}$ should be a space of sections of a vector bundle (it could be some more general manifold of maps, as in the nonlinear sigma models). All that we need to construct ω are the functions S_D on $\mathcal{F}$, and the foliations P_Σ. These are constrained only by the conditions that the action should generate hyperbolic field equations, so that it is possible to distinguish Cauchy surfaces, and that the locality property should hold, as it does when the action is the integral of a Lagrangian density of which the value at x depends only on $\phi(x)$ and a finite number of derivatives of ϕ at x. All that we need to construct h is a way of lifting the flow of v to $\mathcal{F}$ so that (7.3.7) holds.

The general theory is illustrated by the various forms of the Hilbert action for Einstein's equations.[6]

Example. *Einstein's equations.* The Einstein vacuum equations $R_{ab} = 0$ for the space-time metric are derived from the *Hilbert action*

$$S_D = \int_D R\,\epsilon$$

where R is the scalar curvature and ϵ is the metric volume element. We take $\mathcal{F}$ to be the space of all symmetric tensor fields g_{ab} of signature $+---$ and represent a tangent vector to $\mathcal{F}$ by an arbitrary symmetric tensor field X_{ab}.

To the first order in t, the connection of the metric $\hat{g}_{ab} = g_{ab} + tX_{ab}$ is given by

$$\hat{\nabla}_a v^b = \nabla_a v^b + \frac{t}{2}\left(\nabla_a X^b{}_c + \nabla_c X^b{}_a - \nabla^b X_{ac}\right) v^c, \tag{7.3.9}$$

where ∇ is the connection of g, and g is used to raise and lower indices. With the curvature conventions of §A.1, therefore,

$$\hat{R} = R + t\nabla_b\left(\nabla_a X^{ab} - \nabla^b x\right) - tX^{ab}R_{ab} + O(t^2),$$

where $x = X^a{}_a$. The volume element of $\hat{g}$ is $(1 + \frac{1}{2}tx)\epsilon$, again to the first order. It follows that

$$X \lrcorner\, \mathrm{d}S_D = \int_{\partial D} y \lrcorner\, \epsilon - \int_D X_{ab}(R^{ab} - \tfrac{1}{2}Rg^{ab})\,\epsilon,$$

where ∂D is the boundary of D and $y^a = \nabla_b X^{ab} - \nabla^a x$.

By varying the action with g and its first derivatives held fixed on ∂D, we obtain the field equation $R_{ab} = 0$. The space of solutions is denoted by $\mathcal{M}$. If Σ is a Cauchy surface in Q, then the corresponding 1-form $\theta \in \Omega^1(\mathcal{M})$ is given by

$$X \lrcorner \theta = \int_\Sigma y \lrcorner \epsilon.$$

Even though the action integral contains second derivatives of the metric, the field equations are only second order. At the cost of violating covariance, it is possible to remove the second derivative terms by adding a divergence to $R\epsilon$. This leaves the field equations and $\omega = \mathrm{d}\theta$ unchanged, but adds a gradient to θ.

Somewhat simpler potentials for ω can be obtained by replacing the action by

$$S'_D = \int_D R\epsilon + \int_{\partial D} 2k\, n \lrcorner \epsilon$$

where n is the outward-pointing normal, scaled so that $n^a n_a = 1$, and $k = \kappa^a{}_a$, where κ_{ab} is the extrinsic curvature (to simplify calculations, we shall extend n off Σ by making it geodesic; then $n^a \nabla_a n^b = 0$, $\nabla_a n_b = \nabla_b n_a$, and $\kappa_{ab} = \nabla_a n_b$). The additional term has no effect on the field equations since the variation of the boundary term vanishes if g and its first derivatives are held fixed on ∂D; and it also leaves ω unchanged. There is, however, an assumption that the boundary of D is spacelike. This is not a problem in the derivation of the symplectic potentials since we need to know S'_D only for a region bounded by two Cauchy surfaces. It does, on the other hand, raise the issue of boundary conditions at spatial infinity. To avoid this complication, we shall assume that Σ is compact, so that the solutions are spatially closed.

To calculate the potential θ' associated with S' and the Cauchy surface Σ, we must find the variation of the boundary term for a general X_{ab}. The unit normal to ∂D with respect to $\hat{g}$ is

$$\hat{n}^a = n^a + \frac{t}{2} X_{bc} n^b n^c n^a - t X^{ab} n_b + O(t^2).$$

By using (7.3.9), the trace of the extrinsic curvature with respect to $\hat{g}$ is

$$2\hat{k} = 2k + t\nabla_a (X_{bc} n^b n^c n^a) - 2t n^a \nabla^b X_{ab} + t n^a \nabla_a x - 2t X^{ab} \kappa_{ab}$$

plus higher order terms in t. The variation in ϵ is $\frac{1}{2} t x \epsilon$. Hence, since $y \lrcorner \epsilon|_{\partial D} = (y^a n_a) n \lrcorner \epsilon|_{\partial D}$,

$$X \lrcorner \mathrm{d}S'_D = \int_D X_{ab}(R^{ab} - \tfrac{1}{2} R g^{ab})\epsilon + \int_{\partial D} \left(X_{ab} p^{ab} - \nabla_a u^a\right) n \lrcorner \epsilon,$$

where $p^{ab} = -\kappa^{ab} + kg^{ab} - kn^a n^b$ and $u^a = X^{ab}n_b - n^a n^b n^c X_{bc}$.

Now if w is a vector field on Q tangent to Σ, then its divergence with respect to the induced metric on Σ is $\nabla_a w^a - n^a n^b \nabla_a w_b$. Therefore, since $n \lrcorner \epsilon$ is the intrinsic volume element and since Σ is compact,

$$\int_\Sigma (\nabla_a w^a - n^a n^b \nabla_a w_b)\, n \lrcorner \epsilon = 0.$$

by the divergence theorem. But $n^a n^b \nabla_a u_b = 0$, so the integral over Σ of $\nabla_a u^a$ vanishes. We conclude that

$$X \lrcorner \theta' = \int_\Sigma p^{ab} X_{ab}\, n \lrcorner \epsilon.$$

Because $p^{ab}n_a = 0$, $X \lrcorner \theta'$ depends only on the restriction of X_{ab} to Σ; that is, only on the variation of the induced metric on Σ. We have derived the standard result that $p^{ab} n \lrcorner \epsilon$ is 'canonically conjugate' to the intrinsic metric on Σ.

Both S and S' are invariant under general diffeomorphisms of Q. Consequently any (complete) vector field v determines a symmetry of S and S' by Lie dragging. The corresponding vector field on $\mathcal{M}$ is $X_{ab} = -\mathcal{L}_v g_{ab} = -2\nabla_{(a}v_{b)}$. If v vanishes in a neighbourhood of the Cauchy surface that we use to evaluate ω, then $X \lrcorner \omega = 0$. However any $v \in V(Q)$ can be written $v = v' + v''$, where v' vanishes in a neighbourhood of a Cauchy surface Σ', and v'' vanishes in a neighbourhood of a disjoint Cauchy surface Σ''. The corresponding vector fields on $\mathcal{M}$ satisfy $X = X' + X''$, with $X' \lrcorner \omega = X'' \lrcorner \omega = 0$. It follows that $X \lrcorner \omega = 0$ for any vector field on Q.

Thus vector fields on $\mathcal{M}$ corresponding to infinitesimal diffeomorphisms of Q are tangent to the characteristic distribution of ω. The reduction of $\mathcal{M}$ identifies solutions that differ by diffeomorphisms of Q.

When Σ is not compact, one must take careful account of the boundary conditions. The argument still works for vector fields on Q with compact support, but fails for vector fields that look like translations at spatial infinity: the corresponding Xs cannot be decomposed within the class for which ω is defined. The Hamiltonians of these Xs are given by 2-surface integrals at infinity, which represent the energy and momentum of the gravitational field.

The Hilbert form of the action (with the boundary term) leads to a parametrization of $\mathcal{M}$ by the values of the intrinsic metric q and the tensor-valued 3-form $(n \lrcorner \epsilon)p$ on Σ. These can be regarded as canonical coordinates on $T^*\mathcal{Q}$, where $\mathcal{Q}$ is the space of metrics on the 3-manifold. However, $\mathcal{M}$ is not the whole of $T^*\mathcal{Q}$ because p and q cannot be specified arbitrarily: they must satisfy certain differential constraint equations. There is also the problem that not all of $\mathcal{M}$ is contained in $T^*\mathcal{Q}$ because a fixed hypersurface in Q will not be a Cauchy surface for every vacuum metric on Q.

Consequently even a formal canonical treatment of quantum gravity is very hard. From the present point of view, one of the difficulties is that the real foliation of $\mathcal{M}$ given by fixing the intrinsic metric on Σ does not descend to the reduced phase space.

Ashtekar (1988) has found another form of the action, by a variant on the Palatini trick of treating the metric and connection as independent variables, which greatly simplifies the form of the constraints. In the version described by Mason and Frauendiener (1990), the variables are an $\mathrm{sl}(2,\mathbb{C})$-valued 1-form Γ and an $\mathrm{sl}(2,\mathbb{C})$-valued 2-form ζ; Γ is to be thought of as the local representative of the connection on the bundle $\mathcal{S} \to Q$ of 2-component unprimed spinors; and ζ as the map that assigns anti-self-dual 2-forms to symmetric spinors. That is, if ϕ^{AB} is a symmetric spinor field, then $\zeta(\phi)$ is the tensor equivalent of

$$\phi_{AB}\epsilon_{A'B'}.$$

However, these interpretations are not imposed in advance. The action is the integral of the 4-form

$$\tfrac{1}{2}\int \mathrm{tr}(F \wedge \zeta),$$

where F is the curvature of Γ. The connection is unconstrained initially, but the ζs must satisfy algebraic constraints that allow the determination of a real Lorentzian metric. The field equations obtained by varying Γ force the connection to be compatible with the metric; and the field equations obtained by varying ζ are Einstein's equations. The new 1-form θ is given by

$$X \lrcorner\, \theta = \int_\Sigma \mathrm{tr}\,(\delta\Gamma \wedge \zeta),$$

where $\delta\Gamma$ is the variation in Γ. Again the 2-form $\mathrm{d}\theta$ is degenerate, and reduction removes the 'gauge freedom'. This suggests that one might be able to construct a complex polarization by using the Γs as holomorphic coordinates. Unfortunately, in spite of some remarkable simplifications in Ashtekar's approach, this idea does not work because the polarization again fails to survive reduction.

7.4 Polarizations

It is generally hard to find polarizations in the phase space of a classical field, particularly if one requires invariance under the symmetries of the field equations.

When the space of solutions is symplectic, the foliation P_Σ (p. 141) may restrict to a real polarization of $\mathcal{M} \subset \mathcal{F}$. In the case of the Klein-Gordon field, for example, the solutions that have the same values at points of Σ are the leaves of a real polarization. In fact one can regard the values of the Cauchy data $(\phi, \dot{\phi})$ at each point of Σ as canonical coordinates. But a different choice of Σ gives a different foliation: in flat space-time, P_Σ is not invariant under the action of the Poincaré group on $\mathcal{M}$. When ω is degenerate, P_Σ may not survive reduction.

The linear relativistic wave equations are exceptional in that their phase spaces have invariant Kähler polarizations given by the decomposition into positive and negative frequency parts. These polarizations are central to the particle interpretation of relativistic quantum field theories.

Scalar equations

Any real solution of the Klein-Gordon equation $\Box\phi + \mu^2\phi = 0$ in Minkowski space can be written as a superposition of plane waves of the form

$$\mathrm{e}^{-\mathrm{i}k_a x^a} \quad \text{where} \quad k_a k^a = \mu^2.$$

In fact by taking the Fourier transforms of ϕ and $\dot{\phi} = \partial\phi/\partial t$ at $t = x^0 = 0$, we have

$$\phi(x) = \left(\frac{1}{2\pi}\right)^{3/2} \int E^{-1}(\psi^+ \mathrm{e}^{-\mathrm{i}Et} + \psi^- \mathrm{e}^{\mathrm{i}Et}) \exp(\mathrm{i}\mathbf{k}\cdot\mathbf{x})\, \mathrm{d}^3 k \qquad (7.4.1)$$

where $\mathbf{x} = (x^1, x^2, x^3)$, $\mathbf{k} = (k_1, k_2, k_3)$, E is the positive square root of $\mu^2 + \mathbf{k}\cdot\mathbf{k}$, and ψ^+ and ψ^- are the functions of $\mathbf{k}$ defined by

$$\psi^\pm = \left(\frac{1}{2\pi}\right)^{3/2} \int_{t=0} \tfrac{1}{2}(E\phi \pm \mathrm{i}\dot{\phi}) \exp(-\mathrm{i}\mathbf{k}\cdot\mathbf{x})\, \mathrm{d}^3 x.$$

The right-hand side of (7.4.1) satisfies the Klein-Gordon equation and has the same Cauchy data as ϕ on $t = 0$. It is therefore equal to ϕ everywhere.

Equation (7.4.1) takes on an invariant form when written as an integral over the hyperboloid $H_\mu = \{k_a k^a = \mu^2\}$ in the space of four-vectors. There are two sheets, H_μ^+, on which k is future pointing, and H_μ^-, on which k is past pointing; and there is an invariant volume element $\mathrm{d}\tau$ induced by the Lorentz metric. With k_1, k_2, k_3 as coordinates on H_μ,

$$\mathrm{d}\tau = \frac{\mathrm{d}k_1 \wedge \mathrm{d}k_2 \wedge \mathrm{d}k_3}{|k_0|}.$$

If we define $\psi : H_\mu \to \mathbb{C}$ by $\psi = \psi^+$ on H_μ^+ and $\psi = \psi^-$ on H_μ^-, then

$$\phi(x) = \left(\frac{1}{2\pi}\right)^{3/2} \int_{H_\mu} \psi(k)\mathrm{e}^{-\mathrm{i}k_a x^a}\, \mathrm{d}\tau.$$

The reality condition on ϕ is $\psi(-k) = \overline{\psi}(k)$. A solution is said to be positive (negative) frequency whenever $\psi^- = 0$ ($\psi^+ = 0$). The negative frequency solutions form a Lagrangian subspace of $\mathcal{M}_\mathbb{C}$, for which the corresponding

Kähler polarization of $\mathcal{M}$ is positive. In fact, if we use it to turn $\mathcal{M}$ into a complex inner product space, then the Fourier transform identifies $\mathcal{M}$ with the space of complex-valued functions on H_μ^-, with the inner product

$$\langle\psi, \psi'\rangle = 4\int_{H_\mu^-} \overline{\psi}\psi' \,\mathrm{d}\tau.$$

The function ψ is not invariantly associated with ϕ: if we change the origin in Minkowski space, then we must also change the phase of ψ. We shall see later that this phase change can be accommodated in a natural way by interpreting ψ as a section of a line bundle over H_μ. On the other hand, the decomposition of ϕ into its positive and negative frequency parts by writing it as the sum of intergals over H_μ^+ and H_μ^- is invariant under the identity component of the Poincaré group. The polarization is also invariant.

In keeping with the intention to avoid technical issues, we have not considered the precise definition of $\mathcal{M}$. In quantum field theory, the inner product space associated with $\mathcal{M}$ and the positive-negative frequency decomposition is interpreted as the space of one particle states. We shall want this to be a Hilbert space, which it will be if we *define* $\mathcal{M}_{\mathbb{C}}$ to be the space of square-integrable functions on H_μ; this is equivalent to taking an appropriate completion of the space of smooth solutions of the Klein-Gordon equation with compactly supported Cauchy data. We have also not considered the meaning of 'polarization' in the context of infinite-dimensional manifolds. In informal terms, polarizations are defined in the same way as in finite dimensions, with the Lagrangian condition interpreted as meaning 'maximal isotropic'. This causes no problem in the case of massive fields, but is more awkward in the case of massless fields.

The Fourier decomposition of solutions of the wave equation $\Box\phi = 0$ is the same as in the massive case, except that the hyperboloid is replaced by the light-cone N, which is the set of nonzero null four-vectors. Again there are future and past components, N^+ and N^-, and so we can use the decomposition to define a complex structure in exactly the same way as in the massive case. The volume element $\mathrm{d}\tau$ is given by the same formula, and it is still invariant.

The awkward issue is the 'infra-red problem': because k_0 vanishes at the vertex of the light-cone, some care must be taken over the boundary conditions on ψ as $k \to 0$. In particular, if the Fourier decomposition is to be well-defined, then we must have $\psi(0) = 0$, and therefore

$$\int_{t=0} \dot{\phi} \,\mathrm{d}\sigma = 0.$$

This is not special to four dimensions.

Example. Consider the solutions of the two-dimensional wave equation of the form $\phi(t,x) = f(t+x)$, where f is a function of one variable. Restricted to this subspace, the symplectic structure is

$$\omega(f,g) = \tfrac{1}{2}\int_{-\infty}^{\infty} [f'(u)g(u) - g'(u)f(u)]\,du.$$

If $f \not\to 0$ as $u \to \pm\infty$, then ϕ does not have a well-defined decomposition into positive and negative frequency parts.

Suppose that, in addition to solutions for which the decomposition is well-defined, $\mathcal{M}$ contains solutions of the form $\phi = p + q\,\theta(t+x)$, where p and q are constant and $\theta(u)$ is the unit step function; that is $\theta = 0$ when $u < 0$ and $\theta = 1$ when $u > 0$. The positive frequency solutions and the constants then span a complex Lagrangian subspace $P \subset \mathcal{M}$. The corresponding polarization has one real direction. An example of this sort of structure occurs in the quantization of the radiative modes of the gravitational field (Ashtekar 1981, 1987).

The leaves of the coisotropic foliation E associated with P are labelled by the 'topological charge' $q = f(\infty) - f(-\infty)$. For each leaf Λ, P induces a positive polarization of Λ/D. The reduced phase spaces Λ/D are the classical analogues of the 'non-Fock' representation spaces in Streater and Wilde (1970).

Spinor wave equations

The fundamental example is the Dirac equation, which governs the wave function of an electron. The field variables are a pair of spinor fields $\alpha^A, \beta^{A'}$ on Minkowski space, which are usually combined to form a single four-component spinor[7] $\phi = (\alpha, \beta)$. The Lagrangian density L_D is the imaginary part of

$$\alpha^A \nabla_{AA'} \overline{\alpha}^{A'} + \beta^{A'} \nabla_{AA'} \overline{\beta}^A + \sqrt{2}\mu \alpha_A \overline{\beta}^A, \tag{7.4.2}$$

which generates the field equations

$$\nabla_{AA'}\alpha^A = \frac{\mu}{\sqrt{2}}\beta_{A'}, \quad \nabla_{AA'}\beta^{A'} = \frac{\mu}{\sqrt{2}}\alpha_A \tag{7.4.3}$$

and the symplectic form

$$\omega(\phi_1,\phi_2) = -\frac{\mathrm{i}}{2}\int_\Sigma (\alpha_1^A\overline{\alpha}_2^{A'} + \beta_1^{A'}\overline{\beta}_2^A - \alpha_2^A\overline{\alpha}_1^{A'} - \beta_2^{A'}\overline{\beta}_1^A) n_{AA'}\, d\sigma,$$

where n^a is the unit normal to Σ (we are regarding the space of solutions as a real vector space). The equations (7.4.3) are the two-component form of the Dirac equation.

An immediate consequence of (7.4.2) is that $\beta^{A'}$ (or equally α^A) satisfies the Klein-Gordon equation

$$\Box\beta^{A'} + \mu^2\beta^{A'} = 0.$$

Nothing is lost by taking this as fundamental and treating the second equation in (7.4.2) as the *definition* of α.

In the same way as a scalar field, we can write β as a superposition of plane waves of the form

$$\beta^{A'} = \left(\frac{1}{2\pi}\right)^{3/2} \int_{H_\mu} \psi^{A'} \mathrm{e}^{-\mathrm{i}k_a x^a} \, \mathrm{d}\tau$$

where $\psi^{A'}$ is a spinor-valued function on H_μ. Then the symplectic form becomes

$$\omega(\phi_1, \phi_2) = \mathrm{i} \int_{H_\mu} \mu^{-2} k_{AA'} (\overline{\psi}_1^A \psi_2^{A'} - \overline{\psi}_2^A \psi_1^{A'}) \zeta \, \mathrm{d}\tau.$$

where $\zeta = 1$ on H_μ^+ and $\zeta = -1$ on H_μ^-.

The general massive wave equation is

$$(\Box + \mu^2)\beta^{A'B'\ldots C'} = 0, \tag{7.4.4}$$

where $\beta^{A'\ldots C'}$ is symmetric, with n indices (n is twice the spin of the field). This reduces to the complex Klein-Gordon equation when $n = 0$ and to the Dirac equation when $n = 1$. The general Fourier decomposition is

$$\beta^{A'B'\ldots C'} = \left(\frac{1}{2\pi}\right)^{3/2} \int_{H_\mu} \psi^{A'B'\ldots C'} \mathrm{e}^{-\mathrm{i}k_a x^a} \, \mathrm{d}\tau$$

and, with appropriate scaling, the symplectic form on the space of solutions is the imaginary part of

$$-\int_{H_\mu} k_{AA'} \ldots k_{CC'} \overline{\psi}_1^{A\ldots C} \psi_2^{A'\ldots C'} \zeta \, \mathrm{d}\tau. \tag{7.4.5}$$

When $n = 0$, the integrand is $\zeta\overline{\psi}_1\psi_2\mathrm{d}\tau$ and the symplectic form coincides with (7.2.6). In the general case, it can be written as an integral over a spacelike hyperplane of a divergence-free vector field constructed from β and its derivatives.

We have seen (§6.7) that the energy in an inertial coordinate system is the generator of the one-parameter family of time translations $\rho_{t'}$: $(t, x, y, z) \mapsto (t - t', x, y, z)$. These act on the spinor fields $\beta^{A'B'\ldots C'}$ by the push-forward map $\rho_{t'*}$, and hence on the ψs by

$$\psi_{A'B'\ldots C'} \mapsto \mathrm{e}^{-\mathrm{i}k_0 t'} \psi_{A'B'\ldots C'}.$$

The energy is therefore

$$\frac{\mathrm{d}}{\mathrm{d}t'}\left(\omega(\beta, \rho_{t'*}\beta)\right)_{t'=0} = \int_{H_\mu} \zeta k_0 k_{AA'} \ldots k_{CC'} \overline{\psi}^{A\ldots C} \psi^{A'\ldots C'} \, \mathrm{d}\tau.$$

On H_μ^+, $k_0 > 0$ and k_a is future-pointing; on H_μ^-, $k_0 < 0$ and k_a is past-pointing. Thus the integrand is positive on H_μ^+ and H_μ^- whenever n is even, but is negative on H_μ^- when n is odd.

This is the 'classical' form of the spin-statistics problem (Pauli 1940, Streater and Wightman 1964). When n is even (the bosonic case) it is possible to find an invariant positive complex structure on the solution space such that the energy becomes a positive operator on the corresponding quantum Hilbert space. The complex structure is defined not the obvious way by $\beta \mapsto \mathrm{i}\beta$, which is not positive, but by multipling the positive frequency part by $-\mathrm{i}$ and the negative frequency part by i.

In the odd case, the positive complex structure multiplies both parts of the field by $-\mathrm{i}$, but the energy operator on the corresponding Hilbert space (the generator of time translations) then turns out not to be a positive operator. To obtain a physically sensible theory, we are forced to use fermionic quantization, which takes as its starting point the real part of (7.4.5), which is symmetric.

Massless spinor equations

A fundamental example is Maxwell's equations in spinor form. If we write

$$F_{ab} \sim \phi_{A'B'}\epsilon_{AB} + \overline{\phi}_{AB}\epsilon_{A'B'},$$

then Maxwell's equations reduce to

$$\nabla^{AA'}\phi_{A'B'} = 0.$$

The Fourier decomposition is

$$\phi_{A'B'} = \left(\frac{1}{2\pi}\right)^{3/2} \int_N \psi z_{A'} z_{B'} \mathrm{e}^{-\mathrm{i}k_a x^a}\, \mathrm{d}\tau$$

where N is the light-cone, ψ is a complex function on N and $z_{A'}$ is a spinor-valued function on N such that $k_{AA'} = \zeta \overline{z}_A z_{A'}$, with $\zeta = 1$ on the future light-cone and $\zeta = -1$ on the past light-cone. The phase of $z_{A'}$ cannot be made to vary continuously with k, a difficulty that is resolved by interpreting $\psi z^{A'}$ as a section of a nontrivial line bundle over N_0.

In this case the symplectic form is given by

$$\omega(F, F') = \mathrm{i} \int_N (\overline{\psi}\psi' - \overline{\psi}'\psi)\zeta\, \mathrm{d}\tau,$$

We obtain a positive complex structure compatible with ω by $J(\phi) = -\mathrm{i}\phi^+ + \mathrm{i}\phi^-$, where ϕ^+ and ϕ^- are the positive and negative frequency parts of ϕ.

There is a second invariant complex structure given by $\phi \mapsto \mathrm{i}\phi$, which sends F to F^*, but it is not positive.

In spinor form, Maxwell's equations are a special case of the zero-rest mass field equations

$$\nabla^{AA'}\phi_{A'B'\dots C'} = 0,$$

for which the Fourier decomposition is

$$\phi_{A'\dots C'} = \left(\frac{1}{2\pi}\right)^{3/2} \int_N \psi z_{A'} \dots z_{C'} \mathrm{e}^{-\mathrm{i}k_a x^a} \, \mathrm{d}\tau,$$

from which we again obtain a positive polarization.

7.5 Nonhyperbolic examples

In the hyperbolic case, the construction of the closed 2-form on the solution manifold depends on the fact that the boundary of the region on which the fields are defined has two components, the initial and final Cauchy surfaces. There is a corresponding decomposition of the variations (the tangent vectors to the space of sections of $E \to Q$) and of the gradient of the action: each variation is the sum of a variation that vanishes on the initial Cauchy surface and a variation that vanishes on the final Cauchy surface. At a solution, the action is constant along variations that vanish on both hypersurfaces. Consequently its gradient at points of the solution submanifold can be written in a unique way as the difference of two 1-forms, θ_0 and θ_1. The symplectic or presymplectic form is the restriction to the space of solutions of $\mathrm{d}\theta_0$.

This is not the only way in which symplectic structures can arise in field theory, as the following examples illustrate. The first shows how the action in a boundary value problem, together with a complex splitting of the variations at the boundary, can give a symplectic structure on the space of solutions of an elliptic field equation. The second shows how a space of 'fields' can inherit a symplectic structure from the background manifold.

Example. *Elliptic equations.* Let D be the unit disc in the x, y-plane, bounded by the unit circle ∂D. We shall consider the manifold $\mathcal{M}$ of real solutions of the equation $\nabla^2\phi = f_\phi(x, y, \phi)$ on the closure $\overline{D}$, where f is some given function and the subscript denotes the partial derivative with respect to ϕ. The solutions are the critical points of the action

$$S_D = \int_D \left(\tfrac{1}{2}\nabla\phi \cdot \nabla\phi + f(x, y, \phi)\right) \mathrm{d}A,$$

where the variations are made with the values of ϕ kept fixed on ∂D. The space of fields is the space $\mathcal{F}$ of all smooth real functions on $\overline{D}$. A tangent vector X to $\mathcal{F}$ is again a smooth function on $\overline{D}$.

An analogy with the wave equation might suggest

$$\tfrac{1}{2} \int_{\partial D} \mathbf{n} \cdot (X'\nabla X - X\nabla X') \, \mathrm{d}s$$

as a candidate for a symplectic form, where $\mathbf{n}$ is the outward pointing normal. However, this vanishes identically on $\mathcal{M}$, since it is the exterior derivative of $\mathrm{d}S_D|_{\mathcal{M}}$.

As a substitute for splitting a general variation X into parts that vanish on the future and past components of boundary, we shall use the Fourier decomposition on ∂D to to write $X = X^+ + X^-$, where $X^- = \overline{X}^+$, and

$$X^+ = \tfrac{1}{2}a_0 + \sum_{n>0} a_n \mathrm{e}^{\mathrm{i}n\alpha} \quad \text{and} \quad X^- = \tfrac{1}{2}a_0 + \sum_{n<0} a_n \mathrm{e}^{\mathrm{i}n\alpha}$$

on the boundary. Here $\mathrm{e}^{\mathrm{i}\alpha} = x + \mathrm{i}y$ on ∂D. Any $X \in C^\infty(\overline{D})$ can be decomposed in this way, but not, of course, uniquely: there is the freedom to replace X^+ and X^- by $X^+ + Z$ and $X^- - Z$, respectively, where Z is any smooth real function that vanishes on ∂D. However, if Z is such a function, and ϕ is a solution, then $Z \lrcorner \mathrm{d}S_D(\phi) = 0$. Hence at points of the solution space, $\mathrm{d}S_D$ has a well-defined decomposition

$$\mathrm{d}S_D = \mathrm{i}\theta - \mathrm{i}\overline{\theta},$$

where $X^+ \lrcorner \theta = 0 = X^- \lrcorner \overline{\theta}$ for any X. At $\phi \in \mathcal{M}$,

$$X \lrcorner \theta = -\mathrm{i} \int_{\partial D} X^- \mathbf{n} \cdot \nabla\phi \, \mathrm{d}s.$$

The exterior derivative, $\omega = \mathrm{d}\theta|_{\mathcal{M}}$, is a well-defined closed 2-form on $\mathcal{M}$. Because $(\overline{\theta} - \theta)|_{\mathcal{M}}$ is exact, ω is real.

For example, in the case of Poisson's equation, $f(x, y, \phi) = \rho(x, y)\phi$, where ρ is the matter density, and the tangent vectors to $\mathcal{M}$ are harmonic functions on $\overline{D}$. In this case, we can decompose a tangent vector as $X = X^+ + X^-$, where X^+ is a holomorphic function of $x + \mathrm{i}y$ on D and X^- is an antiholomorphic function. The 2-form on $\mathcal{M}$ is given by

$$\begin{aligned}
\omega(X, X') &= \frac{\mathrm{i}}{2} \int_{\partial D} (X^- \mathbf{n} \cdot \nabla X' - X'^- \mathbf{n} \cdot \nabla X) \, \mathrm{d}s \\
&= \frac{\mathrm{i}}{2} \int_{\partial D} (X^- \mathbf{n} \cdot \nabla X'^+ - X'^- \mathbf{n} \cdot \nabla X^+) \, \mathrm{d}s \\
&= \tfrac{1}{2} \int_D (\partial X \wedge \overline{\partial} X' - \partial X' \wedge \overline{\partial} X) \\
&= \tfrac{1}{2} \int_D \mathrm{d}X \wedge \mathrm{d}X'.
\end{aligned}$$

It is real, but degenerate. Reduction identifies the gravitational potentials that differ by a constant. The holomorphic Xs span a polarization. ∎

Example. *Gauge fields on a symplectic manifold* (Donaldson and Kronheimer 1990). Let $B \to M$ be a Hermitian vector bundle over a compact $2n$-dimensional symplectic manifold (M, ω), and let $\mathcal{F}$ denote the space of connections on B (compatible with its Hermitian structure). A tangent vector to $\mathcal{F}$ at a connection ∇ is a 1-form α with values in the Hermitian matrices. The symplectic structure is

$$\omega(\alpha, \alpha') = \int_M \operatorname{tr}(\alpha \wedge \alpha')\, \omega^{n-1}.$$

This makes a notable appearance in the application of geometric quantization in Jones-Witten theory (Axelrod, Della Pietra, and Witten 1991, Atiyah 1990): here M is a Riemann surface and B is trivial. The quantization is of the reduction of the subspace of flat connections: the reduced symplectic structure can also derived from the 'Chern-Simons' action on $M \times \mathbb{R}$.

8

PREQUANTIZATION

8.1 The quantum conditions

THE first problem of quantization concerns the kinematic relationship between the classical and quantum domains. At the quantum level, the states of a physical system are represented by the rays in a Hilbert space $\mathcal{H}$ and the observables by a collection $\mathcal{O}$ of symmetric operators on $\mathcal{H}$, while in the limiting classical description, the state space is a symplectic manifold (M, ω) and the observables are the smooth functions on M. The kinematic problem is: given M and ω, is it possible to to reconstruct $\mathcal{H}$ and $\mathcal{O}$?

According to Dirac's general principles, the canonical transformations of M generated by the classical observables should correspond to the unitary transformations of $\mathcal{H}$ generated by the quantum observables, and Poisson brackets of classical observables should correspond to commutators of quantum observables. Each classical observable $f : M \to \mathbb{R}$ should correspond to an operator $\hat{f} \in \mathcal{O}$ such that

(Q1) the map $f \mapsto \hat{f}$ is linear (over $\mathbb{R}$);

(Q2) if f is constant, then $\hat{f}$ is the corresponding multiplication operator;

(Q3) if $[f_1, f_2] = f_3$, then $\hat{f}_1 \hat{f}_2 - \hat{f}_2 \hat{f}_1 = -i\hbar \hat{f}_3$.

These are Dirac's quantum conditions (Dirac 1925): they imply that the quantum operators form a Hilbert space representation of the classical observables. In his terms, the Poisson bracket is the *classical analogue* of the quantum commutator.

The quantum conditions do not determine the underlying quantum system uniquely. Nor is every representation of the classical observables a physically reasonable quantization. First Q2 must hold. This ensures that a pair of complementary observables, such as a position coordinate and its conjugate momentum, that have a constant Poisson bracket at the classical level, should have the commutator at the quantum level required by the uncertainty principle. Second, some sort of irreducibility conditon is needed to restrict the 'size' of $\mathcal{H}$. This must be such that $f \mapsto \hat{f}$ gives an irreducible representation of the generators of the symmetry group whenever M is a coadjoint orbit, so that there is a correspondence between the notions of an 'elementary system' at the classical and quantum levels.

We shall see that it is not possible to have a one-to-one correspondence between $\mathcal{O}$ and $C^\infty(M)$ without making $\mathcal{H}$ too large.[1] In all quantization schemes, some additional structure in M is needed to pick out a subalgebra of classical observables for which the quantum conditions hold. This can be thought of as the *shadow* of the quantum system in the classical system. In geometric quantization, the structure is a polarization of M, and the freedom in its selection is the second point at which we come across an ambiguity in the passage from the classical to the quantum domain. The first was in the selection of the Lagrangian (§2.6).

A more difficult problem is the dynamical relationship between classical and quantum mechanics. Here we are interested in the connection between the flow generated by the classical Hamiltonian H and the corresponding time evolution in $\mathcal{H}$. This is straightforward enough at the formal level when H is in the subalgebra for which the quantum conditions hold (we shall see that this is when the canonical flow preserves the polarization), but is much harder in the more usual situation in which it is not.

To start with, we shall set aside this problem, and the issue of irreducibility, and look at prequantization. This is a very general and elegant construction, and it is the basic element in all the quantizations we shall look at later on.[2] It is the first step in a two stage procedure, in which we first introduce wave functions on phase space, and then pick out the subspace of physical wave functions (for example, the wave functions that depend on q, but not p) by introducing a polarization. The second stage is the subject of the next chapter.

8.2 Prequantization

A symplectic manifold (M, ω) is oriented and has a natural volume element

$$\epsilon = \left(\frac{1}{2\pi\hbar}\right)^n \mathrm{d}p_1 \wedge \mathrm{d}p_2 \wedge \cdots \wedge \mathrm{d}p_n \wedge \mathrm{d}q^1 \wedge \mathrm{d}q^2 \wedge \cdots \wedge \mathrm{d}q^n,$$

so there is an obvious Hilbert space associated with M: the space $L^2(M)$ of square-integrable complex functions on M, with the inner product

$$\langle \psi, \psi' \rangle = \int_M \overline{\psi}\psi' \epsilon.$$

Apart from a normalization factor, ϵ is equal to ω^n. The normalization will be useful later on—it allows ψ and ψ' to be dimensionless, since $\hbar$ and ω both have dimensions of 'action' ($\mathrm{ML^2T^{-1}}$).

Each classical observable f acts on $L^2(M)$ as a symmetric operator by

$$\psi \mapsto -\mathrm{i}\hbar X_f \psi. \tag{8.2.1}$$

Although this correspondence satisfies Q1 and Q3, it does not satisfy Q2 because the constants in $C^\infty(M)$ are mapped to the zero operator, so that, for example, position and momentum commute. One could try to correct this by replacing the (8.2.1) by $\psi \mapsto -\mathrm{i}\hbar X_f(\psi) + f\psi$. Then the constants would act by multiplication, but Q3 would no longer hold since the commutator of $-\mathrm{i}\hbar X_{f_1} + f_1$ and $-\mathrm{i}\hbar X_{f_2} + f_2$ is the operator

$$-\mathrm{i}\hbar(-\mathrm{i}\hbar X_{f_3} + f_3) - \mathrm{i}\hbar f_3,$$

where $f_3 = [f_1, f_2]$.

The unwanted additional contribution $-\mathrm{i}\hbar f_3$ is removed by adding yet another term and mapping $f \in C^\infty(M)$ to the operator

$$\hat{f}(\psi) = -\mathrm{i}\hbar \left[X_f(\psi) - \frac{\mathrm{i}}{\hbar}(X_f \lrcorner\, \theta)\psi \right] + f\psi, \tag{8.2.2}$$

where θ is a symplectic potential. The constants still act by multiplication, but now the commutator $[\hat{f}_1, \hat{f}_2]$ is $-\mathrm{i}\hbar \hat{f}_3$, since

$$X_{f_1}(X_{f_2} \lrcorner\, \theta) - X_{f_2}(X_{f_1} \lrcorner\, \theta) = X_{[f_1, f_2]} \lrcorner\, \theta + [f_1, f_2]. \tag{8.2.3}$$

There is, however, a price to pay: the definition of $\hat{f}$ depends on the choice of θ. As it stands, the construction only works if ω is exact, and even then it does not look natural. The way out is to allow gauge transformations: if we replace θ by $\theta' = \theta + \mathrm{d}u$, where u is real, then $\hat{f}$ is replaced by $\hat{f}' = \hat{f} - X_f \lrcorner\, \mathrm{d}u$. However,

$$\hat{f}'\left(\mathrm{e}^{\mathrm{i}u/\hbar}\psi\right) = \mathrm{e}^{\mathrm{i}u/\hbar}\left(\hat{f}'(\psi) + (X_f \lrcorner\, \mathrm{d}u)\psi\right) = \mathrm{e}^{\mathrm{i}u/\hbar}\hat{f}(\psi).$$

So if, at the same time, we change the phases of the wave functions by replacing ψ by

$$\psi' = \mathrm{e}^{\mathrm{i}u/\hbar}\psi, \tag{8.2.4}$$

then $\hat{f}$ becomes a well-defined operator, independent of the choice of symplectic potential. The transition amplitudes are unaffected, because the phases at the same point of M are all transformed in the same way; but it is no longer possible to keep track of the phases of wave functions at different points of M—a fact that will be important in the context of the Bohm-Aharonov effect and Berry's phase.

This is the intuitive idea behind prequantization, but it is not the whole story: u is determined by θ and θ' only up to the addition of a constant. Given θ, θ' and ψ, there is an ambiguity in the overall phase of ψ'. To take account of this, and to make the construction consistent when ω is not exact, $\hat{f}$ must be made to act not on functions subject to gauge transformations, but on sections of a Hermitian line bundle over M.

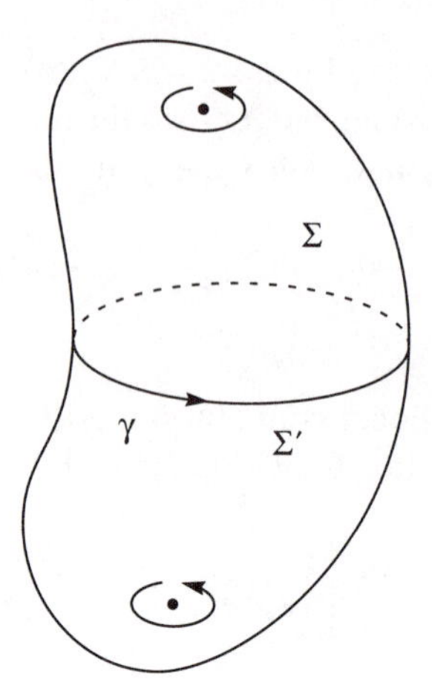

Fig. 8.1. The integrality condition.

8.3 The integrality condition

By comparing (8.2.2) with (A.3.2), it is clear that what is needed is a Hermitian line bundle $B \to M$ and a connection ∇ on B with potentials of the form $\hbar^{-1}\theta$; that is, with curvature $\hbar^{-1}\omega$. Such a bundle and connection exist if and only if ω satisfies Weil's integrality condition (Weil 1958, Kostant 1970a).

The integrality condition is closely related to the quantization rule in the old quantum theory (Messiah 1961, Chapter I, §15, Simms 1973). Its geometric origin can be understood by considering parallel transport with respect to ∇ around a closed curve γ through $m \in M$. Suppose that γ can be spanned by a surface Σ, and that Σ is contained in the domain of θ. Then the result is equivalent to the linear transformation $B_m \to B_m$ given by multiplication by

$$\exp\left(\frac{\mathrm{i}}{\hbar}\oint_\gamma \theta\right) = \exp\left(\frac{\mathrm{i}}{\hbar}\int_\Sigma \omega\right).$$

By dividing Σ into small pieces, it is easy to see that the factor is still given by the expression on the right-hand side even when Σ is not contained in the domain of a single potential.

We must get the same result when we span γ by a second 2-surface Σ'. Hence, taking account of the relationship between the orientations of γ, Σ, and Σ' (Fig. 8.1), the integral of $\hbar^{-1}\omega$ over $\Sigma \cup \Sigma'$ must be an integer multiple of 2π. So if $\hbar^{-1}\omega$ is to be the curvature of a connection, then ω must satisfy the following form of the integrality condition.

IC1. The integral of ω over any closed oriented 2-surface in M is an integral multiple of $2\pi\hbar$.

This is necessary for the existence of B and ∇. When M is simply connected, we can show that it is also sufficient by reconstructing ∇ from its holonomy. Suppose that (IC1) holds. Pick a base point $m_0 \in M$ and consider the set of all triples (m, z, γ), where $m \in M$, $z \in \mathbb{C}$ and γ is a piecewise smooth path from m_0 to m. Two triples (m, z, γ) and (m', z', γ') are defined to be equivalent whenever $m = m'$ and

$$z' = z \exp\left(\frac{\mathrm{i}}{\hbar}\int_\Sigma \omega\right),$$

where Σ is any oriented 2-surface with boundary made up of γ (oriented from m_0 to m) and γ' (oriented from m to m_0). Because M is simply connected, such a surface exists; and because the integrality condition holds, it does not matter which surface is chosen. The set of equivalence classes has the structure of a line bundle $B \to M$.

Addition and scalar multiplication within the fibres are defined by

$$[(m, z, \gamma)] + [(m, z', \gamma)] = [(m, z + z', \gamma)] \quad \text{and} \quad c[(m, z, \gamma)] = [(m, cz, \gamma)],$$

where $c \in \mathbb{C}$ and the square brackets denote equivalence classes. The local trivializations are determined by symplectic potentials. Suppose that θ is a potential on some simply-connected open set $U \subset M$. Pick a point m_1 in U and a curve γ_0 from m_0 to m_1; and define a section s of B in U by taking $s(m)$ to be the class of

$$\left(m, \exp\left(-\frac{\mathrm{i}}{\hbar}\int_{\gamma_1} \theta\right), \gamma\right)$$

where γ_1 is any curve from m_1 to m in U and γ is the curve from m_0 to m made up of γ_0 and γ_1 (Fig. 8.2). Since $\mathrm{d}\theta = \omega$, a different choice of γ_1 gives an equivalent triple and hence the same value of $s(m)$. A different choice of m_1 or γ_0 gives the same section, but multiplied by a constant of modulus one. The effect of replacing θ by $\theta' = \theta + \mathrm{d}u$, where $u(m_1) = 0$, is to replace s by $s' = \mathrm{e}^{-\mathrm{i}u/\hbar}s$. We can therefore define a global connection and a Hermitian structure on B by

$$\nabla s = -\frac{\mathrm{i}}{\hbar}\theta s \quad \text{and} \quad (s, s) = 1,$$

independently of the local choices of m_1, γ_1 and θ. The curvature is $\hbar^{-1}\omega$.

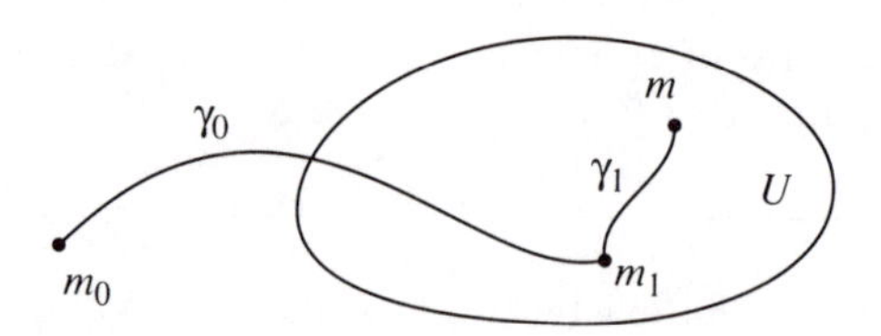

Fig. 8.2. The construction of B.

It is harder to deal in the same direct way with the general case in which M is not simply connected since it turns out that B and ∇ are not unique—there are different possibilities for the holonomy of ∇ around closed loops that cannot be contracted to points. Instead, we shall take as our starting point a different form of the integrality condition. See Weil (1958, p. 87) and, for example, Vaisman (1973).

> **IC2.** The class of $(2\pi\hbar)^{-1}\omega$ in $H^2(M, \mathbb{R})$ lies in the image of $H^2(M, \mathbb{Z})$.

In other words, there is an open cover $U = \{U_j\}$ such that the class of $(2\pi\hbar)^{-1}\omega$ in $H^2(U, \mathbb{R})$ contains a cocycle z in which all the z_{ijk}s are integers (see §A.6).

> **Proposition (8.3.1).** There exists a Hermitian line bundle $B \to M$ and a connection ∇ on B with curvature $\hbar^{-1}\omega$ if and only if IC2 holds. When IC2 holds, the inequivalent choices of B and ∇ are parametrized by $H^1(M, T)$, where $T \subset \mathbb{C}$ is the circle group.

Proof. Suppose that the condition holds. Then there is a contractible open cover $\{U_j\}$ of M, a collection of symplectic potentials $\theta_j \in \Omega^1(U_j)$, and a collection of functions $u_{jk} \in C^\infty(U_j \cap U_k)$ such that

(1) $\mathrm{d}u_{jk} = \theta_j - \theta_k$ whenever $U_j \cap U_k \neq \emptyset$; and

(2) $(2\pi\hbar)^{-1}(u_{jk} + u_{kl} + u_{lj}) \in \mathbb{Z}$ whenever $U_j \cap U_k \cap U_l \neq \emptyset$.

Put $c_{jk} = \exp(\mathrm{i}u_{jk}/\hbar)$. Then

$$\frac{\mathrm{d}c_{jk}}{c_{jk}} = \frac{\mathrm{i}}{\hbar}(\theta_j - \theta_k),$$

whenever $U_j \cap U_k \neq \emptyset$; and

$$c_{jk}c_{kl}c_{lj} = \exp\left(\frac{2\pi\mathrm{i}}{\hbar}(u_{jk} + u_{kl} + u_{lj})\right) = 1$$

whenever $U_j \cap U_k \cap U_l \neq \emptyset$ (there is no summation over the repeated indices). It follows that the cs are the transition functions of a line bundle $B \to M$ and that the $\hbar^{-1}\theta$s determine a connection on B with curvature $\hbar^{-1}\omega$ (see §A.3). Since the potentials are real and the transition functions are of unit modulus, there is also a compatible Hermitian structure $(\cdot, \cdot)$.

Suppose, conversely, that we are given B and a connection on B with curvature $\hbar^{-1}\omega$. Let $\{c_{jk}\}$ be the transition functions of B relative to some open cover. For each nonempty triple intersection, put

$$z_{jkl} = \frac{1}{2\pi\mathrm{i}}(\log c_{jk} + \log c_{kl} + \log c_{lj}).$$

Then, by (A.3.4), z_{jkl} is an integer, and hence constant. The zs satisfy the cocycle condition. There is an ambiguity in the definition of the logarithms since $(2\pi\mathrm{i})^{-1}\log c_{jk}$ is defined only up to the addition of an integer x_{jk}. However, the cohomology class $[z]$ of z in $H^2(M, \mathbb{Z})$ is independent of the choice of branches. It is called the *Chern class* of B. It follows from (A.3.6) that $2\pi\hbar z$ is a representative cocycle of the class in $H^2(M, \mathbb{R})$ determined by ω. Hence IC2 is also a necessary condition.

There is freedom in the construction of B and ∇ from ω since we can replace u_{jk} by $u_{jk} + y_{jk}$, where the ys are real constants chosen so that

$$y_{jk} = -y_{kj} \quad \text{and} \quad \frac{1}{2\pi\hbar}(y_{jk} + y_{kl} + y_{lj}) \in \mathbb{Z},$$

whenever $U_j \cap U_k \cap U_l \neq \emptyset$. The effect is to replace B by $B \otimes F$, where F is the Hermitian line bundle with transition functions[3]

$$t_{jk} = \exp(\mathrm{i}y_{jk}/\hbar) \in T.$$

Since these are constant, F has a connection with vanishing curvature, and so $B \otimes F$ has a connection with the same curvature as ∇.

Conversely, if (B, ∇) and (B', ∇') both have curvature $\hbar^{-1}\omega$, then $F = B^{-1} \otimes B'$ is a Hermitian line bundle with flat connection. Hermitian line bundles with flat connections are labelled by elements of $H^1(M, T)$, where T is the circle group: the correspondence is given by the transition functions of local trivializations in which the connection forms vanish. Thus the various choices for B and ∇ are also parametrized by $H^1(M, T)$. ∎

Note that when M is simply connected, $H^1(M, T) = 0$, and B and ∇ are unique up to equivalence.

We end this section with some remarks about trivializations of B and complex symplectic potentials. Any non-vanishing local section s determines a local trivialization and also a symplectic potential by $\theta s = \mathrm{i}\hbar\nabla s$. If $(s,s) = 1$, then θ is real. In general, however, θ is complex, with imaginary part equal to $\mathrm{d}(\frac{1}{2}\hbar\log(s,s))$. This follows from

$$\mathrm{d}(s,s) = (\nabla s, s) + (s, \nabla s) = \frac{\mathrm{i}}{\hbar}(\bar{\theta} - \theta)(s,s).$$

Conversely, if we are given a (real or complex) symplectic potential θ on a simply-connected neighbourhood $U \subset M$, then we can define a non-vanishing section $s : U \to B$ by picking any $m_0 \in U$ and $b_0 \in B_{m_0}$ and putting

$$s(m) = b\exp\left(-\frac{\mathrm{i}}{\hbar}\int_\gamma \theta\right),$$

where γ is a curve from m_0 to m and b is obtained from b_0 by parallel transport along γ. The value of $s(m)$ is independent of γ. Moreover, $\mathrm{i}\hbar\nabla s = \theta s$, so s determines a local trivialization in which the connection form is $\hbar^{-1}\theta$.

8.4 Prequantization of canonical transformations

A symplectic manifold (M,ω) is said to be *quantizable* whenever ω satisfies the integrality condition. There then exists a Hermitian line bundle $B \to M$ and a connection ∇ on B with curvature $\hbar^{-1}\omega$. We shall call B the *prequantum bundle*.

The Hilbert space of prequantization is the space $\mathcal{H}$ of square integrable sections $s : M \to B$, with the inner product

$$\langle s, s'\rangle = \int_M (s, s')\,\epsilon.$$

The operator $\hat{f}$ corresponding to $f \in C^\infty(M)$ is

$$\hat{f}s = -\mathrm{i}\hbar\nabla_{X_f}s + fs,$$

where s lies in some appropriate subset of $\mathcal{H}$. It is always symmetric; and when X_f is complete, it is actually self-adjoint. The definition also makes sense for complex f, but $\hat{f}$ is then no longer symmetric.

There is another way of looking at the 'quantum observable' $\hat{f}$ that may make the definition appear more natural. Let V_f be the vector field on B given in a local trivialization by[4]

$$V_f = X_f + \hbar^{-1}L\frac{\partial}{\partial\phi}$$

where $z = re^{\mathrm{i}\phi}$ is the coordinate on the fibre of B and

$$L = X_f \lrcorner \theta - f$$

is the Lagrangian of f. A direct calculation shows that V_f is invariant under gauge transformations, something that also follows from its intrinsic characterization by

$$\mathrm{pr}_* V_f = X_f \quad \text{and} \quad \hbar V_f \lrcorner \alpha = \hbar V_f \lrcorner \overline{\alpha} = f \circ \mathrm{pr},$$

where pr is the projection onto M and α is the connection form on the complement of the zero section in B (§A.3). In the local trivialization,

$$\alpha = \frac{1}{\hbar}\theta + \mathrm{i}\frac{\mathrm{d}z}{z}.$$

If g is another real function on M, then $[V_f, V_g] = V_{[f,g]}$, by making use, once again, of (8.2.3). The map $f \mapsto V_f$ is therefore a Lie algebra isomorphism from $C^\infty(M)$ to a subalgebra of the vector fields on B; Kostant (1970a) gives an intrinsic description of this subalgebra.

Let ξ_t denote the flow of V_f. This preserves the fibres of B and projects onto the canonical flow ρ_t of X_f. It also preserves the Hermitian inner product and the connection, since

$$\mathcal{L}_{V_f}\alpha = \mathrm{d}(V_f \lrcorner \alpha) + V_f \lrcorner \mathrm{d}\alpha = 0,$$

by using $\hbar \mathrm{d}\alpha = \mathrm{pr}^*\omega$.

The flow induces a linear 'pull-back' action $s \mapsto \hat{\rho}_t s$ on the sections of B, where

$$\xi_t\big(\hat{\rho}_t s(m)\big) = s(\rho_t m) \tag{8.4.1}$$

(Fig. 8.3). When X_f is not complete, $\hat{\rho}_t s(m)$ does not exist for all $(m,t) \in M \times \mathbb{R}$; but when it is complete, so is V_f, and $\hat{\rho}_t$ is a well defined linear transformation of the sections of B for any $t \in \mathbb{R}$. Its restriction to $\mathcal{H}$ is a one-parameter unitary group, which preserves the action of the quantum observables in the sense that, for any $h \in C^\infty(M)$, $\hat{\rho}_t \hat{h} = \hat{h}_t \hat{\rho}_t$, where $h_t = h \circ \rho_t$ (this follows from the fact that ξ_t preserves all the ingredients of prequantization—the bundle, the Hermitian structure, and the connection).

In a local trivialization, s and $\hat{\rho}_t s$ are represented by complex functions ψ and $\hat{\rho}_t \psi$, where

$$\hat{\rho}_t \psi(m) = \psi(\rho_t m) \exp\left(-\frac{\mathrm{i}}{\hbar}\int_0^t L \, \mathrm{d}t'\right). \tag{8.4.2}$$

The integration is along the integral curve of X_f from m to $\rho_t(m)$; that is, the curve $t' \mapsto \rho_{t'}(m)$. The reappearance of the action integral is significant. By taking the derivative, we find that

$$\frac{\mathrm{d}\hat{\rho}_t}{\mathrm{d}t} = \frac{\mathrm{i}}{\hbar}\hat{\rho}_t \hat{f}.$$

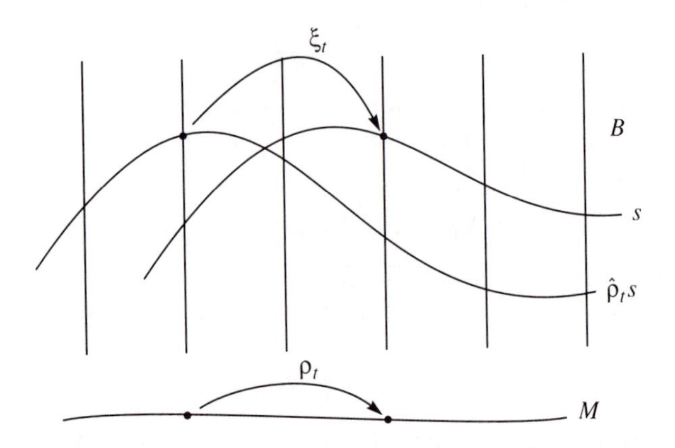

Fig. 8.3. The definition of $\hat{\rho}_t$.

It follows that $\hat{\rho}_t$ is generated by $\hat{f}$; and hence, by Stone's theorem, that $\hat{f}$ is self-adjoint whenever X_f is complete.

This extends in a straightforward way to the flow $\rho_{tt'} : M \to M$ of a time-dependent Hamiltonian vector field X_f, where $f \in C^\infty(M \times \mathbb{R})$. The formula for $\hat{f}$ is unchanged, but $\hat{f}$ now becomes a time-dependent operator on $\mathcal{H}$. The pull-back under the classical flow is a family of unitary operators $\hat{\rho}_{tt'} : \mathcal{H} \to \mathcal{H}$, such that $\hat{\rho}_{tt''} = \hat{\rho}_{tt'}\hat{\rho}_{t't''}$. They are defined by

$$\frac{\mathrm{d}\hat{\rho}_{tt'}}{\mathrm{d}t'} = \frac{\mathrm{i}}{\hbar}\hat{\rho}_{tt'}\hat{f}_{t'}, \quad \text{with initial condition} \quad \hat{\rho}_{tt} = 1,$$

where $\hat{f}_{t'}$ is the value of $\hat{f}$ at time t'. Eqn (8.4.2) becomes

$$\hat{\rho}_{tt'}\psi(m) = \psi(\rho_{tt'}m)\exp\left(-\frac{\mathrm{i}}{\hbar}\int_t^{t'} L\,\mathrm{d}t''\right). \tag{8.4.3}$$

The real functions on M are the generators of Hamiltonian vector fields. Under prequantization, they determine symmetric operators on $\mathcal{H}$. One might expect, by analogy, that canonical transformations of M should generate unitary transformations of $\mathcal{H}$. In certain circumstances, this is what happens, but the issue is not entirely straightforward: in general, there are problems over both existence and uniqueness.

One case can be dealt with very simply. Let G be a simply-connected Lie group with Lie algebra $\mathcal{G}$, and suppose that G has a Hamiltonian action on M. The moment map then picks out a Hamiltonian h_A for each $A \in \mathcal{G}$, and the corresponding operator $\hat{h}_A$ generates a one-parameter group of unitary transformations of $\mathcal{H}$. Since the commutator of $\hat{h}_A$ and $\hat{h}_B$ is $-\mathrm{i}\hbar\hat{h}_{[A,B]}$, and since G is simply connected, these unitary transformations extend to a unitary representation of G on $\mathcal{H}$ (Chevalley 1946). In this case the unitary action of G on $\mathcal{H}$ exists and is uniquely determined by the moment map.[5] If G is connected, but not simply connected—for example, if it is the circle group—then the exponentiation of the $\hat{h}$s may only give a representation of the covering group.

The general case is more awkward. Suppose that $\rho : M \to M$ is canonical. Then $\rho^* B$ is a Hermitian line bundle and $\rho^* \nabla$ is a connection on $\rho^* B$ with curvature $\hbar^{-1}\omega$. When $H^1(M, T) \neq 0$, $(\rho^* B, \rho^* \nabla)$ and (B, ∇) may be inequivalent, in which case we can get no further. To eliminate this possibility, we shall assume that M is simply connected. By (8.3.1), B and ∇ are then uniquely determined by ω, so there must exist a line bundle equivalence $\iota : \rho^*(B) \to B$ which takes $\rho^* \nabla$ to ∇. We can therefore define a unitary transformation $\hat{\rho} : \mathcal{H} \to \mathcal{H}$ by

$$\hat{\rho} s(m) = \iota\big(\rho^* s(m)\big)$$

(ρ^* is the pull-back of s—it is a section of $\rho^* B$; see §A.2). The difficulty is that ι is determined by ρ only up to a constant phase factor, so a general group of canonical transformations will only have a projective action on $\mathcal{H}$. This is not surprising since the 'quantum states' are the rays in $\mathcal{H}$ rather than the individual sections of B.

We remark finally that anticanonical transformations give rise to antiunitary operators. By the same argument, if M is simply connected and $\rho^* \omega = -\omega$ then $\rho^* B$ is equivalent to $\overline{B}$. The antiunitary map $\hat{\rho} : \mathcal{H} \to \mathcal{H}$ is defined by

$$\hat{\rho} s(m) = \overline{\iota(\rho^* s(m))},$$

where $\iota : \rho^* B \to \overline{B}$ is the equivalence map. Again it is determined only up to a constant phase factor.

Example. *Cotangent bundles.* Let $M = T^* Q$, where Q is a configuration space. There is then a natural choice for B and ∇, namely

$$B = M \times \mathbb{C} \quad \text{and} \quad \nabla = \mathrm{d} - \mathrm{i}\hbar^{-1}\theta, \tag{8.4.4}$$

where θ is the canonical 1-form.

A coordinate system q^a on Q determines a canonical coordinate system p_a, q^b on M. The corresponding operators on $\mathcal{H}$ are

$$\hat{p}_a = -\mathrm{i}\hbar \frac{\partial}{\partial q^a} \quad \text{and} \quad \hat{q}^a = \mathrm{i}\hbar \frac{\partial}{\partial p_a} + q^a.$$

Thus prequantization does not, by itself, reproduce the canonical quantization rules. The wave functions depend on momentum as well as position, and can be localized in arbitrarily small regions of phase space. The full quantization construction will restrict the size of $\mathcal{H}$ by admitting only those wave functions that are constant along the vertical polarization; i.e. that are functions of q alone.

When Q is not simply connected, there are other possible choices for B and ∇. The physical significance of these will be considered in the context of the Aharonov-Bohm experiment.

Example. *The Heisenberg group.* Let (V, ω) be a $2n$-dimensional symplectic vector space. We shall think of V as a symplectic manifold. It has a unique prequantization.

Pick a symplectic frame and put

$$\theta_0 = \tfrac{1}{2}(p_a \mathrm{d}q^a - q^a \mathrm{d}p_a) \tag{8.4.5}$$

in the corresponding linear canonical coordinates. Then θ_0 is a symplectic potential.[6] It is invariant under $SP(V, \omega)$ since it can be characterized without coordinates by $(X \lrcorner \theta_0)(Y) = -\omega(X, Y)$ where $X, Y \in V$. In the trivialization determined by θ_0 the sections B are represented by complex functions $\psi = \psi(p, q)$.

Consider a real linear function on V of the form

$$f(p, q) = v^a p_a - u_a q^a = 2\omega(X, W)$$

where p and q are the coordinates of X, u and v are constant, and

$$W = u_a \frac{\partial}{\partial p_a} + v^a \frac{\partial}{\partial q^a}.$$

The corresponding quantum observable is

$$\hat{f} = -\mathrm{i}\hbar \left(u_a \frac{\partial}{\partial p_a} + v^a \frac{\partial}{\partial q^a} \right) + \tfrac{1}{2}(v^a p_a - u_a q^a).$$

Let ρ_t be the canonical flow generated by f and put $\rho_W = \rho_1$; then ρ_W is translation through W. It follows from (8.4.2) that the corresponding unitary operator $\hat{W} = \hat{\rho}_1$ is

$$(\hat{W}\psi)(p, q) = \psi(p + u, q + v)\mathrm{e}^{-\mathrm{i}(u_a q^a - v^a p_a)/2\hbar}, \tag{8.4.6}$$

since in this case $L = W \lrcorner \theta_0 - f = -\frac{1}{2}f$. That is, in coordinate-free notation,

$$(\hat{W}\psi)(X) = \psi(X + W)\mathrm{e}^{-\mathrm{i}\omega(W, X)/\hbar}, \tag{8.4.7}$$

for any $W, X \in V$.

The association of f with W does not define a moment for the abelian group of translations, and $W \mapsto \hat{W}$ is not a representation. In fact,

$$\hat{W}\hat{Z} = \mathrm{e}^{\mathrm{i}\omega(W,Z)/\hbar}\,(\widehat{W + Z}), \tag{8.4.8}$$

so what we have is a (reducible) representation of the Heisenberg group—the group $V \times T$, where $T \subset \mathbb{C}$ is the circle group, with multiplication

$$(W, w) \cdot (Z, z) = (W + Z, wze^{i\omega(W,Z)/\hbar}),$$

where $W, Z \in V$, $w, z \in T$ (see Pressley and Segal 1986, p. 188). ∎

The following proposition is useful for showing that symplectic manifolds are quantizable.[7]

Proposition (8.4.9). Let (M', ω') be a reducible presymplectic manifold (§1.7). Let K be the characteristic foliation and let (M, ω) be the reduced phase space. Suppose that $\omega' = d\theta'$ for some $\theta' \in \Omega^1(M')$. Then a sufficient condition for (M, ω) to be quantizable is that θ' should satisfy the integrality condition

$$\frac{1}{2\pi\hbar} \int_\gamma \theta' \in \mathbb{Z}$$

whenever γ is a closed curve in a leaf of K. If M' is simply connected, then it is also necessary.

The integrality condition on θ' is the Bohr-Sommerfeld condition of the old quantum theory: this is the first of several appearances.

Proof. Suppose that the condition is satisfied. Then we can construct a line bundle $B \to M$ and a connection as follows. Let $m \in M$ and let Λ be the corresponding leaf of K. We define B_m to be the space of smooth functions $\psi : \Lambda \to \mathbb{C}$ such that for any $m_1, m_2 \in \Lambda$,

$$\psi(m_2) = \psi(m_1) \exp\left(\frac{i}{\hbar} \int_{m_1}^{m_2} \theta'\right), \tag{8.4.10}$$

where the integral is along any path from m_1 to m_2 in Λ. The choice of path does not matter because the integral of θ' around a closed path in Λ is an integer multiple of $2\pi\hbar$. The space B_m is one-dimensional because ψ is uniquely determined by its value at any point of Λ.

As m varies, the Bs fit togther to form a line bundle over M. A smooth section $s : M \to B$ is simply a smooth complex function ψ on M' such that, $X(\psi) = i\hbar^{-1}(X \lrcorner \theta')\psi$, for every $X \in V_K(M')$. In other words, $\nabla'_X \psi = 0$ where ∇' is the connection on $M' \times \mathbb{C}$ with potential $\hbar^{-1}\theta'$.

We define the connection ∇ on B as follows. Let s be a section of B and let ψ be the corresponding function on M'. Let $Y \in V(M)$ and let Z be any vector field on M' that projects onto Y (Fig. 8.4). The covariant derivative $\nabla_Y s$ is the section determined by $\nabla'_Z \psi$. Since Z is unique up

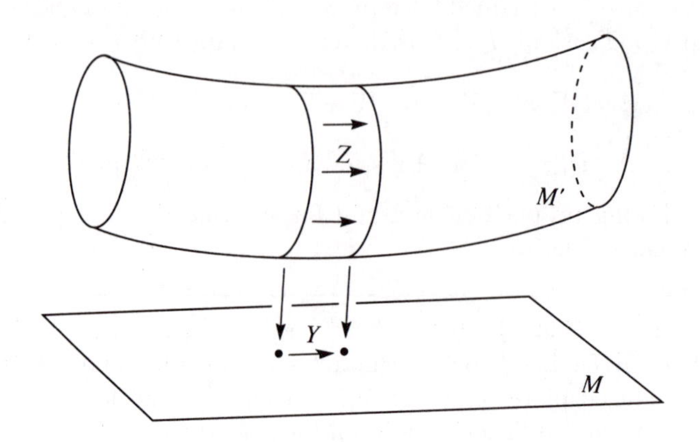

Fig. 8.4. The construction of B.

to the addition of a vector field tangent to K, $\nabla'_Z\psi$ is independent of the choice of Z. Moreover if X is tangent to K, then $[X, Z]$ is tangent to K and

$$\nabla'_X(\nabla'_Z\psi) = \nabla'_Z(\nabla'_X\psi) + \nabla'_{[X,Z]}\psi - \frac{2\mathrm{i}}{\hbar}\omega'(X, Z) = 0$$

since $X \lrcorner\, \omega' = 0$, $\nabla'_X\psi = 0$, and $\nabla'_{[X,Z]}\psi = 0$. Therefore $\nabla'_Z\psi$ is also covariantly constant along K. A similar calculation shows that the curvature of ∇ is ω. Hence ω satisfies the integrality condition.

To prove the partial converse, suppose that M' is simply connected and that ω satisfies IC1. Let γ be a closed curve in a leaf of K. Since M' is simply connected, γ can be spanned in M' by a 2-surface Σ'. The projection of Σ' into M is a closed 2-surface $\Sigma \subset M$. By Stokes' theorem and the integrality of ω,

$$\frac{1}{2\pi\hbar}\oint_\gamma \theta' = \frac{1}{2\pi\hbar}\int_{\Sigma'} \omega' = \frac{1}{2\pi\hbar}\int_\Sigma \omega \in \mathbb{Z}.$$

Therefore θ' satisfies the integrality condition. ∎

The proposition can also be proved by showing that ω satisfies IC2. One constructs local symplectic potentials on M and hence a representative cocycle of $[\omega]$ by restricting θ' to local sections of K: the difference between two such potentials is $\mathrm{d}f$, where the value of f at $m \in M$ is the integral of θ' from one section to the other along a curve in the corresponding leaf of K. The integrality of $[\omega]$ is a direct consequence of the fact that the integral of θ' around a closed loop in the leaf is an integer multiple of $2\pi\hbar$.

The advantage of our indirect method of proof is that it constructs a prequantum bundle on the reduced phase space. The method is also useful when it comes to quantizing group actions. If G acts on M' on the right and preserves θ', then it also has a Hamiltonian action on M (see 3.4.1). The quantized action of G on the prequantum Hilbert space of (M, ω) is simply the pull-back action on the functions ψ that satisfy (8.4.10).

Example. *Elementary systems.* Let G be a connected real Lie group with Lie algebra $\mathcal{G}^*$ and let $f \in \mathcal{G}^*$. Then $(G, \mathrm{d}\theta_f)$ is a presymplectic manifold (§3.5), with characteristic foliation spanned by the right-invariant vector fields R_A, $A \in \mathcal{G}_f$. Its reduction M is the symplectic manifold of right cosets Hg, where $H = (G_f)_0$. The integrality condition on θ_f is that

$$\chi_f(h) = \exp\left(\frac{\mathrm{i}}{\hbar}\int_e^h \theta_f\right)$$

should be a single-valued function on H.

Now if $A \in \mathcal{G}_f$, then $R_A \lrcorner\, \theta_f = f(A)$ and $R_A \lrcorner\, \mathrm{d}\chi_f = \mathrm{i}\hbar^{-1} f(A)\chi_f$. It follows that

$$\chi_f(\mathrm{e}^{tA}h) = \chi_f(\mathrm{e}^{tA})\chi_f(h),$$

and hence that $\chi_f(h'h) = \chi_f(h')\chi_f(h)$ for every $h, h' \in H$. So the integrality condition on θ_f is the condition that $\mathrm{i}\hbar^{-1}f$ should be the gradient at e of a homomorphism $\chi_f : H \to T$, where T is the circle group (Kostant 1970a). When the condition holds, the sections of $B \to M$ are identified with functions $\psi : G \to \mathbb{C}$ such that

$$\psi(hg) = \chi_f(h)\psi(g) \tag{8.4.11}$$

for every $g \in G$, $h \in H$. By pulling back the ψs under right translations, we obtain a reducible representation of G on the sections of B.

If θ_f satisfies the integrality condition, then $f(A)$ is an integer multiple of $\hbar$ for every A in the Lie algebra of H such that $\exp(2\pi A)$ is the identity.

Example. *Angular momentum.* Let (M, ω) be the coadjoint orbit of $SU(2)$ representing the phase space of a particle of spin s. Then M is a sphere of radius s and ω is the area element divided by s (p. 54). It is quantizable whenever $4\pi s$ is an integral multiple of $2\pi\hbar$. That is, whenever s is an integral multiple of $\frac{1}{2}\hbar$.

Another way to see this is to use the fact that M is the reduction of the presymplectic manifold $(M', \mathrm{d}\theta')$, where M' is the 3-sphere

$$S^3 = \{z^0\bar{z}^1 + z^1\bar{z}^1\} = 1$$

in $(\mathbb{C}^2, \mathrm{d}\theta')$, and

$$\theta' = \mathrm{i}s(z^0\mathrm{d}\bar{z}^0 + z^1\mathrm{d}\bar{z}^1 - \bar{z}^0\mathrm{d}z^0 - \bar{z}^1\mathrm{d}z^1)$$

(restricted to M'). The leaves of K are the circles $\{e^{i\phi}(z^0, z^1) \mid \phi \in [0, 2\pi]\}$, and the integrality condition on θ' is

$$\frac{1}{2\pi\hbar}\oint \theta' = \frac{1}{2\pi\hbar}\int_0^{2\pi} 2s\, d\phi = \frac{2s}{\hbar} \in \mathbb{Z},$$

which gives the same quantization condition on s. Since M' is simply connected, integrality of θ' is necessary as well as sufficient.

The sections of $B \to M$ are functions $\psi : S^3 \to \mathbb{C}$ such that

$$\psi(e^{i\phi}z, e^{-i\phi}\overline{z}) = e^{2is\phi/\hbar}\psi(z, \overline{z}).$$

The Hilbert space of square-integrable sections is, of course, infinite dimensional. But if (as we shall do later) we admit only the ψs which are the restrictions to S^3 of holomorphic functions of z^0, z^1, then we obtain the finite-dimensional space of homogeneous polynomials $\psi = \psi_{AB\ldots C'}z^A z^B \ldots z^{C'}$, and we can represent the states by symmetric n-index spinors $\psi_{AB\ldots C'}$, where $n = 2s/\hbar$.

We can also see this phase space as a special case of the previous example, by representing (M, ω) as the reduction of the presymplectic form $d\theta_f$ on $SO(3)$, where $f = (0, 0, s)$ in the notation of §3.5, p. 54. In this case, H is the group of rotations about the z-axis, which is generated by Z. Since $f(Z) = s$, θ_f satisfies the integrality condition only when s is an integer multiple of $\hbar$. There is no contradiction here: $SO(3)$ is not simply connected, so integrality of θ_f is not a necessary condition for quantization.

9

QUANTIZATION

9.1 Polarizations and quantization

THE problem with prequantization is that the Hilbert space it constructs is too large to represent the phase space of a physically reasonable quantum system. In the first example in §8.4 (a cotangent bundle), the Hilbert space contained wave functions that depended on momentum as well as position; and in the third (a coadjoint orbit), the group representation was reducible. We have already touched on a possible solution: we should admit only sections of B that are parallel along a polarization. We shall now look in detail at how this works. It should be stressed, however, that the physical justification is not based on general mathematical results (such as the Borel-Weil theorem), but on the way in which the construction works in particular examples. It generalizes and unifies a number of quantization techniques that, in the past, have not appeared to have any obvious connection with each other and that have sometimes seemed overspecialized with applications only to particular physical systems. Also it should be noted that not all symplectic manifolds admit polarizations (Gotay 1987).

Let P be a strongly integrable polarization of a symplectic manifold (M, ω) and let B be a prequantum bundle over M.

Definition (9.1.1). A smooth section $s : M \to B$ is said to be polarized if $\nabla_{\overline{X}} s = 0$ for every $X \in V_P(M)$.

Local polarized sections exist because the curvature of ∇ vanishes on restriction to directions in P. We can represent them explicitly by recalling (5.4.7) that, locally, there are coordinates p_a, q^b, z^α (p, q real, z complex) and a real function K such that P is spanned by $\partial/\partial p_a$ and $\partial/\partial z^\alpha$, and such that

$$\theta = p_a \mathrm{d}q^a - \mathrm{i}\frac{\partial K}{\partial z^\alpha}\mathrm{d}z^\alpha - \frac{\mathrm{i}}{2}\frac{\partial K}{\partial q^a}\mathrm{d}q^a \tag{9.1.2}$$

is a symplectic potential adapted to P. In a local trvialization in which the connection form is $\hbar^{-1}\theta$, the polarized sections are the functions of the form $\phi(q, z)$, holomorphic in z; that is, $\phi \in C^\infty_P(M)$. In the Kähler case, $\theta = -\mathrm{i}\partial K$, K is a Kähler scalar and ϕ is a holomorphic function of z.

By adding a constant to K, we can arrange that the Hermitian structure on B is given by

$$(\phi, \phi) = \overline{\phi}\phi \mathrm{e}^{-K/\hbar}.$$

If $\phi \in C^\infty_P(M)$, then (ϕ, ϕ) is independent of p. So if s is a polarized section, then (s, s) is constant on the leaves of $D = P \cap \overline{P} \cap TM$ (the isotropic foliation determined by P).

If s is the section $\phi = 1$, then $K = -\hbar \log(s, s)$. Conversely, given any local polarized section s, we can choose the coordinates p_a, q^b, z^α so that (9.1.2) is an adapted potential with $K = -\hbar \log(s, s)$. For suppose that s is represented by $\phi(q, z)$ with respect to some initial choice of the coordinates and trivialization. We can replace θ, K, and p_a by

$$\begin{aligned} \theta' &= \theta + \mathrm{i}\hbar \mathrm{d}(\log \phi), \\ K' &= K - \hbar \log(\overline{\phi}\phi), \\ p'_a &= p_a + \frac{\mathrm{i}\hbar}{2}\frac{\partial}{\partial q^a}\Big(\log \phi - \log \overline{\phi}\Big). \end{aligned}$$

Then $K' = -\hbar \log(s, s)$ and

$$\theta' = p'_a \mathrm{d}q^a - \mathrm{i}\frac{\partial K'}{\partial z^\alpha}\mathrm{d}z^\alpha - \frac{\mathrm{i}}{2}\frac{\partial K'}{\partial q^a}\mathrm{d}q^a.$$

In the trivialization in which $\hbar^{-1}\theta'$ is the connection form, s is represented by $\phi' = 1$.

We should like to quantize M by replacing the Hilbert space $\mathcal{H}$ of prequantization by the subspace of square-integrable polarized sections of B. There are two difficulties. First, the operator $\hat{f}$ coresponding to $f \in C^\infty(M)$ maps local polarized sections to polarized sections only if the flow of X_f preserves P. This follows from

$$\nabla_{\overline{X}}(\hat{f}s) = \hat{f}(\nabla_{\overline{X}}s) - \mathrm{i}\hbar\nabla_{[\overline{X}, X_f]}s;$$

so if $\hat{f}s$ is polarized whenever s is polarized, then $[X, X_f] \in V_P(M)$ whenever $X \in V_P(M)$. Only a limited class of observables can be quantized directly. In the real case, f must be of the form (4.8.1); in the Kähler case, X_f must be a Killing vector.

Second, there may not be any nonzero square-integrable polarized sections. When P contains real directions, the difficulty is that (s, s) is constant on the leaves of D whenever s is polarized; so if the leaves are complete and noncompact, the integral of $(s, s)\epsilon$ over M does not converge (for $s \neq 0$). If the leaves are compact, then there are no nonzero polarized sections (§10.6).

There is also a problem if M is compact and P is Kähler, but not positive (Bott 1957). Let $s \neq 0$ be a polarized section and suppose that (s, s) achieves its maximum at a point $m \in M$. Then the Hessian of $-\log(s, s)$ at m is nonnegative definite. On the other hand, away from the zeros of s,

$K = -\hbar \log(s, s)$ is a Kähler scalar. Consequently, the matrix

$$\left(\frac{\partial^2 K}{\partial z^\alpha \partial \overline{z}^\beta} \right)$$

has the same signature as the Hermitian metric on M, which is positive definite. But if this matrix has negative eigenvalues, then so does the Hessian. So if P is not positive, then there are no nonzero polarized sections.

We shall think again about how to deal with general polarizations, but in the case of positive Kähler polarizations, the quantization can be applied directly.

9.2 Holomorphic quantization

Suppose that P is Kähler and positive. Any two nonvanishing polarized sections of B are related by $s = \phi s'$, where ϕ is a holomorphic function. If we use only polarized sections to define the local trivializations of B, then the transition functions are holomorphic. Therefore P gives B the structure of a holomorphic line bundle. Let $\mathcal{H}_P \subset \mathcal{H}$ denote the set of square-integrable polarized sections. Then $\mathcal{H}_P$ is a closed subspace of $\mathcal{H}$, and is therefore a Hilbert space.[1]

When M is compact, $\mathcal{H}_P$ is finite dimensional (see, for example, Field 1982, p. 41), and its dimension is constrained by the Riemann-Roch formula. In the semiclassical limit $\hbar \to 0$,

$$\dim \mathcal{H}_P \sim \int_M \epsilon.$$

(Simms 1974, Guillemin and Sternberg 1982b). There is therefore an analogy between the volume of the classical phase space and the dimension of the quantum Hilbert space.[2]

The examples below suggest that when $s \in \mathcal{H}_P$ has unit norm, it is reasonable to interpret $(s, s)\epsilon$ as a probability measure on phase space.

Example. *Fock space.* Let (V, ω) be the symplectic vector space discussed in §8.4, p. 166, and let J be a compatible complex structure on V. We shall think of V as a flat Kähler manifold.

Pick a symplectic frame such that (5.2.1) holds and let p_a, q^b be the corresponding linear canonical coordinates on V. Put $z^a = g^{ab}p_b + \mathrm{i}q^a$. Then the zs are holomorphic coordinates, $K = \frac{1}{2} g_{ab} z^a \overline{z}^b$ is a Kähler scalar, and

$$\omega = \mathrm{d}p_a \wedge \mathrm{d}q^a = \tfrac{1}{2}\mathrm{i} g_{ab} \mathrm{d}z^a \wedge \mathrm{d}\overline{z}^b = \mathrm{i}\partial\overline{\partial}K.$$

The symplectic potential $\theta = -\frac{1}{2}\mathrm{i} g_{ab}\overline{z}^b \mathrm{d}z^a$ is adapted to P, and determines a global trivialization of B in which the polarized sections are represented by entire holomorphic functions $\phi(z)$. It is related to θ_0 (8.4.5) by

$$\theta_0 = \tfrac{1}{2}(p_a \mathrm{d}q^a - q^a \mathrm{d}p_a) = \theta + \tfrac{1}{2}\mathrm{i}\,\mathrm{d}K. \qquad (9.2.1)$$

So in the θ_0 gauge, the polarized wave functions are of the form

$$\psi(z,\overline{z}) = \phi(z)\exp\left(-\frac{1}{4\hbar}g_{ab}z^a\overline{z}^b\right).$$

The inner product is

$$\langle\psi,\psi'\rangle = \int_M \overline{\phi}\phi'\exp\left(-\frac{1}{2\hbar}g_{ab}z^a\overline{z}^b\right)\epsilon.$$

This can converge for nonzero holomorphic ϕ only if g is positive definite; that is, only if J is positive, which we shall assume to be the case.

The only real functions on V that generate canonical flows preserving P are of the form

$$f = \tfrac{1}{2}\mathrm{i}\overline{w}_a z^a - \tfrac{1}{2}\mathrm{i}w_a\overline{z}^a + \tfrac{1}{2}U_{ab}z^a\overline{z}^b + c,$$

where w_a, U_{ab} and c are constants, and $\overline{U}_{ab} = U_{ba}$. The linear terms generate translations, and the quadratic term generates unitary transformations of $V_{(J)}$. The corresponding self-adjoint operator on $\mathcal{H}_P$ is

$$\hat{f}\left(\phi\mathrm{e}^{-\overline{z}_a z^a/4\hbar}\right) = \left(-\mathrm{i}\hbar w^a\frac{\partial\phi}{\partial z^a} + \frac{\mathrm{i}}{2}\overline{w}_a z^a\phi + \hbar U_a{}^b z^a\frac{\partial\phi}{\partial z^b} + c\phi\right)\mathrm{e}^{-\overline{z}_a z^a/4\hbar} \tag{9.2.2}$$

(indices are raised and lowered by g^{ab} and g_{ab}).

Now consider the operator $\hat{W}:\mathcal{H}\to\mathcal{H}$ given by translation through $W\in V$ (8.4.6). Because translation preserves the polarization, $\hat{W}(\mathcal{H}_P)\subset\mathcal{H}_P$. If W has holomorphic coordinates w^a, then $\hat{W}$ acts on $\mathcal{H}_P$ by

$$\hat{W}\left(\phi(z)\mathrm{e}^{-\overline{z}_a z^a/4\hbar}\right) = \phi(z+w)\exp\left(-\frac{1}{4\hbar}(2\overline{w}_a z^a + \overline{w}_a w^a + \overline{z}_a z^a)\right). \tag{9.2.3}$$

The resulting representation of the Heisenberg group on $\mathcal{H}_P$ is irreducible.

In the physical examples, the state in which ϕ is constant is the vacuum or ground state. The translations of the ground state under the representation of the Heisenberg group are called *coherent states.*

Let ψ_0 be the state for which $\phi = 1$ and let $\psi_W = \widehat{(-W)}\psi_0$ be the coherent state based on W. In the θ_0-gauge $\psi_W = \phi_W\mathrm{e}^{-\overline{z}_a z^a/4\hbar}$, where[3]

$$\phi_W(Z) = \exp\left(\frac{1}{4\hbar}(2\overline{w}_a z^a - \overline{w}_a w^a)\right). \tag{9.2.4}$$

It is 'the best approximation' in $\mathcal{H}_P$ to the classical state W. Its probability measure is the normal distribution

$$\overline{\psi}_W\psi_W\epsilon = \exp\left(-\frac{1}{2\hbar}(\overline{z}_a - \overline{w}_a)(z^a - w^a)\right)\epsilon.$$

'Best' is in the sense of the maximum likelihood property: ψ_W maximizes $\overline{\psi}_W\psi_W$ at W, subject to $\psi \in \mathcal{H}_P$ and $\langle\psi,\psi\rangle = 1$, so ψ_W is the 'quantum state that most closely approximates the classical state W'. 'Coherent' refers to the fact that the state is invariant under the circle action $z^a - w^a \mapsto e^{i\alpha}(z^a - w^a)$ and retains its form under general unitary transformations of $V_{(J)}$.

The coherent states have the useful 'reproducing' property that

$$\langle\psi_W,\psi\rangle = \psi(W) \tag{9.2.5}$$

for any $\psi \in \mathcal{H}_P$. To prove this, we consider first the case $W = 0$. When $n = 1$, we can choose the coordinate z so that $g = 1$. Then, if $\psi = \phi e^{-\overline{z}z/4\hbar}$,

$$\langle\psi_0,\psi\rangle = \frac{1}{2\pi\hbar}\int_0^\infty\int_0^{2\pi} \phi e^{-r^2/2\hbar} r\, dr d\alpha = \frac{1}{\hbar}\int_0^\infty \phi(0)e^{-r^2/2\hbar}\, dr = \psi(0),$$

where $z = re^{i\alpha}$ and the α integral has been evaluated by Cauchy's theorem. The extension to a general value of n follows by choosing coordinates so that g is the identity matrix (this is possible since J is positive) and applying the same argument n times over. The result when $W \neq 0$ follows from $\langle\psi_W,\psi\rangle = \langle\psi_0,\hat{W}\psi\rangle$. In particular, we deduce that the coherent states span $\mathcal{H}_P$.

More than this is true. If we put $\psi'(W) = \langle\psi_W,\psi\rangle$ where ψ is any element of $\mathcal{H}_P$, then ψ' is the image of ψ under the orthogonal projection $\mathcal{H} \to \mathcal{H}_P$.

An expression for the inner product in $\mathcal{H}_P$ can be deduced from (9.2.5) by expanding ϕ and ϕ_W as Taylor series. If

$$\phi(z) = \phi^{(0)} + \phi^{(1)}_a z^a + \phi^{(2)}_{ab} z^a z^b + \cdots + \phi^{(k)}_{ab\ldots c} z^a z^b \ldots z^c + \cdots$$

and $\psi = \phi e^{-K/2\hbar}$, then

$$\langle\psi,\psi\rangle = \overline{\phi}^{(0)}\phi^{(0)} + 2\hbar\overline{\phi}^{(1)a}\phi^{(1)}_a + \cdots + (2\hbar)^k k!\overline{\phi}^{(k)ab\ldots c}\phi^{(k)}_{ab\ldots c} + \cdots.$$

Let $\mathcal{H}_1$ denote the dual space to $V_{(J)}$ (as complex vector space) and let $\mathcal{H}_k$ be the k-fold symmetric tensor product of $\mathcal{H}_1$. Appropriately scaled, the coefficients in the kth term in the Taylor expansion can be regarded as the components of a tensor in $\mathcal{H}_k$, and so we have a Hilbert space isomorphism from $\mathcal{H}_P$ to the symmetric Fock space

$$\mathcal{F} = \mathbb{C} \oplus \mathcal{H}_1 \oplus \mathcal{H}_2 \oplus \cdots.$$

Example. *The harmonic oscillator.* When $n = 1$, the symplectic vector space (V,ω) is the phase space of the harmonic oscillator, which has Hamiltonian

$$H = \tfrac{1}{2}(p^2 + q^2) = \tfrac{1}{2}z\overline{z}.$$

The powers of z in $\mathcal{H}_P$ are the energy eigenstates and the operators

$$\phi \mapsto 2\hbar \frac{\partial \phi}{\partial z} \quad \text{and} \quad \phi \mapsto z\phi$$

corresponding to the complex functions $\overline{z}$ and z are the lowering and raising operators.

The Hamiltonian flow preserves J, and H is quantized as the operator

$$\hat{H} : \phi \mapsto \hbar z \frac{\partial \phi}{\partial z},$$

which is, of course, incorrect: there should be an additional constant term $\frac{1}{2}\hbar$. From the point of view of geometric quantization, the extra term arises from the metaplectic correction (Chapter 8).

Example. *Free bosons.* When V is infinite-dimensional, the construction of $\mathcal{H}_P$ still makes sense at a formal level. It is made rigorous by *defining* $\mathcal{H}_P$ to be the symmetric Fock space $\mathcal{F}$ and by using the inner product in $\mathcal{F}$ to define the various integrals over V. As an example, let V be the space of real solutions of the Klein-Gordon equation $\Box u + \mu^2 u = 0$ and let J be given by $Ju = -\mathrm{i}u^+ + \mathrm{i}u^-$. Then $V_{(J)}$ is the Hilbert space of square-integrable functions on H_μ^- (p. 148) and $\mathcal{H}_1 = \overline{V}_{(J)}$ is the space of positive-frequency solutions, which are interpreted as single particle states. Elements of $\mathcal{H}_k$ are states consisting k indistinguishable particles.

A real test function f on Minkowski space determines a linear function $f \in V^*$ by

$$f(u) = \int f u \, \epsilon,$$

and hence, by (9.2.2), an operator $\hat{f}$ on $\mathcal{H}_P$, which is the sum of an annihilation operator (the $\partial/\partial z$ term) and a creation operator (the z term). The $\hat{f}$s are the field operators, and the 'quantum field' is the distribution-valued operator $f \mapsto \hat{f}$.

Example. *Compact groups.* Let G be a simply-connected compact group, let $f \in \mathcal{G}^*$, and let (M, ω) be the reduction (G, ω_f). Then M has an invariant positive Kähler polarization (§5.4). When ω satisfies the integrality condition, the quantization of M is a finite-dimensional unitary representation of G. The Borel-Weil theorem states that it is irreducible and that every finite-dimensional irreducible representation arises in this way (Bott 1957, Baston and Eastwood 1989).

The f that generates a given irreducible unitary representation on a Hilbert space V can be any element of the coadjoint orbit of the 'highest weight'. This is usually defined by picking a maximal torus T and looking at the eigenvalues and eigenvectors of its generators as operators on V. The highest weight orbit can also be characterized in a direct geometric way as follows (Kostant 1973, Atiyah 1982).

Each $v \in V$ determines an $f_v \in \mathcal{G}^*$ by $f(A) = -\mathrm{i}\hbar\langle v, Av\rangle/\langle v, v\rangle$, where $A \in \mathcal{G}$ and $v \mapsto Av$ is the infinitesimal action of $\mathcal{G}$ on V. Note that f_v depends only on the image of v in the projective space $\mathbb{P}V$. The map $v \mapsto f_v$ is the moment of the action of G on $\mathbb{P}V$ with respect to the Fubini-Study symplectic structure (p. 95). The 'highest weight orbit' is the set of f_vs for which f_v has the maximum possible distance from the origin in $\mathcal{G}^*$ with respect to any invariant metric on $\mathcal{G}^*$. Another way to say this is in terms of the action of T on $\mathbb{P}V$. This has a moment that maps $\mathbb{P}V$ onto a convex polytope in $\mathcal{T}^*$ (p. 48). The vertices of the polytope are the images of a collection of 'highest weight points' in $\mathbb{P}V$ (they are in fact the projection into $\mathbb{P}V$ of the orbit of the highest weight vector in V under the Weyl group). The highest weight orbit is singled out by the fact that it contains the corresponding f_vs.

The vectors in V determine functions on G satisfying (8.4.11), and hence sections of a prequantum bundle on M. If $f = f_v$, then the function $\psi \in C^\infty(G)$ determined by $u \in V$ is $\psi(g) = \langle v, gu\rangle$. The corresponding section is polarized, and every polarized section arises in this way.

Example. *Angular momentum.* Consider again the orbit of $SU(2)$ with spin $s = N\hbar/2$. This has a positive polarization P, which is the reduction of the holomorphic polarization of $\mathbb{C}^2$ (p. 55). If we use $z = z^1/z^0$ as holomorphic coordinate, then

$$\omega = \frac{\mathrm{i}N\hbar \mathrm{d}z \wedge \mathrm{d}\overline{z}}{(1+z\overline{z})^2},$$

and $K = N\hbar\log(1+z\overline{z})$. The coordinate patch covers the whole orbit, apart from the point at infinity, $z^0 = 0$. In the gauge of the potential $-\mathrm{i}\partial K$, the elements of $\mathcal{H}_P$ are polynomials of the form

$$\psi(z) = \sum_{k=0}^{N} \begin{pmatrix} N \\ k \end{pmatrix} \psi_k z^k$$

where the ψ_ks are constants. In terms of our previous description of the sections of the prequantum bundle as functions on S^3 (p. 169), ψ corresponds to the homogeneous polynomial $\psi_{AB\ldots C} z^A z^B \ldots z^C$, where $\psi_{AB\ldots C}$ is symmetric, with $\psi_{1\ldots10\ldots0} = \psi_k$ (with k ones and $N-k$ zeros). The inner product is

$$\langle\psi, \psi\rangle = \frac{1}{2\pi}\int \frac{\mathrm{i}\overline{\psi}\psi\, \mathrm{d}z \wedge \mathrm{d}\overline{z}}{(1+z\overline{z})^{N+2}} = \frac{N}{N+1}\overline{\psi_{AB\ldots C}}\psi_{AB\ldots C},$$

with summation over $A, B, \ldots$. Under the action of $SU(2) \subset SL(2,\mathbb{C})$, the coefficients $\psi_{AB\ldots C}$ behave as the components of a symmetric spinor.

Example. *The determinant bundle.* Let (V, ω) be a $2n$-dimensional real symplectic vector space (V, ω) and let L^+V be the positive Lagrangian Grassmannian. The symplectic structure Ω and the complex structure on L^+V were defined in §5.4 (p. 93).

For each $J \in L^+V$, let

$$L_J = \{\xi \in \Lambda^n V_{\mathbb{C}} \mid X \wedge \xi = 0 \, \forall \, X \in P_J\}.$$

Then $L_J = \bigwedge^n P_J$ is one dimensional. As J varies, the Ls form a holomorphic line bundle $L \to L^+V$, called the determinant bundle. It will be important later on in the construction of the metaplectic representation, so we shall look at its structure in detail.

The determinant bundle has a Hermitian structure and consequently a connection. It is, in fact, the unique line bundle over L^+V with connection with curvature Ω, as the following argument establishes.

Define $\Xi \in \bigwedge^{2n} V$ by $\Xi = X^1 \wedge \ldots \wedge X^n \wedge Y_1 \wedge \ldots \wedge Y_n$, where X^a, Y_b is any symplectic frame. Then the Hermitian structure on L is $(\cdot, \cdot)$, where

$$(\xi, \xi')\Xi = \mathrm{i}^n \overline{\xi} \wedge \xi'.$$

For each J, let $Z^a = X^a - z^{ab} Y_b$ be the basis for P_J introduced in (5.4.4), and let $\xi = Z^1 \wedge \cdots \wedge Z^n$. Then ξ is a holomorphic section of L, and

$$(\xi, \xi) = \det \mathrm{i}(\overline{z} - z) = \det\, 2y,$$

where $z = x + \mathrm{i}y$. Therefore, the curvature of the connection is $\Omega = \mathrm{i}\partial\overline{\partial}(-\log \det y)$. There is no other Hermitian line bundle with connection with this curvature because L^+V is simply connected.

Example. *Coherent states.* The coherent states in Fock space (p. 174) have a number of interesting properties, which suggest that we might look in a more general context for a way of assigning quantum states $s_m \in \mathcal{H}_P$ to classical states $m \in M$ so that one or more of the following holds:

(CS1) s_m has the 'coherence property' that it is invariant under an action of the circle group of which m is a fixed point;

(CS2) s_m has the 'reproducing property' that for every $s' \in \mathcal{H}_P$, $\langle s_m, s' \rangle = k(s_m(m), s'(m))$ for some constant k independent of s' (the first bracket is the inner product in $\mathcal{H}_P$ and the second is the Hermitian structure on the fibre of the prequantum bundle over m: this is a gauge-invariant form of 9.2.5); and

(CS3) s_m has the 'maximum likelihood property' that it maximizes $(s(m), s(m))$ over all $s \in \mathcal{H}_P$ such that $\langle s, s \rangle = 1$.

Any of these can be taken as the definition of 'the coherent state based on m' in a more general context (Simpson 1984). Rawnsley (1977b), for

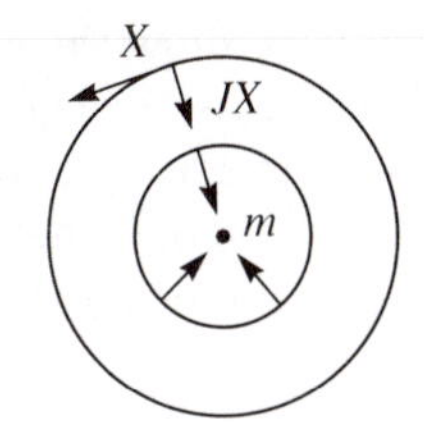

Fig. 9.1. The vector field JX.

example, uses CS2 to define the coherent states on a quantizable symplectic manifold with a positive polarization.

One can see some general relationships between CS1–3. First it is clear that CS3 implies CS2 since CS3 implies that s_m is an eigenvector of the Hermitian form $(s(m), s(m))$ on $\mathcal{H}_P$.

Second, consider CS1 in more detail. Let (M, ω) be a compact connected quantizable symplectic manifold with a positive polarization P. Let J be the corresponding complex structure on M. Suppose that $f \in C^\infty(M)$ generates circle actions on M and $\mathcal{H}$, so that $\rho_{t+2\pi} = \rho_t$ and $\hat{\rho}_{t+2\pi} = \hat{\rho}_t$, where ρ_t is the canonical flow of f (the second equation follows from the first if one adds an appropriate constant to f). We shall assume that ρ_t preserves P and that f has only nondegenerate critical points. We shall also assume that for any $m \in M$, there is an $s \in \mathcal{H}_P$ such that $s(m) \neq 0$.

We can construct an invariant state $s \in \mathcal{H}_P$ by 'averaging over the circle'; that is by taking any $s_0 \in \mathcal{H}_P$ and defining

$$s = \frac{1}{2\pi} \int_0^{2\pi} \hat{\rho}_t(s_0)\, \mathrm{d}t.$$

Then $\hat{\rho}_t s = s$. If m is a fixed point of the circle action, then by (8.4.2), $\hat{\rho}_t s(m) = \mathrm{e}^{\mathrm{i}tf(m)/\hbar} s(m)$ for all t. So either $s(m) = 0$ or $f(m) = 0$.

Since M is compact, f has a unique minimum (p. 48). Take this point to be m and choose $s_0 \in \mathcal{H}_P$ such that $s_0(m) \neq 0$. Then $s(m) = s_0(m) \neq 0$, so $f(m) = 0$, but s vanishes at the other critical points of f. Moreover,

$$-\mathrm{i}\hbar \nabla_X s + fs = 0 \quad \text{and} \quad \nabla_{JX} s = \mathrm{i}\nabla_X s,$$

where $X = X_f$ (the first equation holds since s is invariant, the second because it is polarized). Therefore

$$\nabla_{JX} s = \frac{f}{\hbar} s. \tag{9.2.6}$$

Now JX is a 'radial' vector field at m (Fig. 9.1). It is the gradient vector field of $-f$ with respect to the metric $g(\cdot, \cdot) = 2\omega(\cdot, J\cdot)$ and all its integral curves in a neighbourhood of m have m as their limit points. Consequently (9.2.6) determines s uniquely in terms of $s(m)$ in a neighbourhood of m. However, two polarized sections that are equal on an open set are equal everywhere. Up to proportionality, therefore, there is only one invariant state which does not vanish at m (Guillemin and Sternberg 1982b). Moreover (9.2.6) also implies

$$JX \lrcorner \, \mathrm{d}(s, s) = \frac{2f}{\hbar}(s, s)$$

and hence that (s, s) can have critical points only where $(s, s) = 0$ or $f = 0$. Consequently m must be the unique point at which (s, s) achieves its maximum.

There is therefore a unique invariant state which is concentrated around the minimum of f. It necessarily has the reproducing property since, for any $s' \in \mathcal{H}_P$,

$$\frac{1}{2\pi}\int_0^{2\pi} \hat{\rho}_t(s')\, \mathrm{d}t = \frac{(s(m), s'(m))}{(s(m), s(m))} s$$

and therefore,

$$\langle s, s' \rangle = \frac{1}{2\pi}\int_0^{2\pi} \langle \hat{\rho}_t s, \hat{\rho}_t s' \rangle \, \mathrm{d}t = \frac{1}{2\pi}\int_0^{2\pi} \langle s, \hat{\rho}_t s' \rangle \, \mathrm{d}t = k(s(m), s'(m)),$$

where $k = \langle s, s \rangle / (s(m), s(m))$.

Example. *Eigenfunctions.* Suppose that the canonical flow of a classical observable f preserves P. Then $\hat{f} : \mathcal{H}_P \to \mathcal{H}_P$. What is the relationship between the eigenfunctions of $\hat{f}$ and their classical counterparts, the hypersurfaces of constant f?

Since adding a constant to f also adds the same constant to $\hat{f}$, it is enough to consider the hypersurface C on which $f = 0$. Let $X = X_f$ and let $Y = JX$, where J is the complex structure on M determined by P. Then $-Y$ is the gradient vector field of f with respect to the Riemannian metric $g(\cdot, \cdot) = 2\omega(\cdot, J\cdot)$. We shall assume that $X \neq 0$ on C.

Because the canonical flow of f preserves P, X is the real part of a holomorphic vector field. Therefore we can introduce local holomorphic coordinates z^a, such that

$$X = \frac{\partial}{\partial x} \quad \text{and} \quad Y = \frac{\partial}{\partial y},$$

where $x + \mathrm{i}y = z^1$. In these coordinates,

$$\omega = \mathrm{i}\frac{\partial^2 K}{\partial z^a \partial \bar{z}^b} \mathrm{d}z^a \wedge \mathrm{d}\bar{z}^b,$$

where the Kähler scalar can be chosen to be independent of x. Since $X = X_f$,

$$\partial f + \overline{\partial} f = \mathrm{d}f = -X \lrcorner\, \omega = -\mathrm{i}\overline{\partial}\left(\frac{\partial K}{\partial z^1}\right) + \mathrm{i}\partial\left(\frac{\partial K}{\partial \overline{z}^1}\right).$$

Therefore $f + \mathrm{i}\partial K/\partial z^1$ is holomorphic, and independent of z^1. We are free to add the real part of a holomorphic function to K, so we can assume without loss of generality that $f + \mathrm{i}\partial K/\partial z^1 = 0$. Then

$$\frac{\partial K}{\partial y} = 2\mathrm{i}\frac{\partial K}{\partial z^1} = -2f. \tag{9.2.7}$$

In the trivialization determined by the potential $-\mathrm{i}\partial K$, the elements of $\mathcal{H}_P$ are holomorphic functions of the zs and $\hat{f}$ is the operator

$$\hat{f} = -\mathrm{i}\hbar\frac{\partial}{\partial z^1}. \tag{9.2.8}$$

The elements of $\mathcal{E}_0 = \operatorname{Ker}\hat{f} \subset \mathcal{H}_P$ are the holomorphic functions of $z^2, \ldots, z^n$. If $s \in \mathcal{E}_0$ is a normalized section represented by such a function, then

$$(s, s) = \overline{\phi}\phi\,\mathrm{e}^{-K/\hbar}.$$

Now, by (9.2.7), K is increasing along the integral curves of Y when $f < 0$, and decreasing when $f > 0$; and since $\overline{\phi}\phi$ is independent of z^1, it is constant along Y. Therefore, certainly in a neighbourhood of C, the probability density (s, s) is maximal on C. In fact it follows from $\hbar\nabla_X s + \mathrm{i}fs = 0$ that $Y \lrcorner\, \mathrm{d}(s, s) = 2f(s, s)/\hbar$ everywhere on M and hence that, when M is compact, the global maximum of (s, s) occurs on C, since at the critical points of (s, s), either $f = 0$ or $(s, s) = 0$.

By differentiating (9.2.7) along Y,

$$\frac{\partial^2 K}{\partial y^2} = -2\frac{\partial f}{\partial y} = 4\omega(X, Y) = 2g(X, X) \tag{9.2.9}$$

since $X \lrcorner\, \omega = -\mathrm{d}f$. Therefore, near C, the y-dependence of (s, s) is that of a normal distribution with variance

$$\sigma_y^2 = \frac{\hbar}{2g(X, X)}.$$

However, $g(X, X) = g(Y, Y)$; and for small y, the distance d from C with respect to g is $\sqrt{g(Y, Y)}y$. Therefore, for small $\hbar$, d is normally distributed about $d = 0$ with variance $\frac{1}{2}\hbar$: in this sense, the elements of $\mathcal{E}_0$ are concentrated on C in the semiclassical limit.

For example, in the case of the harmonic oscillator, $\phi = z^n$ is an eigenfunction of $\hat{H}$ with eigenvalue $n\hbar$, where $H = \frac{1}{2}z\overline{z}$, and

$$(s, s) = (z\overline{z})^n \mathrm{e}^{-z\overline{z}/2\hbar},$$

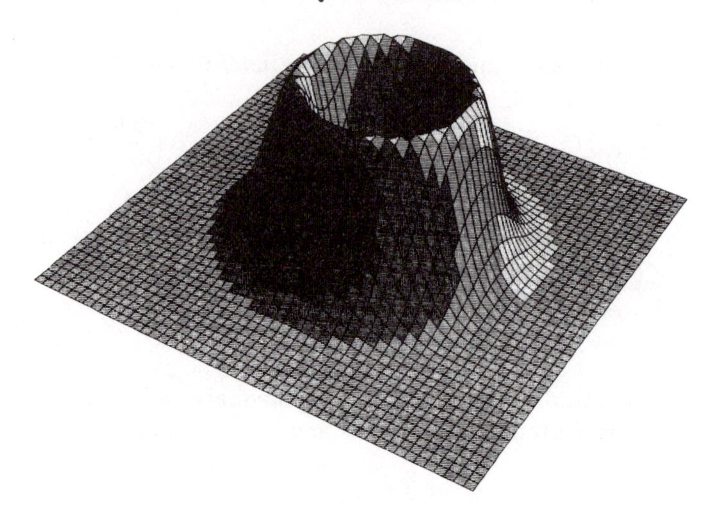

Fig. 9.2. The distribution of (s, s) in the harmonic oscillator.

which is concentrated around the circle $z\bar{z} = n\hbar$. Fig. 9.2 shows the form of the graph of (s, s) over the complex z-plane.

Example. *Quantization of constrained systems.* A general issue that arises is whether constraints should be imposed before or after quantization. This has been addressed in the context of geometric quantization by Guillemin and Sternberg (1982b), Śniatycki (1983), Ashtekar and Stillerman (1986), and Gotay (1986); see also Blau (1988). Some features of the problem are illustrated by taking C and f as in the previous example, and by treating C as a constraint.

We shall suppose that C is reducible and that the leaves of the characteristic foliation are circles of circumference λ with respect to the Riemannian metric g, where λ is a function on the reduced phase space M_0. We shall also assume that the holonomy of ∇ around the leaves of the characteristic foliation is trivial. Then M_0 is quantizable and P projects onto a positive polarization P_0 of M_0. By an extension of the construction in (8.4.9), the sections of the prequantum bundle $B_0 \to M_0$ are represented by sections of B_C such that $\nabla_X s = 0$. The polarized sections of B_0 correspond to sections of B_C such that

$$\nabla_X s = 0 \quad \text{and} \quad \nabla_{\overline{Z}} s = 0 \tag{9.2.10}$$

for every complex vector Z on C tangent to $P \cap T_{\mathbb{C}}C$.

The issue in this case is the following. Should we identify the states of the constrained system with the elements of $\mathcal{H}_0$, the Hilbert space constructed from M_0 and P_0, or with the elements of

$$\mathcal{E}_0 = \operatorname{Ker} \hat{f} \subset \mathcal{H}_P,$$

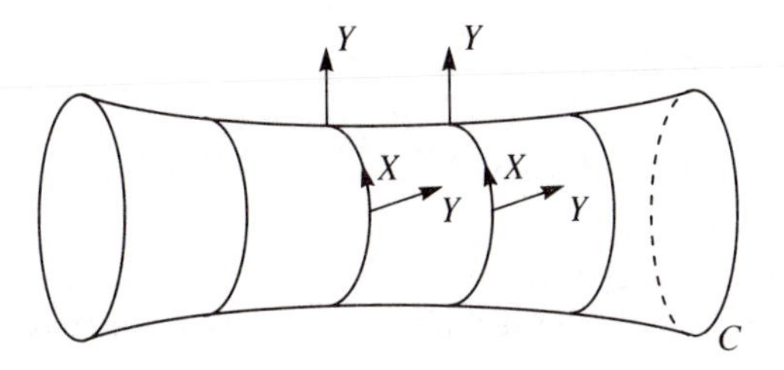

Fig. 9.3. The vector fields X and Y.

where $\mathcal{H}_P$ is the Hilbert space constructed from M and P? If $s \in \mathcal{E}_0$, then the restriction of s to C determines an element of $\mathcal{H}_0$, so we have a map in one direction, $\mathcal{E}_0 \to \mathcal{H}_0$. Conversely, given a section s of B_C such that (9.2.10) holds, we can extend s to a neighbourhood of C in M by imposing

$$\nabla_Y s = \frac{f}{\hbar} s.$$

The integral curves of Y are transverse to C, so this determines s near C (Fig. 9.3). The result is a polarized section s of B, which is defined on a neighbourhood of C, and which satisfies $\hat{f}s = 0$. Therefore, since two polarized sections of B coincide if they coincide on some open subset, the map $\mathcal{E}_0 \to \mathcal{H}_0$ is injective. If we could show that

(E1) s can be extended to the whole of M; and

(E2) the inner products in $\mathcal{E}_0$ and $\mathcal{H}_0$ coincide,

then we could identify $\mathcal{E}_0$ with $\mathcal{H}_0$, and we could make the pleasing deduction that it makes no difference whether the constraint is imposed before or after quantization. In the case that M is compact and f generates a circle action on M, the first point is cleared up by Guillemin and Sternberg (1982b): they show that the map is bijective in a general context in which C is the zero-set of the moment of the Hamiltonian action of a compact group on M which preserves P.

The second is a more serious problem. In the local representation introduced in the last example, the elements of $\mathcal{E}_0$ are holomorphic functions $\phi(z^2, \dots, z^n)$. If ϕ is such a function, then $\overline{\phi}\phi e^{-K/\hbar}|_C$ is independent of x, and so can be regarded as a function on M_0. The two inner products we must compare are

$$I_1 = \int_{M_0} \overline{\phi}\phi e^{-K/\hbar}\, \epsilon_0 \quad \text{and} \quad I_2 = \int_M \overline{\phi}\phi e^{-K/\hbar}\, \epsilon,$$

where ϵ_0 and ϵ are the respective volume elements on M_0 and M. Since x runs from 0 to $\lambda g(X, X)^{-1/2}$ around the leaves of the characteristic foliation of C, we can rewrite the first integral as

$$I_1 = \int_C \overline{\phi}\phi \mathrm{e}^{-K/\hbar} \frac{\sqrt{g(X,X)}}{\lambda} \mathrm{d}x \wedge \tau,$$

where τ is the pull-back of ϵ_0 to C.

To compare this with the second integral, we note that at points of C,

$$\epsilon = \frac{\omega(X,Y)}{\pi\hbar} \mathrm{d}x \wedge \mathrm{d}y \wedge \tau.$$

If we extend τ away from C so that this holds in a neighbourhood of C, then the second integral is

$$I_2 = \frac{1}{2\pi\hbar} \int_M \overline{\phi}\phi \mathrm{e}^{-K/\hbar} g(X,X)\, \mathrm{d}x \wedge \mathrm{d}y \wedge \tau,$$

since $g(X,X) = 2\omega(X,Y)$. These expressions involve our special choice of local coordinates and local trivialization, but the integrands are coordinate and gauge-invariant.

In general, it is hard to say very much about the relationship between the two integrals. However, as a consequence of (9.2.7), K is minimal as a function of y on the constraint hypersurface $f = 0$. So in the semiclassical limit, the principal contribution to the second integral comes from the immediate neighbourhood of C. In fact, if we use Laplace's method to determine the leading term in the asymptotic expansion of the y integral in powers of $\hbar$, then as a consequence of (9.2.9) the second integral reduces to

$$I_2 = c \int_C \overline{\phi}\phi \mathrm{e}^{-K/\hbar} \sqrt{g(X,X)}\, \mathrm{d}x \wedge \tau,$$

where c is independent of ϕ. Therefore the two inner products are the same (up to a constant factor) in the semiclassical limit if the circles foliating C all have the same circumference. It is not hard to derive in the same way an analogous condition on the volumes of the leaves of the characteristic foliation in the case of a general coisotropic constraint.

A special case is when C is a sphere centred on the origin in a symplectic vector space with the constant positive polarization (see the example of Fock space above). Then M_0 is complex projective space. In this case, not only are the circles of constant length throughout C, but the inner products coincide exactly (up to a constant factor).

9.3 Nonnegative polarizations

We shall now look at the modifications that are needed to extend the quantization scheme to nonnegative polarizations with real directions. First, we shall consider how to deal with a real polarization P. We shall assume that the leaves of P are complete and simply-connected, and that $Q = M/P$ is an oriented Hausdorff manifold.

A section s of the prequantum bundle which is constant along P depends only on the position coordinates and looks like a wave function in the 'configuration space' Q; but such sections are not square-integrable on M, and so P does not pick out a subspace of $\mathcal{H}$ in the same way as a Kähler polarization. We shall modify the holomorphic construction to make (s, s) come out not as a function, but as an n-form on Q. We shall then define an inner product by integration over Q.

Denote by $\Delta_Q \to Q$ the line bundle $\bigwedge^n T^*_{\mathbb{C}} Q$, where, as usual, $n = \frac{1}{2}\dim M$. Its sections are the complex volume forms on Q. The first step is to represent sections of Δ_Q by objects on M. Let α be a section of Δ_Q and let $\beta = \mathrm{pr}^*\alpha \in \Omega^n_{\mathbb{C}}(M)$, where $\mathrm{pr} : M \to Q$ is the projection. Then, for every $X \in V_P(M)$,

$$X \lrcorner \beta = 0 \quad \text{and} \quad X \lrcorner \mathrm{d}\beta = 0. \tag{9.3.1}$$

Conversely, suppose that these hold for some $\beta \in \Omega^n_{\mathbb{C}}(M)$. Then β is the pull-back of a section of Δ_Q; moreover, if Z is a vector field on M whose flow preserves P, then $\mathcal{L}_Z\beta$ also satisfies (9.3.1). In fact, if $\beta = \mathrm{pr}^*\alpha$ and $Y = \mathrm{pr}_* Z$, then $\mathcal{L}_Z\beta = \mathrm{pr}^*(\mathcal{L}_Y\alpha)$.

The complex n-forms on M such that $X \lrcorner \beta = 0$ for every $X \in V_P(M)$ are the sections of the canonical line bundle $K_P \subset \bigwedge^n T_{\mathbb{C}} M$ (§A.4). The correspondence between α and β identifies K_P with $\mathrm{pr}^*\Delta_P$. Since Q is oriented, the transition functions of Δ_P and K_P can all be made real and positive; so K_P is trivial, although it does not have a natural trivialization. We can, however, pick out the real sections of K_P and distinguish amongst them between positive and negative sections.

There are two types of derivative operator acting on the sections of K_P.

The partial connection. The covariant derivative ∇_X is defined for $X \in V_P(M)$ by

$$\nabla_X\beta = X \lrcorner \mathrm{d}\beta.$$

Apart from the fact that it is restricted to vector fields tangent to P, it has all the properties of a flat connection. In particular, for $X, Y \in V_P(M)$,

$$\nabla_X\nabla_Y - \nabla_Y\nabla_X = \nabla_{[X,Y]}.$$

The sections which are covariantly constant along P are the pull-backs of n-forms on Q.

The Lie derivative. If the flow of Z preserves P, then $\mathcal{L}_Z$ (acting on n-forms) maps sections of K_P to sections of K_P. If Z is tangent to P, then $\nabla_Z = \mathcal{L}_Z$.

Since K_P is trivial with positive transition functions, we can take its square root by taking the square roots of the transition functions (§A.3). The two derivative operators pass to the resulting line bundle $\delta_P = \sqrt{K_P}$, where they are determined by

$$2(\nabla_X \nu)\nu = \nabla_X \nu^2 \quad \text{and} \quad 2(\mathcal{L}_Z \nu)\nu = \mathcal{L}_Z \nu^2.$$

Here ν is a section of δ_P: if ν and ν' are sections of δ_P, then $\nu\nu'$ is a section of K_P. Again δ_P is a trivial bundle.

The principal modification is to replace the prequantum bundle B by $B_P = B \otimes \delta_P$. The sections of B_P are of the form $\tilde{s} = s\nu$, where s is a section of B and ν is a section of δ_P. The smooth sections such that

$$\nabla_X \tilde{s} = (\nabla_X s)\nu + s\nabla_X \nu = 0$$

for every $X \in V_P(M)$ are called P *wave functions*; they are the basic elements in the construction of $\mathcal{H}_P$.

If $\tilde{s} = s\nu$ and $\tilde{s}' = s'\nu'$ are P wave functions, then $(\tilde{s}, \tilde{s}') = (s, s')\bar{\nu}\nu'$ is a section of K_P; and, for $X \in V_P(M)$,

$$\nabla_X(\tilde{s}, \tilde{s}') = (\nabla_X \tilde{s}, \tilde{s}') + (\tilde{s}, \nabla_X \tilde{s}') = 0.$$

Hence we can identify $(\tilde{s}, \tilde{s}')$ with an n-form on Q, and define an inner product on P wave functions by

$$\langle \tilde{s}, \tilde{s}' \rangle = \int_Q (\tilde{s}, \tilde{s}').$$

The square-integrable P wave functions form a pre-Hilbert space, and its completion $\mathcal{H}_P$, which is our model for the quantum phase space, is a Hilbert space.[4] We interpret $(\tilde{s}, \tilde{s})$ as a probability measure in configuration space.

As in the Kähler case, the only classical observables that directly determine operators on $\mathcal{H}_P$ are those whose flows preserve P. If f is such an observable, then the corresponding quantum observable is the operator $\tilde{f}$ that acts on the P wave functions by

$$\tilde{f}\tilde{s} = \hat{f}(s)\nu - i\hbar s\mathcal{L}_{X_f}\nu, \quad \text{where} \quad \tilde{s} = s\nu.$$

The quantum conditions hold; and if X_f is complete, then $\tilde{f}$ is self-adjoint, and generates the one-parameter unitary group $\tilde{\rho}_t : \mathcal{H}_P \to \mathcal{H}_P$ that acts on P wave functions by

$$\tilde{\rho}_t \tilde{s} = \hat{\rho}_t(s)\rho_t^*(\nu),$$

where $(\rho_t^*\nu)^2 = \rho_t^*(\nu^2)$. The square root is continuous in t, and defined so that $\rho_0^*\nu = \nu$.

Example. *Canonical quantization.* Let $M = T^*Q$ and let μ be a volume element on Q. Then μ picks out a trivialization of K_P, and hence of δ_P. We shall choose B and ∇ as in (8.4.4).

Let q^a be a coordinate system on Q and let q^a, p_b be the corresponding canonical coordinate system on M. The sections of B_P are represented by complex functions $\psi(p, q)$, and the P wave functions by functions of q alone, with the inner product

$$\langle \psi, \psi' \rangle = \int_Q \overline{\psi} \psi' \, \mu.$$

A classical observable that generates a flow preserving P is of the form

$$f = v^a(q) p_a + u(q),$$

where $v \in V(Q)$ and $u \in C^\infty(Q)$. The corresponding $\tilde{f}$ acts on P wave functions by

$$\tilde{f}\psi = -\mathrm{i}\hbar v(\psi) + (u + \tfrac{1}{2}\operatorname{div} v)\psi,$$

where the divergence is defined by $\mathcal{L}_v \mu = (\operatorname{div} v)\mu$. In particular, if the coordinates are chosen so that $\mu = \mathrm{d}^n q$, then

$$\tilde{q}^a \psi = q^a \psi \quad \text{and} \quad \tilde{p}_a \psi = -\mathrm{i}\hbar \frac{\partial \psi}{\partial q^a}, \tag{9.3.2}$$

which are the familar rules of canonical quantization.

Example. *Charge quantization and minimal coupling.* (Śniatycki 1974, Souriau 1970). Suppose that we replace the canonical 2-form ω on T^*Q by the charged symplectic structure

$$\omega_F = \mathrm{d}p_a \wedge \mathrm{d}q^a + \frac{e}{2} F_{ab} \mathrm{d}q^a \wedge \mathrm{d}q^b,$$

where F is a closed 2-form on Q and e is the charge. Then M is quantizable if and only if eF satisfies the integrality condition on Q. This is a quantization condition on e: in Dirac's (1931) original example, Q was Euclidean space, less a number of points at which there were monopole singularities in an otherwise well-behaved magnetic field, represented by the 2-form (2.6.3). The integrality condition restricts the ratios of e and the monopole strengths.

The local symplectic potentials are now of the form $p_a \mathrm{d}q^a + eA_a \mathrm{d}q^a$, where $F = 2\mathrm{d}A$. In the corresponding local trivializations of B, eqns (9.3.2) are replaced by

$$\tilde{q}^a = q^a \quad \text{and} \quad \tilde{p}_a = -\mathrm{i}\hbar \frac{\partial}{\partial q^a} - eA_a,$$

which are the conventional 'minimally coupled' operators.

Example. *The Aharonov-Bohm effect.* When Q is not simply connected, the prequantization of T^*Q, with the canonical symplectic structure, is not unique: the various possibilities are parametrized by $H^1(Q, T)$, the elements of which measure the phase change induced by parallel propagation around closed loops in Q. The natural choice (8.4.4), for which there is no phase change, is not necessarily physically correct.

For example, consider a charge e moving in the region outside a long cylinder, in which there is a nonvanishing magnetic field. The space Q outside the cylinder is then not simply connected. Even when the magnetic field vanishes in Q, there is no gauge in which the magnetic vector potential A also vanishes since the integral

$$\oint A_a \mathrm{d}q^a$$

around a closed curve surrounding the cylinder is nonzero and gauge invariant.

Aharonov and Bohm's (1959) suggestion, which was confirmed by measuring the interference between two beams of electrons passing on each side of a magnetized iron cylinder, was that the phase change of the wave function of the charge around a closed loop surrounding the cylinder is not zero when the field vanishes outside Q, but is given by

$$\exp\left(\frac{\mathrm{i}e}{\hbar}\oint A_a \mathrm{d}q^a\right).$$

Thus the prequantum bundle in this case is not $Q \times \mathbb{C}$, but the bundle labelled by the corresponding element of $H^1(Q, T)$. ∎

A similar modification of the quantization construction is needed in the case of a general nonnegative polarization P. Let the corresponding isotropic and coisoptropic foliations be D and E. The most straightforward possibility is that the leaves of D are complete and simply-connected, that E is integrable, that $Q = M/E$ is a Hausdorff manifold, and that for each leaf Λ of E, Λ/D is also a Hausdorff manifold. Then each Λ/D is a Kähler manifold, and it follows from our local representation of the polarized sections of B as functions of q and z (p. 171) that $B|_\Lambda$ is the pull-back of a holomorphic line bundle $B' \to \Lambda/D$, and that the restrictions to Λ of the local polarized sections of B are the pull-backs of the holomorphic sections of B'.

The square-integrable polarized sections of B' form a Hilbert space $\mathcal{H}'_\Lambda$. As Λ varies, the $\mathcal{H}'$s fit together to form a Hermitian vector bundle $\mathcal{H}' \to Q = M/E$. The polarized sections of B are the smooth sections of this bundle.[5] As in the case of a real polarization, the difficulty is that there is no natural inner product since the norm of a polarized section of B is constant along D and so its integral over M diverges.

We get around this by including the 'square root of a volume element on Q' in the definition of the wave functions on M by tensoring B with the $\delta_D = \sqrt{K_D}$, where K_D is the canonical bundle of D. The space of sections of $\mathcal{H}'$ then has an inner product, given by integration over Q. The details are straightforward, but we shall not develop this modified construction here since it is superseded by the metaplectic theory in Chapter 10, and because in many applications Q carries a volume element that can be used to define the inner product more simply.

Example. *Vector bundles.* Every Hermitian vector bundle arises in this way: given a Hermitian vector bundle $V \to Q$, we construct a symplectic manifold M and nonnegative polarization P from the (complex) dual bundle $V^* \to Q$, as follows. First pick a connection on V^* (and hence on V), which is given locally by

$$\nabla = \mathrm{d} - \mathrm{i}\Theta,$$

where $\Theta = (\Theta^\alpha{}_\beta)$ is a 1-form with values the Hermitian matrices. By pulling it back by the projection $V^* \to Q$, we can also regard the Θ as a 1-form on V^*.

Now consider the total space of $T^*Q \oplus V^*$, the bundle whose fibre over $q \in Q$ is the sum of T^*_qQ and V^*_q as real vector spaces. This has coordinates q^a, p_b and (the real and imaginary parts of) z^α, where the qs are coordinates on Q, the ps are the conjugate momenta on T^*Q, and the zs are complex linear coordinates on the fibres of V^*.

Define a real 1-form θ on $T^*Q \oplus V^*$ by

$$\theta = p_a \mathrm{d}q^a + \frac{\mathrm{i}\hbar}{2} g_{\alpha\beta} \left[z^\alpha (\mathrm{d}\bar{z}^\beta - \mathrm{i}\bar{\Theta}^\beta{}_{\gamma a} \bar{z}^\gamma \mathrm{d}q^a) - \bar{z}^\beta (\mathrm{d}z^\alpha + \mathrm{i}\Theta^\alpha{}_{\gamma a} z^\gamma \mathrm{d}q^a) \right],$$

where g is the Hermitian metric on the fibres of V^*. Then θ is independent of the choice of coordinates: it is the sum of the canonical 1-form $p_a \mathrm{d}q^a$ and the vertical part of $\frac{1}{2}\mathrm{i}\hbar g_{\alpha\beta}(z^\alpha \mathrm{d}\bar{z}^\beta - \bar{z}^\beta \mathrm{d}z^\alpha)$.

The 2-form $\omega = \mathrm{d}\theta$ is a symplectic structure on $T^*Q \oplus V^*$. We define M to be the reduction of the constraint $g_{\alpha\beta} z^\alpha \bar{z}^\beta = 1$, and P to be the reduction of the polarization of $T^*Q \oplus V^*$ spanned by $\partial/\partial p_a$ and $\partial/\partial z^\alpha$. The Kähler manifolds Λ/D are the projective spaces $\mathbb{P}V^*_q$.

Because ω is exact, we can quantize M by using Proposition (8.4.9). Apart from the inclusion of the real variables p, q, the construction works in the same way as the quantization of angular momentum (p. 177). The polarized sections of the prequantum bundle are represented by the functions $\psi(q, z)$ on the constraint manifold that are independent of p, and holomorphic and homogeneous of degree one in z. They are therefore of the form

$$\psi(q, z) = s_\alpha(q) z^\alpha$$

where s is a section of $V^{**} = V$. Thus quantization of M recovers V (see Sternberg 1977, Guillemin and Sternberg 1978, Weinstein 1978).

A special case is when Q is a single point. The construction then makes precise the analogy between a complex vector space V with a Hermitian inner product and the Fubini-Study symplectic structure on $\mathbb{P}V^*$. ∎

9.4 Relativistic quantization

Further illustrations are provided by the quantization of elementary relativistic systems, by using the invariant polarizations constructed in Chapter 5.

The phase space of a massive particle of spin s and mass m and its associated antiparticle is $M_{sm} = M_{sm}^+ \cup \overline{M}_{sm}^-$. It is the reduction of the two components of the coisotropic constraint

$$C_{sm} = \left\{ p_a p^a = 2\left(p_{AA'} z^A \overline{z}^{A'}\right)^2 = m^2 \right\}$$

in $(T^*\mathbb{M} \times S, \omega')$, where

$$\omega' = \frac{\sqrt{2}\,\mathrm{i}s}{m}(p_{AA'}\mathrm{d}z^A \wedge \mathrm{d}\overline{z}^{A'} + \overline{z}^{A'}\mathrm{d}z^A \wedge \mathrm{d}p_{AA'}) + \mathrm{cc} - \zeta \mathrm{d}p_a \wedge \mathrm{d}q^a,$$

where $\zeta = 1$ when $p_0 > 0$ and $\zeta = -1$ when $p_0 < 0$. There is a nonnegative invariant polarization of M_{sm}. On M_{sm}^+, it is the reduction of the polarization of $T^*\mathbb{M} \times S$ spanned by $\partial/\partial q^a$ and $\partial/\partial z^A$; on the other component, $\overline{M}_{sm}^-$, it is the reduction of the conjugate polarization spanned by $\partial/\partial q^a$ and $\partial/\partial \overline{z}^{A'}$. The leaves of E are the surfaces of constant p, and the leaves of D are the surfaces of constant p and z.

Now $\omega' = \mathrm{d}\theta'$, where[6]

$$\theta' = \frac{\sqrt{2}\,\mathrm{i}s}{m}\left(p_{AA'} z^A \mathrm{d}\overline{z}^{A'} - p_{AA'}\overline{z}^{A'}\mathrm{d}z^A\right) + \zeta q^a \mathrm{d}p_a.$$

Since ω' is exact, M_{sm} is quantizable whenever θ' satisfies the integrality condition of Proposition (8.4.9). That is, whenever $s = \frac{1}{2}n\hbar$ for some positive integer n. A section of the prequantum bundle is represented by a function $\psi \in C_{\mathbb{C}}^\infty(C_{sm})$ which satisfies the differential equations

$$p^a \frac{\partial \psi}{\partial q^a} = 0 \quad \text{and} \quad z^A \frac{\partial \psi}{\partial z^A} - \overline{z}^{A'} \frac{\partial \psi}{\partial \overline{z}^{A'}} = n\psi.$$

For a polarized section on M_{sm}^+, we must have, in addition,

$$\frac{\partial \psi}{\partial q^a} = 0, \qquad \frac{\partial \psi}{\partial \overline{z}^{A'}} = 0,$$

so that ψ is independent of q, and holomorphic and homogeneous of degree n in z. It must therefore be of the form $\psi(p, z) = \psi_{AB\ldots C} z^A z^B \ldots z^C$, where $\psi_{AB\ldots C}$ is an n-index symmetric spinor field on the mass hyperboloid $H_m^+ = \{p_a p^a = m^2\}$. If we look at just M_{sm}^+, then $Q = M_{sm}^+/E$ is the hyperboloid H_m^+ and the Hermitian vector bundle over $\mathcal{H}' \to Q$ is the bundle of symmetric spinors. Up to a constant, the Hermitian structure is

$$(\psi, \psi) = p^{AA'} p^{BB'} \ldots p^{CC'} \overline{\psi}_{A'B'\ldots C'} \psi_{AB\ldots C}, \tag{9.4.1}$$

by the same calculation as in the quantization of angular momentum (p. 177). We can therefore construct a Hilbert space $\mathcal{H}_{sm}^+$ from the sections by integration over H_{sm}^+ against the invariant volume element $\mathrm{d}\tau$.

The quantization of the other half of the classical phase space is similar: the polarized sections are of the form

$$\psi = \overline{\psi}_{A'B'\ldots C'}(p) \overline{z}^{A'} \overline{z}^{B'} \ldots \overline{z}^{C'}, \tag{9.4.2}$$

where $\psi_{AB\ldots C}$ is a symmetric spinor field on the past sheet of the mass hyperboloid H_m^-. The Hermitian structure is given by the right-hand side of (9.4.1) when n is even, and by minus this expression when n is odd.

The potential θ' is not invariant under translations in Minkowski space, so when we move the origin, we must also make a gauge transformation of the ψs. When $q^a \mapsto q^a + t^a$, $\theta' \mapsto \theta' + \mathrm{d}(\zeta t^a p_a)$ and $\psi \mapsto e^{i\zeta t^a p_a/\hbar}\psi$. When $\zeta > 0$, this is the same as the behaviour of the coefficients in the Fourier decomposition of a positive frequency solution of the massive wave equation (7.4.4), provided that we take $\mu = m/\hbar$. Thus the Fourier transform

$$\phi_{AB\ldots C} = \left(\frac{1}{2\pi}\right)^{3/2} \int_{H_m^+} \psi_{AB\ldots C} e^{-ip_a x^a/\hbar} \mathrm{d}\tau$$

identifies in an invariant way the polarized sections of the prequantum bundle over M_{sm}^+ with the positive frequency solutions of the massive wave equation

$$\hbar^2 \Box \phi_{AB\ldots C} + m^2 \phi_{AB\ldots C} = 0. \tag{9.4.3}$$

Similarly, polarized sections over H_{sm}^- are identified with negative frequency solutions of (9.4.3) by

$$\phi_{AB\ldots C} = \left(\frac{1}{2\pi}\right)^{3/2} \int_{H_m^-} \psi_{AB\ldots C} e^{-ip_a x^a/\hbar} \mathrm{d}\tau. \tag{9.4.4}$$

Without the complex conjugate in (9.4.2), this would not be gauge invariant.

The entire Hilbert space $\mathcal{H}_{sm} = \mathcal{H}_{sm}^+ \oplus \mathcal{H}_{sm}^-$ is the same as the space V of all solutions of (9.4.3) with appropriately behaved Fourier transforms. The correspondence is linear if we regard both $\mathcal{H}_{sm}$ and V as real vector spaces, but because of the complex conjugate in (9.4.2), it is not linear over

$\mathbb{C}$ with respect to the obvious complex structure on the V, $\phi \mapsto \mathrm{i}\phi$. It is, however, *antilinear* with respect to the complex structure

$$J : V \to V : \phi \mapsto -\mathrm{i}\phi^+ + \mathrm{i}\phi^- .$$

Therefore the natural complex linear identification is between $\overline{V}_{(J)}$ and $\mathcal{H}_{sm}$. This is consistent with the Fock space quantization of V, in which $\mathcal{H}_1 = \overline{V}_{(J)}$ is interpreted as the space of one particle states.

It is straightforward to read off how the Poincaré group and charge conjugation act at the quantum level, either on $\mathcal{H}_{sm}$ or on the corresponding space-time fields. The identity component acts on the spinor fields $\psi_{AB\ldots C}$ in the natural way, with the usual sign ambiguity. When $n = 1$, the time reversal $T : x^a \mapsto x^a - t^a x^b t_b$, where $t^a t_a = 2$, becomes, up to an undetermined phase, the antiunitary transformation $\psi_A \mapsto \tilde{\psi}_A$, where

$$\tilde{\psi}^A(p_a) = t^{AA'}\overline{\psi}_{A'}(-p_a + t_a p_b t^b).$$

At the field level, this is $\tilde{\phi}^A(x) = t^{AA'}\overline{\phi}_{A'}(Tx)$, with the obvious generalization to higher values of n.

The parity map $P : x^a \mapsto -x^a + t^a x^b t_b$ becomes the unitary transformation $\psi_A \mapsto \tilde{\psi}_A$, where now

$$\tilde{\psi}_A(p_a) = -\frac{\sqrt{2}}{m} p_{AA'} t^{BA'} \psi_B(-p_a + t^a p_b t^b),$$

or, at the field level,

$$\tilde{\phi}(x)_A = -\frac{2\mathrm{i}\hbar}{m} \nabla_{AA'} \left(t^{BA'} \phi_B(Px) \right).$$

Finally charge conjugation becomes the unitary transformation $\psi_A \mapsto \tilde{\psi}_A$, where this time

$$\tilde{\psi}^A(p) = \frac{\sqrt{2}}{m} p^{AA'} \overline{\psi}_{A'}(-p); \quad \text{that is,} \quad \tilde{\phi}^A = \frac{\sqrt{2}\mathrm{i}\hbar}{m} \nabla^{AA'} \overline{\phi}_{A'}.$$

Only charge conjugation interchanges the positive and negative frequency parts of the field.

The quantization of massless particles is very similar if we use the real polarization P_1 (p. 126). Given an origin in Q, there is a natural symplectic potential

$$\theta' = -\mathrm{i}\omega^A \mathrm{d}\overline{\pi}_A + \mathrm{i}\overline{\omega}^{A'} \mathrm{d}\pi_{A'}$$

on $\mathbb{T}$. Again the orbit M^+_{s0} is quantizable whenever the restriction of θ' to the constraint manifold C_{s0} satisfies the integrality condition of Proposition (8.4.9), which is whenever $s = \frac{1}{2}n\hbar$ for some positive or negative integer n.

The polarized sections of the prequantum bundle are then the functions on C_{s0} of the form $\psi(\pi_{A'}, \overline{\pi}_A)$ such that

$$\pi_{A'} \frac{\partial \psi}{\partial \pi_{A'}} - \overline{\pi}_A \frac{\partial \psi}{\partial \overline{\pi}_A} = -\frac{2n}{\hbar}\psi.$$

For $n > 0$, these are mapped in a gauge-invariant way to the positive frequency solutions of the right-handed massless field equation by the Fourier transform

$$\phi_{A'B'\ldots C'} = \left(\frac{1}{2\pi}\right)^{3/2} \int_{N+} \psi \pi_{A'} \pi_{B'} \ldots \pi_{C'} \mathrm{e}^{-\mathrm{i}p_a x^a/\hbar}\, \mathrm{d}\tau,$$

where there are n πs, $p_{AA'} = \overline{\pi}_A \pi_{A'}$, and the integration is over all p on the future light-cone N^+ (p. 151). The integrand is constant along the characteristic distribution on C_{s0}, and so is well defined as a function on M^+_{s0}. Similarly, for $n < 0$,

$$\phi_{A'B'\ldots C'} = \left(\frac{1}{2\pi}\right)^{3/2} \int_{N-} \overline{\psi} \pi_{A'} \pi_{B'} \ldots \pi_{C'} \mathrm{e}^{-\mathrm{i}p_a x^a/\hbar}\, \mathrm{d}\tau,$$

where now $p_a = -\overline{\pi}_A \pi_{A'}$, gives an *antilinear* correspondence with the *negative* frequency solutions.

The Hilbert space obtained from $M_{s0} = M^+_{s0} \cup M^+_{-s0}$ is the same as the Hilbert space of one particle states constructed from the solution space of the massless field equation by using the complex structure $J : \phi \mapsto -\mathrm{i}\phi^+ + \mathrm{i}\phi^-$.

Quantization by the twistor polarization P_2, which is Kähler but non-positive, introduces new issues which we shall consider briefly at the end of the next chapter.

9.5 Pairing

If a symplectic manifold has one Kähler polarization, then it has many. For example, there are canonical transformations which preserve ω, but change the polarization.[7] In this section, we shall consider the relationship between the Hilbert spaces of different polarizations.

Let $\mathcal{H}_{P'}$ and $\mathcal{H}_P$ be the Hilbert spaces constructed from two positive polarizations P and P' by using the same prequantum bundle $B \to M$. Then $\mathcal{H}_P$ and $\mathcal{H}_{P'}$ are both subspaces of the prequantum Hilbert space $\mathcal{H}$. Let $\pi : \mathcal{H}_{P'} \to \mathcal{H}_P$ be the orthogonal projection. This map is not, of course, unitary, but does have one significant property. If the flow of $f \in C^\infty(M)$ preserves both polarizations, then $\hat{f}$ acts on both subspaces; and since $\hat{f}$ is a symmetric operator on $\mathcal{H}$, $\langle \hat{f}s, s'\rangle = \langle s, \hat{f}s'\rangle$ for every $s' \in \mathcal{H}_{P'}$ and $s \in \mathcal{H}_P$. It follows that

$$\pi \hat{f} = \hat{f} \pi. \tag{9.5.1}$$

We shall look at some examples later on. In particular, when P and P' are determined by two complex structures on a symplectic vector space, it turns out that a multiple of the projection operator is in fact unitary: it is the operator whose existence is implied by the Stone-von Neumann theorem (§9.9).

Although the Hilbert space $\mathcal{H}_P$ of a nonpositive polarization is not a subspace of $\mathcal{H}$, there is still a natural pairing between the Hilbert spaces of different polarizations which enables us to extend the definition of π to nonnegative polarizations with real directions. We shall consider only two special cases here. The full construction will be given in the next chapter, where we shall also look at the 'metaplectic correction', which introduces additional phase factors into the pairing, but does not otherwise alter in any essential way the examples we shall look at now.

Example. *Two real polarizations.* Suppose that P and P' are transverse real polarizations (so that $TM = P \oplus P'$); and that $Q = M/P$ and $Q' = M/P'$ are both Hausdorff manifolds. Then $M = Q' \times Q$ and

$$\omega = \frac{\partial^2 S}{\partial q'^a \partial q^b} \mathrm{d}q'^a \wedge \mathrm{d}q^b,$$

where q'^a is a coordinate system on Q', q^a is a coordinate system on Q, and $S = S(q', q)$ (§4.9). We shall assume that Q is oriented; then, since M is oriented by ω, Q' is also oriented.[8]

Let β and β' be respective sections of the two canonical bundles K_P and $K_{P'}$. Since P and P' are transverse, we can define a nonsingular pairing $(\beta, \beta') \in C^\infty_{\mathbb{C}}(M)$ by

$$(\beta, \beta')\epsilon = \beta' \wedge \overline{\beta}.$$

From this we obtain a pairing between sections of δ_P and $\delta_{P'}$ by putting $(\nu, \nu') = \sqrt{(\nu^2, \nu'^2)}$; and hence a pairing between $\mathcal{H}_P$ and $\mathcal{H}_{P'}$ by putting

$$\langle \tilde{s}, \tilde{s}' \rangle = \int_M (s, s')(\nu, \nu')\, \epsilon,$$

where $\tilde{s} = s\nu \in \mathcal{H}_P$ and $\tilde{s}' = s'\nu' \in \mathcal{H}_{P'}$. The sign of the square root in the definition of (ν, ν') will be a central issue later on; but we can set it aside for the moment by arranging that ν and ν' should both be real and positive (since Q and Q' are oriented, the transition functions of δ_P and $\delta_{P'}$ are real and positive, so this condition makes sense).

The pairing induces a linear map $\pi : \mathcal{H}_{P'} \to \mathcal{H}_P$. To find an explicit representation of π, we shall use the local trivialization of B determined by the symplectic potential

$$\frac{\partial S}{\partial q^a} \mathrm{d}q^a;$$

and we shall use $\sqrt{\mathrm{d}^n q}$ and $\sqrt{\mathrm{d}^n q'}$ to trivialize δ_P and $\delta_{P'}$. Then the P wave functions and the P' wave functions become complex functions of the form

$$\psi(q) \quad \text{and} \quad \psi(q')\mathrm{e}^{\mathrm{i}S/\hbar}$$

respectively (the asymmetry arises because the potential vanishes on the leaves of P, but restricts to $\mathrm{d}S$ on the leaves of P'). Since

$$\mathrm{d}q'^1 \wedge \cdots \wedge \mathrm{d}q'^n \wedge \mathrm{d}q^1 \wedge \cdots \wedge \mathrm{d}q^n = \frac{(2\pi\hbar)^n}{D}\epsilon \quad \text{where} \quad D = \det\left(\frac{\partial^2 S}{\partial q \partial q'}\right),$$

the pairing is

$$\langle \psi, \psi' \rangle = \left(\frac{1}{2\pi\hbar}\right)^{n/2} \int \overline{\psi}\psi' \mathrm{e}^{\mathrm{i}S/\hbar} \sqrt{D}\, \mathrm{d}^n q \mathrm{d}^n q'.$$

It is assumed that the coordinate systems on Q and Q' are both right-handed, and so D is positive.

A significant example arises when Q is n-dimensional Euclidean space, and the leaves of P' are the surfaces of constant p (in linear coordinates). Then we can replace q' by p and S by $p_a q^a$. In this case, the pairing map $\pi : \mathcal{H}_{P'} \to \mathcal{H}_P$ is the Fourier transform

$$(\pi\psi')(q) = \left(\frac{1}{2\pi\hbar}\right)^{n/2} \int \psi'(p)\mathrm{e}^{\mathrm{i}p_a q^a/\hbar}\, \mathrm{d}^n p.$$

We have recovered the usual unitary correspondence between the quantizations in position and momentum space.[9]

Example. *A real polarization and a Kähler polarization.* Let P be a real polarization, as before, and let P' be the Kähler polarization determined by a positive complex structure J' on M. At each point of M, there is a symplectic frame X^a, Y_b such that the Xs span P and Ys project onto a right-handed frame on Q; and in which J' takes the standard form

$$J' = \begin{pmatrix} 0 & -1 \\ 1 & 0 \end{pmatrix}. \tag{9.5.2}$$

This frame is unique up to proper orthogonal transformations of the Xs and Ys. We can therefore pick out a positive section of K_P by putting

$$\beta(Y^1, \ldots, Y^n) = \frac{1}{n!}.$$

Let $\nu = \sqrt{\beta}$. Then the wave functions in $\mathcal{H}_P$ are uniquely of the form $\tilde{s} = s\nu$, where s is a section of B. The pairing with $\mathcal{H}_{P'}$ is defined by

$$\langle \tilde{s}, s' \rangle = c \int (s, s')\epsilon$$

where $s' \in \mathcal{H}_{P'} \subset \mathcal{H}$ and c is a constant that we shall leave unspecified for the time being. The pairing determines linear maps $\pi : \mathcal{H}_{P'} \to \mathcal{H}_P$ and $\pi' : \mathcal{H}_P \to \mathcal{H}_{P'}$, but again these need not be unitary.

The standard example is Bargmann's transform (Bargmann 1961). The following treatment is based on notes by J. H. Rawnsley.

In the case $n = 1$, M is a 2-dimensional symplectic vector space. In linear canonical coordinates, J' is given by (9.5.2). The leaves of P are the lines of constant q, and $\nu = \sqrt{\mathrm{d}q}$. The symplectic potential is $\frac{1}{2}(p\mathrm{d}q - q\mathrm{d}p)$ and the prequantum bundle is the trivial bundle.

The wave functions in $\mathcal{H}_P$ and $\mathcal{H}_{P'}$ are represented by complex functions of the respective forms

$$\psi(q)\mathrm{e}^{-\mathrm{i}pq/2\hbar} \quad \text{and} \quad \phi(z)\mathrm{e}^{-\bar{z}z/4\hbar},$$

where ϕ is a holomorphic function of $z = p + \mathrm{i}q$. The pairing is

$$\langle \psi, \phi \rangle = \frac{c}{2\pi\hbar} \int \overline{\psi}(q)\phi(z)\mathrm{e}^{(2\mathrm{i}pq - \bar{z}z)/4\hbar}\, \mathrm{d}p\mathrm{d}q. \tag{9.5.3}$$

By the reproducing property of coherent states,

$$\phi(z) = \frac{1}{2\pi\hbar} \int \phi(w)\mathrm{e}^{(\overline{w}z - \overline{w}w)/2\hbar}\, \mathrm{d}x\mathrm{d}y$$

where $w = x + \mathrm{i}y$. By substituting into (9.5.3), and integrating over p,

$$\langle \psi, \phi \rangle = \frac{1}{2\pi\hbar} \int \overline{\psi}(q)\phi(w)\overline{F(q,w)}\mathrm{e}^{-w\overline{w}/2\hbar}\, \mathrm{d}q\mathrm{d}x\mathrm{d}y,$$

where

$$F(q,w) = \frac{\bar{c}}{\sqrt{\pi\hbar}} \exp\left(\frac{1}{4\hbar}(w^2 - 4\mathrm{i}qw - 2q^2) \right).$$

Hence

$$\begin{aligned} (\pi\phi)(q) &= \frac{1}{2\pi\hbar} \int \phi(w)\overline{F(q,w)}\mathrm{e}^{-w\overline{w}/2\hbar}\, \mathrm{d}x\mathrm{d}y \\ (\pi'\psi)(w) &= \int \psi(q)F(q,w)\, \mathrm{d}q. \end{aligned}$$

The energy eigenfunctions z^N in $\mathcal{H}_{P'}$ are mapped to the Hermite functions (i.e. the eigenfunctions of the harmonic oscillator). In particular, the ground state $\phi = 1$ is mapped to

$$\psi = \frac{c}{\sqrt{\pi\hbar}}\mathrm{e}^{-q^2/2\hbar}.$$

In this case π is unitary, provided that we make the apparently eccentric choice $c = (\pi\hbar)^{1/4}$, a choice that will appear more natural in the context of the full pairing construction.

9.6 The WKB approximation

So far we have concentrated on the reconstruction of quantum systems from their classical counterparts. Many of the same elements from symplectic geometry and prequantization also appear naturally in the reverse process of passing to the classical limit. The key construction here is the WKB approximation.[10]

A simple example is provided by the time-independent Schrödinger equation

$$\tfrac{1}{2}\hbar^2\nabla^2\psi + E\psi = 0$$

for the energy eigenfunctions of a free particle of unit mass in $\mathbb{E}$ (three-dimensional Euclidean space). This is solved asymptotically as $\hbar \to 0$ by substituting $\psi = A\mathrm{e}^{\mathrm{i}w/\hbar}$ and collecting the various powers of $\hbar$. The terms of order $\hbar^0$ and $\hbar^1$ give equations for the phase function w and the amplitude A,

$$\tfrac{1}{2}\nabla w \cdot \nabla w = E \quad \text{and} \quad \nabla w \cdot \nabla A + \tfrac{1}{2}A\nabla^2 w = 0.$$

The first is the time-independent form of the Hamilton-Jacobi equation of the Hamiltonian $H = \frac{1}{2}\mathbf{p}\cdot\mathbf{p}$; the second is the 'transport equation', which determines the behaviour of the amplitude A along the vector field $X = \nabla w$. It can be written in the more suggestive form

$$\mathcal{L}_X(A^2\,\mathrm{d}V) = 0,$$

where $\mathrm{d}V$ is the Euclidean volume element. We should thus think of A as the square root of a volume form—an interpretation that is entirely natural in the context of the quantization of $T^*\mathbb{E}$ by the vertical polarization. (One can, of course, take the asymptotic analysis further by expanding A in powers of $\hbar$; but we shall be concerned only with the first two terms.)

The same equations arise for any quantum Hamiltonian constructed by the Schrödinger prescription. Let Q be an n-dimensional configuration space. Consider first a homogeneous classical Hamiltonian on T^*Q of the form

$$H(p,q) = H_k^{ab\ldots c}p_a p_b \ldots p_c$$

where H_k is a symmetric tensor field on Q, with k indices. The Schrödinger prescription

$$p_a \mapsto -\mathrm{i}\hbar\partial_a = \frac{\hbar}{\mathrm{i}}\frac{\partial}{\partial q^a} \tag{9.6.1}$$

transforms H into a differential operator on the wave functions on Q, given by

$$\hat{H}(\psi) = (-\mathrm{i}\hbar)^k\left(H_k^{ab\ldots c}\partial_a\partial_b\ldots\partial_c + \tfrac{1}{2}k\partial_a(H_k^{ab\ldots c})\partial_b\ldots\partial_c + \cdots\right)\psi,$$

where the terms omitted are operators of order $k-2$ or less. This does not determine $\hat{H}$ unambiguously because of the factor-ordering problem: if we write the products of ps and qs in the coordinate expression for H in a different order, then the substitution (9.6.1) gives a different operator. Only the leading term (the operator of order k) will be unchanged by a general reordering. However the next term (the operator of order $k-1$) is fixed by requiring formal self-adjointness with respect to the inner product

$$\langle \psi, \psi \rangle = \int \overline{\psi}\psi \, \mathrm{d}^n q.$$

The lower-order terms cannot be determined uniquely without more detailed 'factor-ordering' rules.

The two leading terms in $\hat{H}$ are invariant under coordinate transformations in Q provided that we think of $\hat{H}$ as acting on sections of $\sqrt{\Delta_Q}$ rather than on functions (as before, Δ_Q is the line bundle of complex n-forms on Q; since Q is oriented, its transition functions are real and positive, so $\sqrt{\Delta_Q}$ is unambiguously defined). That is, we think of $\hat{H}$ as the operator $\psi\sqrt{\mathrm{d}^n q} \mapsto \hat{H}(\psi)\sqrt{\mathrm{d}^n q}$. Under an orientation-preserving coordinate transformation $q^a \mapsto \tilde{q}^a$, we replace ψ by

$$\tilde{\psi} = \psi \sqrt{\det\left(\frac{\partial \tilde{q}}{\partial q}\right)}.$$

The square-integrable sections of $\sqrt{\Delta_Q}$ are precisely the elements of $\mathcal{H}_P$, where P is the vertical polarization of T^*Q. Thus the Schrödinger prescription determines $\hat{H}$ invariantly as an operator on $\mathcal{H}_P$ up to the addition of an operator of order $k-2$. (If $k<2$, there is no ambiguity; in particular, when $k=0$, $\hat{H}$ simply acts on $\mathcal{H}_P$ by multiplication by H.)

By expressing a general polynomial in the ps as a sum of homogeneous polynomials, we can construct a class of symmetric operators $\hat{H}$ for any H of the form

$$H = H_0 + H_1^a p_a + \cdots + H_k^{ab\ldots c} p_a p_b \ldots p_c,$$

by summing the operators constructed from the individual homogeneous terms. Whatever choice we make for $\hat{H}$ within this class, substituting $\psi = A\mathrm{e}^{\mathrm{i}w/\hbar}$ into the eigenvalue equation $(\hat{H}-E)\psi\sqrt{\mathrm{d}^n q} = 0$, and picking out the two leading terms in $\hbar$, results in the Hamilton-Jacobi and transport equations

$$H(q, \mathrm{d}w) = E \quad \text{and} \quad \mathcal{L}_X(A^2 \mathrm{d}^n q) = 0,$$

where

$$X = \sum k H_k^{ab\ldots c} \partial_b w \ldots \partial_c w \frac{\partial}{\partial q^a}.$$

The phase w will always be real, but it will be useful later on to allow the amplitude A to be complex. If we denote by Λ the Lagrangian submanifold of T^*Q generated by w, then X is the projection into Q of $X_H|_\Lambda$.

The solutions of these equations are naturally represented by objects on Λ. The pull-back of the phase factor $e^{iw/\hbar}$ under the projection $\Lambda \to Q$ is a function ψ on Λ such that

$$X(\psi) - \frac{i}{\hbar}(Z \lrcorner \theta)\psi = 0$$

for every Z tangent to Λ, where θ is the canonical 1-form; that is, ψ is the representation in the θ gauge of a covariantly constant section s of B_Λ, the restriction of B to Λ, where B is the product prequantum bundle with connection $d - i\hbar^{-1}\theta$. The pull-back of $A^2 d^n q$ is a section κ of Δ_Λ such that $\mathcal{L}_{X_H}\kappa = 0$, where Δ_Λ is the line bundle of complex volume elements (n-forms) on Λ. Up to sign, we can recover A and $e^{iw/\hbar}$ from s and κ, by introducing coordinates q^a and by using θ to trivialize B.

This representation does not involve the polarization. A 'WKB wave function' is simply a triple (Λ, s, κ) consisting of a Lagrangian submanifold Λ of T^*Q, a covariantly constant section s of B_Λ such that $(s, s) = 1$ (the 'phase function'), and a section κ of Δ_Λ (the 'amplitude squared'). Covariantly constant sections of B_Λ certainly exist locally because the restriction to Λ of the curvature 2-form vanishes. The triple is a 'WKB eigenfunction' of $\hat{H}$ if H is constant on Λ and $\mathcal{L}_{X_H}\kappa = 0$.

Given any real polarization P' transverse to Λ, we recover, up to sign, an element $s'\nu'$ of $\mathcal{H}_{P'}$ from (Λ, s, κ) by extending s from Λ to T^*Q by making it covariantly constant along P', which gives a section s' of B; and by pulling κ back to T^*Q by the projection along P' and then taking the square root, which gives first a section of $K_{P'}$ and then a section of $\delta_{P'}$. To obtain an explicit representation of $s'\nu'$ by a wave function on $Q' = T^*Q/P'$, it is necessary to introduce a potential θ' adapted to P' and coordinates q'^a on $Q' = T^*Q/P'$. Then κ is the pull-back under $\Lambda \to Q'$ of an n-form $A'^2 d^n q'$ on Q' and, in the θ' gauge, s is represented by $e^{iw'/\hbar}$ for some real function w'. The wave function on Q' is $A' e^{iw'/\hbar}$.

A number of points have been glossed over: for example, the wave function on Q' may not be square-integrable. More serious is the sign ambiguity, and the related issue of the relationship between the orientations on Q and Q' (we have to assume that Q' is oriented to define $\delta_{P'}$). We shall set these aside for the moment: their resolution involves the metaplectic construction, which is the subject of the next chapter.

The construction gives us a direct geometric procedure for associating a wave function $A' e^{iw'/\hbar}$ on Q' with a given wave function $A e^{iw/\hbar}$ on Q. We can also construct a wave function on Q' indirectly by pairing. If we substitute $\psi = A e^{iw/\hbar}$ into (9.5), and use the method of stationary phase to evaluate the leading term in $\hbar$ of the integral over q, then we obtain

$$\begin{aligned}\langle\psi,\psi'\rangle &= \left(\frac{1}{2\pi\hbar}\right)^{n/2}\int\overline{A}(q)\psi'(q')\mathrm{e}^{\mathrm{i}[S(q,q')-w(q)]/\hbar}\sqrt{D}\,\mathrm{d}^nq\mathrm{d}^nq' \\ &= \int\overline{A}\psi'\sqrt{\frac{D}{|\det g|}}\,\mathrm{e}^{-\mathrm{i}\pi\sigma/4}\mathrm{e}^{\mathrm{i}[S(q,q')-w(q)]/\hbar}\,\mathrm{d}^nq' + O(\hbar)\end{aligned} \tag{9.6.2}$$

where ψ' is an arbitrary function of q',

$$g_{ab} = \frac{\partial^2 w}{\partial q^a\partial q^b} - \frac{\partial^2 S}{\partial q^a\partial q^b},$$

σ is the signature of g_{ab} (the number of positive eigenvalues less the number of negative eigenvalues), and, in the second integrand, q is expressed as a function of q' by solving

$$\frac{\partial S}{\partial q^a} = \frac{\partial w}{\partial q^a}.$$

This relates the pull-backs under the projections $\Lambda\to Q$ and $\Lambda\to Q'$ of the coordinate systems q^a and q'^a on Q and Q'. By differentiating with respect to q^a, we obtain

$$\frac{\partial^2 w}{\partial q^a\partial q^b} - \frac{\partial^2 S}{\partial q^a\partial q^b} = \frac{\partial^2 S}{\partial q^a\partial q'^c}\frac{\partial q'^c}{\partial q^b}.$$

Hence, by taking determinants,

$$\det g = D\det\left(\frac{\partial q'}{\partial q}\right).$$

But $A^2\mathrm{d}^nq = A'^2\mathrm{d}^nq' = \kappa$. Therefore $A' = A\sqrt{D/\det g}$. Moreover

$$\theta' = \theta - \mathrm{d}S$$

is a symplectic potential adapted to P' (p. 78). If we use θ' to trivialize B in the representation of the P' wave functions by functions on Q', then $\mathrm{e}^{\mathrm{i}w'/\hbar} = \mathrm{e}^{\mathrm{i}(w-S)/\hbar}$ (by 8.2.4). Hence

$$\langle\psi,\psi'\rangle = \mathrm{e}^{-\mathrm{i}\pi\sigma/4}\int\overline{A}'\mathrm{e}^{-\mathrm{i}w'/\hbar}\psi'\,\mathrm{d}^nq' \tag{9.6.3}$$

The integral is the inner product of $A'\mathrm{e}^{-\mathrm{i}w'/\hbar}$ with an arbitrary P' wave function ψ'. It follows that, apart from an integral power of $\sqrt{\mathrm{i}}$, the image of $\psi = A\mathrm{e}^{\mathrm{i}w/\hbar}$ in $\mathcal{H}_{P'}$ under the pairing map is the same as the wave function given by the direct geometric construction, to the first order in $\hbar$.

The power of $\sqrt{\mathrm{i}}$ and the ambiguity in the sign of A' are of no physical importance provided that Λ remains transverse to both P and P', but they become more significant when we try to extend the analysis to allow for the possibility of caustic singularities in the vector field X, where nearby classical trajectories cross over. These correspond to points where Λ fails to be transverse to P. Although Λ itself is still a perfectly well-behaved submanifold of T^*Q at such points, the projection $\Lambda \to Q$ becomes singular and the amplitude A blows up. It is not obvious how to carry solutions of the transport equation through such points. In Chapter 8, we shall see that by replacing Δ_Λ by an appropriately defined line bundle $\delta_\Lambda = \sqrt{\Delta_\Lambda}$, we can remove the ambiguity and absorb the factor $\mathrm{e}^{\mathrm{i}\pi\sigma/4}$ in the definition of the pairing, and so obtain a well-defined prescription for continuing WKB solutions through caustics. This is essentially Maslov's (1972) theory: it incorporates the analogue of the classical phenomenon of geometric optics that the phase of a converging beam of light advances by half a period on passing through a focus. It also incorporates Keller's (1958) correction to the Bohr-Sommerfeld condition: if we allow Λ to be a general Lagrangian submanifold of T^*Q, then B_Λ has a global covariantly constant section only if the Bohr-Sommerfeld condition holds; that is

$$\frac{1}{2\pi\hbar}\int \theta \in \mathbb{Z}$$

for every closed curve in Λ. This restricts the possible semiclassical eigenvalues of $\hat{H}$. Keller's correction to this condition takes account of the discontinuous phase changes at the caustics.

9.7 The Schrödinger equation

Suppose that we have quantized a classical system by introducing a real polarization P in the classical space (M, ω). We shall now consider the dynamical problem of determining the evolution of the quantum state from a classical Hamiltonian $H \in C^\infty(M)$, which is essentially the problem of tying down the lower order terms in the operator constructed by the Schrödinger prescription.

Let ρ_t be the canonical flow generated by H. If ρ_t preserves P, then it generates a one-parameter unitary group $\tilde{\rho}_t : \mathcal{H}_P \to \mathcal{H}_P$ by 'pulling back' the wave functions. The evolution of the quantum state is given by the inverse 'push forward' map. In other words, if the state at time t is $\tilde{s}_t \in \mathcal{H}_P$, then $\tilde{\rho}_t \tilde{s}_t = \tilde{s}_0$, so that

$$\mathrm{i}\hbar \frac{\mathrm{d}\tilde{s}_t}{\mathrm{d}t} = \tilde{H}\tilde{s}_t. \tag{9.7.1}$$

This takes care of the rather trivial case in which H depends linearly on the momenta in coordinates adapted to P.

Blattner, Kostant and Sternberg (BKS) introduced a construction for the Schrödinger equation for Hamiltonians that do not preserve the polarization (Blattner 1973, Guillemin and Sternberg 1977, Simms 1978). Although it works only in special cases, these include the familiar examples of elementary wave mechanics, where it reproduces the standard Hamiltonian operators. However, it is not obvious from the construction that the evolution will turn out to be unitary.

The idea is very simple: it is to move the wave function forward for a small time δt along the flow of H, and then use the pairing to project it back into $\mathcal{H}_P$. Thus the evolution of $\tilde{s}_t \in \mathcal{H}_P$ is defined by replacing (9.7.1) by

$$\left\langle \frac{\mathrm{d}\tilde{s}_t}{\mathrm{d}t}, \tilde{s}' \right\rangle = -\frac{\mathrm{d}}{\mathrm{d}t'} \left\langle \tilde{\rho}_{t'} \tilde{s}_t, \tilde{s}' \right\rangle_{t'=0}$$

for every time-independent $\tilde{s}' \in \mathcal{H}_P$. On the right, $\langle \cdot, \cdot \rangle$ is the inner product in $\mathcal{H}_P$; on the left, it is the pairing between $\mathcal{H}_P$ and $\mathcal{H}_{P'}$, where P' is the polarization given by pulling back P by $\rho_{t'}$. Note that $\tilde{\rho}_{t'}\tilde{s}_t \in \mathcal{H}_{P'}$, so the pairing is well defined provided that P and P' are transverse. The minus sign on the right-hand side appears because $\tilde{\rho}$ is a 'pull-back' map, while the evolution is given by dragging the wave functions forward along the classical flow.

Example. *Geodesic flow.* Let (Q, g_{ab}) be a complete n-dimensional Riemannian manifold and let μ be the metric volume form on Q. Put $M = T^*Q$ and $H = \frac{1}{2}g^{ab}p_a p_b$, which generates the geodesic flow, and take P to be the vertical polarization. For example, Q might be a constraint surface in $\mathbb{R}^3$. Then H is the kinetic energy of a unit mass particle constrained to move on the surface.

The wave functions are of the form $\tilde{s} = \psi\nu$, where ψ is a complex function on Q which is constant along P, and $\nu^2 = \mu$ (we shall not distinguish between μ and its pull-back under $\mathrm{pr} : T^*Q \to Q$). We have to calculate the pairing of $\tilde{\rho}_t(\tilde{s}) = \hat{\rho}_t(\psi)\rho_t^*(\nu)$ and $\tilde{s}' = \psi'\nu$ for an arbitrary function ψ' on Q. By (8.4.2),

$$(\hat{\rho}_t\psi)(m) = \psi_t(m) \exp\left(-\frac{\mathrm{i}}{\hbar}\int_0^t L \, \mathrm{d}t'\right) = \psi_t(m)\mathrm{e}^{-\mathrm{i}tH/\hbar},$$

where $\psi_t = \psi \circ \rho_t$, since $L = X_H \lrcorner\, \omega - H = H$ is constant along geodesics. Therefore the pairing is

$$\langle \tilde{\rho}_t(\tilde{s}), \tilde{s}' \rangle = \int_M \overline{\psi}_t \psi' \sqrt{(\rho_t^*\mu, \mu)}\, \mathrm{e}^{\mathrm{i}tH/\hbar}\, \epsilon. \qquad (9.7.2)$$

We shall evaluate the integral by rescaling the momenta. Because of the homogeneity of the Hamiltonian, if $\rho_t(p, q) = (p', q')$, then $\rho_1(tp, q) = (tp', q')$. That is, $\rho_1 \circ \xi_t = \xi_t \circ \rho_t$, where ξ_t is the dilatation $(p, q) \mapsto (tp, q)$ (Fig.

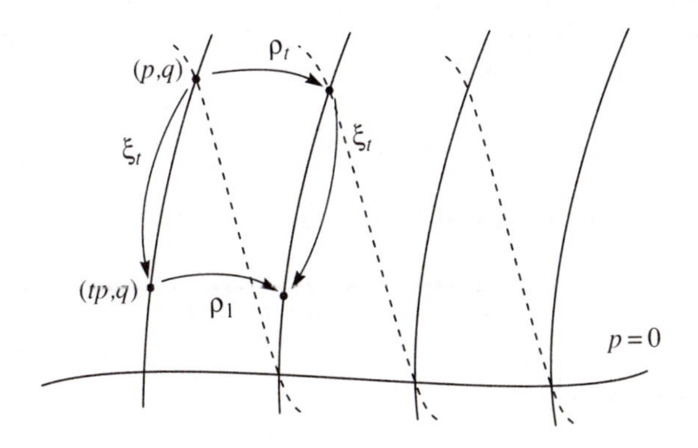

Fig. 9.4. The broken lines are the leaves of $\rho_t^* P$.

9.4).

Now $\xi_t^* \epsilon = t^n \epsilon$, $\xi_t^* \mu = \mu$, and $\psi \circ \xi_t = \psi$. Therefore $\psi \circ \rho_1 \circ \xi_t = \psi \circ \rho_t$ and

$$\begin{aligned}(\rho_t^* \mu, \mu)\epsilon &= \rho_t^*(\mu) \wedge \mu \\ &= \rho_t^*(\xi_t^* \mu) \wedge \xi_t^* \mu \\ &= (\xi_t^* \rho_1^* \mu) \wedge \xi_t^* \mu \\ &= \xi_t^* \big((\rho_1^* \mu, \mu)\epsilon\big) \\ &= t^n \big((\rho_1^* \mu, \mu) \circ \xi_t\big)\epsilon.\end{aligned}$$

That is, $(\rho_t^* \mu, \mu) = t^n (\rho_1^* \mu, \mu) \circ \xi_t$. It follows that

$$\langle \tilde{\rho}_t \tilde{s}, \tilde{s}' \rangle = t^{-n/2} \int_M \overline{\psi}_1 \psi' \sqrt{(\rho_1^* \mu, \mu)}\, \mathrm{e}^{\mathrm{i}H/\hbar t}\, \epsilon.$$

We need the derivative of this expression with respect to t at $t = 0$. By the principle of stationary phase, the integral over the momentum variables is determined in the limit $t \mapsto 0$ by the behaviour of the integrand in a neighbourhood of $p = 0$. The contribution of order t is determined by the second order terms in the Taylor expansion in powers of p.

To find the expansion of $(\rho_1^* \mu, \mu)$, we need the '$\mathrm{d}p$' part of $\rho_1^* \mu$, that is, the restriction of $\rho_1^* \mu$ to the fibres of T^*Q. Now the restriction of $\mathrm{pr} \circ \rho_1$ to T_q^*Q is the exponential map which sends $p \in T^*Q$ to the point $\exp_q(p)$ reached by travelling for unit time from q along a geodesic with velocity $p^a = g^{ab} p_b$. Thus the restriction of $\rho_1^* \mu$ to T^*Q is $\exp_q^* \mu$.

This we can find by using geodesic coordinates (following Guillemin and Sternberg 1977). Pick a frame at q and let p_a be the corresponding linear coordinate system on T_q^*Q. Then q is the origin of the coordinate system in which the point x in some neighbourhood of q is labelled by the coordinates $x^a = g^{ab}(q)p_b$ where $\exp_q(p) = x$. The metric is approximated to the second order in x by[11]

$$g_{ab}(x) = g_{ab}(q) - \tfrac{1}{3}R_{acbd}(q)x^c x^d.$$

Hence the restriction of $\rho_1^*\mu$ to T_q^*Q is approximated to the second order in p by

$$\frac{1}{\sqrt{g(q)}}(1 - \tfrac{1}{6}R_{ab}p^a p^b)\,\mathrm{d}^n p,$$

where $g = \det g_{ab}$. Therefore

$$(\rho_1^*\mu, \mu) = (2\pi\hbar)^n \left(1 - \tfrac{1}{6}R^{ab}p_a p_b + O(p^3)\right).$$

By making a Taylor expansion along the geodesics through q,

$$\psi \circ \rho_1 = \psi(q) + p^a(\nabla_a\psi)(q) + \tfrac{1}{2}p^a p^b(\nabla_a\nabla_b\psi)(q) + O(p^3).$$

We shall use (A.8.1) to expand the p integral in powers of t. Since odd powers of p do not contribute, to the first order in t, we have[12]

$$\begin{aligned}
&\langle\tilde{\rho}_t(\tilde{s}), \tilde{s}'\rangle \\
&\quad= \left(\frac{1}{2\pi\hbar t}\right)^{n/2} \int \left(\overline{\psi} + \tfrac{1}{2}p_a p_b(\nabla^a\nabla^b\overline{\psi} - \tfrac{1}{12}R^{ab}\overline{\psi})\right)\psi' \mathrm{e}^{\mathrm{i}p_c p^c/2\hbar t}\,\mathrm{d}^n p\mathrm{d}^n q \\
&\quad= \mathrm{e}^{n\pi\mathrm{i}/4}\int\left(\overline{\psi} - \frac{\mathrm{i}t}{\hbar}\hat{H}\overline{\psi}\right)\psi'\,\mu,
\end{aligned}$$

where $\hat{H} = -\frac{1}{2}\hbar^2(\nabla^a\nabla_a - R/6)$ and $\mu = \sqrt{g}\,\mathrm{d}^n q$. The phase factor in front of the integral is awkward, and will be absorbed into the pairing by the metaplectic correction. If we ignore it, we conclude that the evolution of $\psi(q,t)\nu \in \mathcal{H}_P$ is given by the Schrödinger equation

$$\mathrm{i}\hbar\frac{\partial\psi}{\partial t} = -\frac{\hbar^2}{2}(\nabla^a\nabla_a - \tfrac{1}{6}R)\psi.$$

It is not hard to generalize to the case in which H includes a potential term and is time-dependent (Śniatycki 1980). However, the analysis depends on the fact that the leading order term in the Hamiltonian is quadratic. The construction fails, for example, if there are cubic terms in the momenta.

For a free particle in flat space, the scalar curvature term is not present, and one can in fact obtain the evolution by dragging $\psi\nu$ forward through any value of t, and then projecting into $\mathcal{H}_P$ by the pairing construction.[13]

9.8 Path integrals and the Feynman expansion

The BKS 'dragging-projection' construction gives a natural approximation to the time evolution by a series of discrete steps. The projection at each stage is an integral over phase space, and the formal limit as the step size goes to zero can be realized as an integral over paths in phase space (Simms 1980a, Gawędzki 1974, 1976, Śniatycki 1980). We shall look here at how this works in the context of Fock space, where it reproduces a form of the path integral similar to that in Klauder (1960, 1979), Faddeev (1976).[14] One remarkable feature is the way in which the construction automatically incorporates the boundary conditions for the Feynman propagator in field theory without the need for 'Euclideanization'. Some of the details are due to Simpson (1984).

Let V, ω, θ_0 and P be as in §9.2 (p. 173). Suppose that $H \in C^\infty(M \times \mathbb{R})$ is a time-dependent Hamiltonian generating a canonical flow $\rho_{tt'} : V \to V$ that does not preserve J. Then for each $t, t' \in \mathbb{R}$, $\hat{\rho}_{tt'}$ maps $\mathcal{H}_P$ to some other subspace of $\mathcal{H}$.

Let the state of the system at time t_0 be $\psi_0 \in \mathcal{H}_P$. According to the BKS construction, to the first order in $\delta\tau$, the state ψ_1 at time $t_1 = t_0 + \delta\tau$ is the image of $\hat{\rho}_{t_1 t_0}\psi_0$ under the orthogonal projection $\mathcal{H} \to \mathcal{H}_P$. The projection into $\mathcal{H}_P$ is given by taking the inner product with the coherent states. To the first order in $\delta\tau$, therefore,

$$\psi_1(z) = \langle \psi_z, \hat{\rho}_{t_1 t_0}\psi_0 \rangle = \langle \hat{\rho}_{t_0 t_1}\psi_z, \psi_0 \rangle, \qquad (9.8.1)$$

where ψ_z is the coherent state based on $z \in V$.

So as not to clutter the argument with notation, we shall take $n = \frac{1}{2}\dim V = 1$. Then $V_{(J)}$ has one complex dimension. The elements of V are labelled by a single complex coordinate z, and

$$\omega = \frac{\mathrm{i}}{2}\mathrm{d}z \wedge \mathrm{d}\overline{z} \quad \text{and} \quad \theta_0 = \frac{\mathrm{i}}{4}\big(z\mathrm{d}\overline{z} - \overline{z}\mathrm{d}z\big).$$

In higher dimensions, the results are the same: the only difference is that z also carries a coordinate index.

In the θ_0-gauge, the coherent state based on the point with coordinate z is the function

$$\psi_z(w) = \exp\frac{1}{4\hbar}\big(2\overline{z}w - \overline{z}z - \overline{w}w\big).$$

Write the time-dependent Hamiltonian vector field generated by H in the form

$$X_H = X\frac{\partial}{\partial z} + \overline{X}\frac{\partial}{\partial \overline{z}},$$

where X is a function of z, $\overline{z}$, and t. Then,

$$\rho_{t_0 t_1}(w) = w + \delta\tau X + O(\delta\tau^2).$$

Hence by putting $L = X_H \lrcorner \theta_0 - H = \frac{1}{4}\mathrm{i}(\overline{X}z - X\overline{z}) - H$ in (8.4.3) and by evaluating ψ_z at $w + \delta\tau X$,

$$(\hat{\rho}_{t_0t_1}\psi_z)(w) = \exp\left(\frac{\delta\tau}{2\hbar}(2\mathrm{i}H + (\overline{z} - \overline{w}))\right)\psi_z(w),$$

to the first order in $\delta\tau$, where H and X are evaluated at w. The evolution from t_0 to t_1 is given by substituting this into (9.8.1) and by calculating the inner product by integration over V.

To find the evolution over $[t_0, t_n]$, we partition $[t_0, t_n]$ into n intervals of length $\delta\tau$, and iterate. The result is an integral over $V^n = V\times V\times\cdots\times V = \{z_0, z_1, \ldots, z_{n-1}\}$. The state at time t_n is ψ_n, where

$$\psi_n(z_n) = \int_{V^n} \psi_0(z_0)\exp\left(\frac{\delta\tau}{2\hbar}\sum_0^{n-1}(-2\mathrm{i}H_k + \overline{X}_k\delta z_k)\right)K\epsilon^n. \qquad (9.8.2)$$

Here $\delta z_k = z_{k+1} - z_k$,

$$K(z_0, z_1, \ldots, z_n) = \overline{\psi}_{z_1}(z_0)\overline{\psi}_{z_2}(z_1)\ldots\overline{\psi}_{z_n}(z_{n-1}),$$

and the subscript k indicates evaluation at $(z, \overline{z}, t) = (z_k, \overline{z}_k, t_0 + k\delta\tau)$.

We now take the formal limit $\delta\tau \to 0$, $n \to \infty$, keeping fixed $t_0 = t$, $t_n = t'$, $z_0 = z$, $z_n = z'$. Then $K\epsilon^n$ becomes a complex 'measure' $\mathrm{d}\kappa$ on the path space $\Gamma = \{\gamma : \tau \mapsto z(\tau)\}$ joining the two fixed points $z = \gamma(0)$ and $z' = \gamma(t')$. The state $\psi_{t'}$ at time t' is given in terms of the state ψ_t at time t by[15]

$$\psi_{t'}(z') = \int_\Gamma \psi_t(z)\exp\left(-\frac{\mathrm{i}}{\hbar}\int_t^{t'} H\,\mathrm{d}\tau\right)\mathrm{d}\kappa,$$

since the $\delta\tau\delta z$ terms are approximated by $\delta\tau^2\dot{z}$ and disappear in the limit; see Fig. 9.5.

The function K has two notable properties. First,

$$K = \exp\left[\frac{\mathrm{i}}{\hbar}\sum_0^{n-1}\left(-\frac{\mathrm{i}}{4}\overline{z}_k\delta z_k + \frac{\mathrm{i}}{4}z_{k+1}\delta\overline{z}_k\right)\right],$$

so in the limit

$$K = \exp\left(\frac{\mathrm{i}}{\hbar}\int_t^{t'}\theta_0\right)$$

since $\theta_0 = \frac{1}{4}\mathrm{i}(z\mathrm{d}\overline{z} - \overline{z}\mathrm{d}z)$. Therefore the path integral can also be written in a more familiar form involving the action

$$\psi_{t'}(z') = \int_\Gamma \psi_t(z)\exp\left(\frac{\mathrm{i}}{\hbar}\int_t^{t'} L\,\mathrm{d}\tau\right)\epsilon^\infty, \qquad (9.8.3)$$

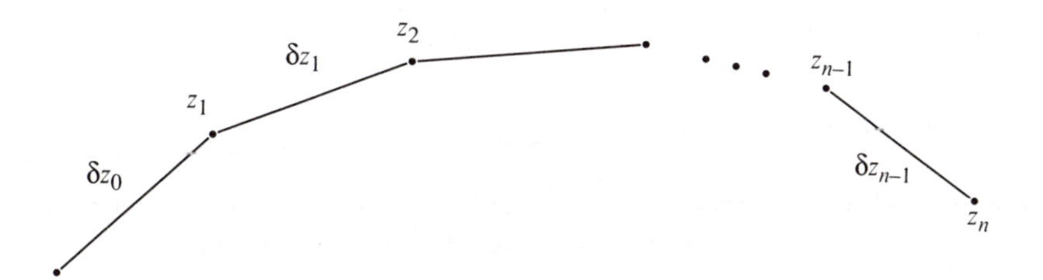

Fig. 9.5. The construction of the path integral.

where, with $\dot{\gamma}$ denoting the tangent to γ, $L = \dot{\gamma} \lrcorner\, \theta_0 - H$. Second,

$$\overline{K}K = \exp\left(-\frac{1}{2\hbar}\sum_{0}^{n-1} \delta\bar{z}_k \delta z_k\right),$$

so $|K|^2$ is normally distributed about the straight path from z_0 to z_n traversed at constant speed. Because of this Gaussian form, integrals involving $\mathrm{d}\kappa$ are more likely to make sense than integrals involving ϵ^{∞}.

In field theory, V is the space of solutions of the free field equations and H is the interaction Hamiltonian. For example, in '$\lambda\phi^4$ theory', V is the space of solutions of $\Box\phi + \mu^2\phi = 0$, J is defined by the positive-negative frequency decomposition (§7.4), and H is a multiple of

$$\int_{y^0=t} \phi^4(y^a)\,\mathrm{d}^3y$$

which depends on $\phi \in V$ and t (the ys are inertial coordinates on Minkowski space). The states are elements of the Fock space of the free field, and the path integral determines the state at time $t' = \infty$ in terms of the state at time $t = -\infty$. One thinks of H as generating a small interaction superimposed on the evolution of the free field, which is determined by the Klein-Gordon equation. The expansion of the path integral expression in powers of H is the 'Feynman expansion'.

This takes us into an area beyond the scope of this book, but it is worth noting one point about the way in which the path integral emerges as a limit of the BKS construction. This is that it automatically incorporates the boundary conditions at infinity: if one starts with the formal expression (9.8.3), then the boundary conditions must be imposed as an additional requirement—for example by 'Wick rotating' in a specified direction from imaginary time. If, on the other hand, one thinks of (9.8.3) as the limit of the BKS construction, then the correct boundary conditions are incorporated automatically. One can see how this works for the Klein-Gordon field by looking at the linear interaction Hamiltonian

$$H(\phi, t) = \int_{x^0=t} f(x)\phi(x)\, \mathrm{d}^3 x$$

where f is some compactly-supported function on Minkowski space, and by calculating the 'vacuum-vacuum' transition amplitude. The value of the time-dependent Hamiltonian vector field X_H at time t is the field $\phi' \in V$, where

$$\phi'(x) = \int_{y^0=t} \Delta(x, y) f(y)\, \mathrm{d}^3 y,$$

by (7.2.5).

First consider the analogous situation in which $n = 1$ and H is a time-dependent linear function on V. That is $H = \frac{1}{2}\mathrm{i}(\overline{X}z - X\overline{z})$, where the X depends on t (but not z). Let $\psi_0(z_0) = \mathrm{e}^{-\overline{z}_0 z_0/4\hbar}$ be the ground state. We are interested in $\langle \psi_n, \psi_0 \rangle$, which we can write in the form

$$\langle \psi_n, \psi_0 \rangle = \int_{V^{n+1}} \exp\left(\frac{\delta\tau}{2\hbar} \sum_0^{n-1} (X_k \overline{z}_{k+1} - \overline{X}_k z_k) \right) K' \epsilon^{n+1} \tag{9.8.4}$$

by using (9.8.2). Here

$$K' = \exp\left(-\frac{1}{2\hbar} \sum_{j,k=0}^{n} Q^{jk} z_j \overline{z}_k \right)$$

with

$$Q^{jk} = \begin{cases} 1 & \text{when } j = k \\ -1 & \text{when } j = k - 1 \\ 0 & \text{otherwise.} \end{cases}$$

Now the inverse of (Q^{jk}) is the matrix (Q_{jk}), where $Q_{jk} = 1$ when $j \leq k$, and $Q_{jk} = 0$ when $j > k$. Therefore the result of the Gaussian integral (9.8.4) is

$$\exp\left(-\frac{1}{2\hbar} \sum_{j,k=0}^{n} \delta\tau^2 Q_{jk} X_{j-1} \overline{X}_k \right),$$

where we define $X_n = X_{-1} = 0$. In the limit $\delta\tau \to 0$, this becomes

$$\exp\left(-\frac{1}{2\hbar}\int_t^{t'}\int_t^{\tau} X(\sigma)\overline{X}(\tau)\,\mathrm{d}\sigma\mathrm{d}\tau\right).$$

The formula is the same for a general value of n, except that there is then summation over the coordinate index a.

When n is infinite, we need to translate the coordinate formula into the notation of field theory. The points of V become solutions of the free field equation. The coordinates $\overline{z}, z$ become, respectively, the positive and negative frequency parts, ϕ^+, ϕ^-, of the field ϕ, and $z'\overline{z}$ becomes $-4\mathrm{i}\omega(\phi'^-, \phi^+)$, where ω is the free-field symplectic structure. Finally in the limit $t \to -\infty$, $t \to \infty$, the double integral over σ and τ becomes

$$\int_{-\infty}^{\infty}\int_{-\infty}^{\tau}\frac{2\mathrm{i}}{\hbar}\omega(X_H^-(\sigma), X_H^+(\tau))\,\mathrm{d}\sigma\mathrm{d}\tau = \int_{\mathsf{M}\times\mathsf{M}}\frac{1}{2\hbar}\Delta_F(x,y)f(x)f(y)\mathrm{d}^4x\mathrm{d}^4y,$$

where Δ_F is the Feynman propagator. So the vacuum-vacuum amplitude is given by the exponential of the Feynman propagator. The key point is that the integral over σ and τ is over the region in $\sigma < \tau$, and involves the negative frequency part of $X_H(\sigma)$ and the positive frequency part of $X_H(\tau)$.

9.9 Projections and Fock spaces

We shall now return to the quantization of a linear system with phase space (V, ω), a $2n$-dimensional symplectic vector space. We have seen (§9.2) that a positive complex structure J determines a positive polarization P_J and hence a Hilbert subspace $\mathcal{H}_{P_J}$ of the prequantum Hilbert space $\mathcal{H}$. To avoid multiple subscripts, we shall denote $\mathcal{H}_{P_J}$ by $\mathcal{F}$ (for 'Fock').

The problem we shall consider now is that of relating the Fock spaces $\mathcal{F}$ and $\mathcal{F}'$ arising from two different positive complex structures, J and J'. We shall see that the orthogonal projection $\pi : \mathcal{F}' \to \mathcal{F}$ in $\mathcal{H}$ is a scalar multiple of the unitary operator that intertwines the representations of the Heisenberg group on $\mathcal{F}$ and $\mathcal{F}'$. The existence and uniqueness up to phase of such a unitary operator is predicted by the Stone-von Neumann theorem. (If it is hard to picture how an orthogonal projection can be proportional to a unitary operator, think of the example of the two two-dimensional subspaces $\{(z_1, z_2, 0, 0)\}$ and $\{(z_1, z_2, z_1, z_2)\}$ of $\mathbb{C}^4$ with the standard Hermitian metric.)

Much of the detail of this application of geometric quantization is due to J. H. Rawnsley (see also Robinson and Rawnsley 1989). The example is not only of interest in its own right—it is a finite-dimensional model of the Bogoliubov transformation in quantum field theory—but it is also important as the basis of the 'metaplectic correction' to the quantization construction.

Proposition (9.9.1). Let (V, ω) be a $2n$-dimensional symplectic vector space and let $\mathcal{H}$ be the Hilbert space constructed from V by prequantization.

(a) Each $X \in V$ generates a translation operator $\hat{X} : \mathcal{H} \to \mathcal{H}$, and if $X, Y \in V$, then

$$\hat{X}\hat{Y} = \mathrm{e}^{\mathrm{i}\omega(X,Y)/\hbar}(\widehat{X+Y}).$$

(b) Each positive complex structure J on (V, ω) picks out a Fock space $\mathcal{F} \subset \mathcal{H}$ such that $\hat{X}\mathcal{F} = \mathcal{F}$ for every $X \in V$.

(c) If J and J' are positive complex structures, and if $\pi : \mathcal{F}' \to \mathcal{F}$ and $\pi' : \mathcal{F} \to \mathcal{F}'$ are the orthogonal projections in $\mathcal{H}$ then $\hat{X}\pi = \pi\hat{X}$ for every $X \in V$.

(d) The rescaled projection map $\sqrt[4]{\det \frac{1}{2}(J + J')}\,\pi$ is unitary.

Proof. The first two statements were established in §§ 8.4 and 9.2. The rest is proved by straightforward, but fairly lengthy calculation.

In §9.2, we used linear holomorphic coordinates z^a that were adapted to J. We need now to treat J and J' even-handedly, so we shall begin by restating our description of $\mathcal{F}$ in a more obviously invariant form. In the θ_0 gauge (9.2.1), the elements of $\mathcal{F}$ are complex functions on V of the form $\psi = \phi \exp(-K/2\hbar)$, where ϕ is holomorphic with respect to J (i.e. $(JX - \mathrm{i}X) \lrcorner \mathrm{d}\phi = 0$ for every $X \in V$). The wave functions in $\mathcal{F}'$ are of the form $\psi' = \phi' \exp(-K'/2\hbar)$, where $K'(X) = \omega(X, J'X)$ and ϕ' is holomorphic with respect to J' (i.e. $(J'X - \mathrm{i}X) \lrcorner \mathrm{d}\phi' = 0$).

The translation operators are given by (8.4.6). They preserve $\mathcal{F}$ and $\mathcal{F}'$, and, on restriction, generate the two representations of the Heisenberg group . If $\psi' \in \mathcal{F}'$, then $\hat{X}\pi\psi' = \pi\hat{X}\psi'$ (as a consequence of 9.5.1), so π maps the action of the translation operators on $\mathcal{F}'$ onto their action on $\mathcal{F}$.

The calculations depend on the properties of the coherent states in $\mathcal{F}$ and $\mathcal{F}'$. Let $\psi_0 = \mathrm{e}^{-K/2\hbar}$ and $\psi'_0 = \mathrm{e}^{-K'/2\hbar}$ be the ground states in $\mathcal{F}$ and $\mathcal{F}'$, so the coherent states in $\mathcal{F}$ and $\mathcal{F}'$ based on $X \in V$ are $\psi_X = (\widehat{-X})\psi_0$ and $\psi'_X = (\widehat{-X})\psi'_0$, respectively. That is

$$\psi_X(Z) = \exp\left(\frac{1}{2\hbar}\Big(2\omega(X, (J + \mathrm{i})Z) - \omega(X, JX) - \omega(Z, JZ)\Big)\right), \quad (9.9.2)$$

with a similar expression for ψ'_X, with J replaced by J'. Note that $\psi_X(Z) =$

$\overline{\psi_Z(X)}$. The orthogonal projection $\pi : \mathcal{F}' \to \mathcal{F}$ is given by

$$(\pi\psi')(X) = \langle \psi_X, \psi' \rangle = \int_V \overline{\psi}_X \psi' \epsilon. \tag{9.9.3}$$

We shall need the following.

Lemma (9.9.4). There is a symplectic frame in which

$$J = \begin{pmatrix} 0 & -1 \\ 1 & 0 \end{pmatrix} \quad \text{and} \quad J' = \begin{pmatrix} 0 & -D^{-1} \\ D & 0 \end{pmatrix}$$

where $D = \mathrm{diag}(\lambda_1^2, \dots, \lambda_n^2)$, with $\lambda_a \in \mathbb{R}$ and $\lambda_a \geq 1$.

Proof. The two positive-definite quadratic forms $\omega(\cdot, J\cdot)$ and $\omega(\cdot, J'\cdot)$ can be diagonalized simultaneously. So there is a basis $Z_1, \dots, Z_{2n}$ in V such that $2\omega(Z_i, JZ_i) = 1$ and $JZ_i = \alpha_i J' Z_i$, with $\alpha_i > 0$ (there is no summation over i).

Now if $J'Z = \alpha JZ$ and $J'Z' = \alpha' JZ'$, then $J'(JZ) = \alpha^{-1} J(JZ)$ and

$$\omega(Z, Z') = \omega(J'Z, J'Z') = \alpha\alpha'\omega(JZ, JZ') = \alpha\alpha'\omega(Z, Z').$$

So either $\omega(Z, Z') = 0$ or $\alpha' = \alpha^{-1}$.

Suppose that $\alpha_i \neq 1$ for all i. Then n of the αs are more than one and n are less than one. We can arrange that $\alpha_i > 1$ for $i = 1, \dots, n$. A symplectic frame with the required property is then given by taking $X^a = Z_a$ and $Y_a = JX_a$ for $a = 1, \dots, n$, with $\lambda_a = \sqrt{\alpha_a}$.

If some of the αs are equal to one, then we first decompose V into the direct sum of two symplectic subspaces $V = V_1 \oplus V_1^\perp$ by putting $V_1 = \{X | JX = J'X\}$. Both V_1 and $V_1^\perp$ are invariant under J and J'. We construct a symplectic frame on $V_1^\perp$, as before, and extend this to a frame for V by picking any symplectic frame in V_1 in which J is in the required standard form. ∎

The projection into $\mathcal{F}$ of the ground state in $\psi'_0 = \mathrm{e}^{-K'/2\hbar}$ in $\mathcal{F}'$ is given by

$$(\pi\psi'_0)(X) = \langle \psi_X, \psi'_0 \rangle = \int_V \overline{\psi}_X \mathrm{e}^{-K'/2\hbar} \epsilon.$$

We shall use the lemma to evaulate the integral.

Let $\Delta = \sqrt[4]{\det \frac{1}{2}(J + J')}$ and $L = (J + J')^{-1}(J - J')$, which is well defined since $J + J'$ is nonsingular. Put

$$\Phi(X) = \exp\left(\frac{1}{2\hbar}\omega(X, JLX - \mathrm{i}LX)\right).$$

Note that $L = (J' - J)(J + J')^{-1}$, and that $JL = -LJ$ since

$$(J + J')J(J' - J) = J + J'JJ' = -(J - J')J(J + J').$$

Therefore $\omega(X, JLX - \mathrm{i}LX)) = \frac{1}{2}\omega(X - \mathrm{i}JX, JL(X - \mathrm{i}JX))$, and so Φ is holomorphic with respect to J. In the quantum field theory interpretation, it is a superposition of states of even particle number.

Lemma (9.9.5).

(i) $\pi\psi_0' = \Delta^{-2}\Phi \mathrm{e}^{-K/2\hbar}$
(ii) $\langle \pi\psi_W', \pi\psi_0' \rangle = \Delta^{-2}\mathrm{e}^{-K'(W)/2\hbar}$ for all $W \in V$.

Proof. Consider, first, the case $n = 1$. Choose a symplectic frame in which

$$J = \begin{pmatrix} 0 & -1 \\ 1 & 0 \end{pmatrix} \quad \text{and} \quad J' = \begin{pmatrix} 0 & -\lambda^{-2} \\ \lambda^2 & 0 \end{pmatrix},$$

and let p, q be the corresponding linear coordinates. We can label the vectors $Z \in V$ either by the complex coordinate $z = p + \mathrm{i}q$, which is holomorphic with respect to J, or by $z' = \lambda p + \mathrm{i}\lambda^{-1}q$, which is holomorphic with respect to J'. In terms of these, $K' = \frac{1}{2}z'\overline{z}'$ and $K = \frac{1}{2}z\overline{z}$. The elements of $\mathcal{F}'$ are of the form $\phi'(z')\mathrm{e}^{-z'\overline{z}'/4\hbar}$, where ϕ' is a holomorphic function of z', and the elements of $\mathcal{F}$ are of the form $\phi(z)\mathrm{e}^{-z\overline{z}/4\hbar}$, where ϕ is a holomorphic function of z. In particular,

$$\psi_0'(Z) = \mathrm{e}^{-z'\overline{z}'/4\hbar} \quad \text{and} \quad \psi_W(Z) = \mathrm{e}^{(2\overline{w}z - z\overline{z} - w\overline{w})/4\hbar},$$

where w is the value of $p + \mathrm{i}q$ at W. Hence $(\psi_W, \psi_0') = \overline{\psi}_W\psi_0'$ is given as a function of p and q by

$$(\psi_W, \psi_0') = \exp\left(\frac{1}{4\hbar}\left(2w(p - \mathrm{i}q) - (\lambda^2 + 1)(p^2 + \lambda^{-2}q^2) - w\overline{w}\right)\right).$$

By integrating over p and q against $\epsilon = (2\pi\hbar)^{-1}\mathrm{d}p \wedge \mathrm{d}q$, we obtain

$$(\pi\psi_0')(W) = \frac{2\lambda}{1 + \lambda^2} \exp\left(\frac{w^2(1 - \lambda^2)}{4\hbar(1 + \lambda^2)} - \frac{w\overline{w}}{4\hbar}\right).$$

However, $\Delta^2 = (1 + \lambda^2)/2\lambda$ and, in this frame,

$$L = \frac{1 - \lambda^2}{1 + \lambda^2}\begin{pmatrix} 1 & 0 \\ 0 & -1 \end{pmatrix}.$$

Therefore

$$\Phi(W) = \exp\left(\frac{w^2(1-\lambda^2)}{4\hbar(1+\lambda^2)}\right),$$

and the first part of the lemma follows.

To prove the second part, note first that $\langle \pi\psi'_W, \pi\psi'_0\rangle = \langle \psi'_W, \pi\psi'_0\rangle$ for any $W \in V$. But

$$\psi'_W(Z) = \mathrm{e}^{(2z'\overline{w}' - w'\overline{w}' - z'\overline{z}')/4\hbar},$$

where w' is the value of $\lambda p + \mathrm{i}\lambda^{-1}q$ at W. Hence, at Z,

$$(\pi\psi'_W, \pi\psi'_0) = (\psi'_W, \pi\psi'_0) = \frac{2\lambda}{1+\lambda^2}\exp\left(\frac{(1-\lambda^2)z^2}{4\hbar(1+\lambda^2)} + \frac{2\overline{z}'w' - w'\overline{w}' - z'\overline{z}' - z\overline{z}}{4\hbar}\right).$$

But

$$(1-\lambda^2)z^2 - (1+\lambda^2)z\overline{z} = (\lambda^2-1)\overline{z}'^2 - (1+\lambda^2)z'\overline{z}'.$$

Therefore

$$\langle \pi\psi'_W, \pi\psi'_0\rangle = \frac{2\lambda}{1+\lambda^2}\left(\int f(\overline{z}')\mathrm{e}^{-z'\overline{z}'/2\hbar}\,\epsilon\right)\mathrm{e}^{-w'\overline{w}'/4\hbar},$$

where

$$f(\overline{z}') = \exp\left(\frac{(\lambda^2-1)\overline{z}'^2 + 2(1+\lambda^2)w'\overline{z}'}{4\hbar(1+\lambda^2)}\right).$$

But f is antiholomorphic in z' and $f(0) = 1$. Therefore

$$\langle \pi\psi'_W, \pi\psi'_0\rangle = \Delta^{-2}\mathrm{e}^{-K'(W)/2\hbar}.$$

With a general value of n, the proof is the same. In the symplectic frame given by lemma (9.9.4), the various integrals split up into products of integrals over the pairs of coordinates p, q of the same form as the $n = 1$ integrals. ∎

Now for any $X, Y \in V$, $\hat{Y}(\psi'_X) = \psi'_{X-Y}\mathrm{e}^{\mathrm{i}\omega(X,Y)/\hbar}$, by (8.4.8). Also $\hat{X}$ is unitary. Therefore

$$\begin{aligned}
\langle \pi\psi'_X, \pi\psi'_Y\rangle &= \langle \hat{Y}\pi\psi'_X, \pi\psi'_0\rangle \\
&= \langle \pi\psi'_{X-Y}, \pi\psi'_0\rangle \mathrm{e}^{-\mathrm{i}\omega(X,Y)/\hbar} \\
&= \Delta^{-2}\exp\left(-\frac{1}{2\hbar}\Big(2\mathrm{i}\omega(X,Y) + K'(X-Y)\Big)\right) \\
&= \Delta^{-2}\psi'_Y(X)
\end{aligned}$$

by using lemma (9.9.5). But $\psi'_Y(X) = \langle \psi'_X, \psi'_Y\rangle$. It follows, since the coherent states span $\mathcal{F}'$, that

$$\langle \pi\psi', \pi\psi'\rangle = \Delta^{-2}\langle \psi', \psi'\rangle$$

for every $\psi' \in \mathcal{F}'$.

It is also true that $\pi(\mathcal{F}') = \mathcal{F}$; for if $\psi \in (\pi\mathcal{F}')^{\perp} \subset \mathcal{F}$, then $\langle \psi', \psi \rangle = 0$ for every $\psi' \in \mathcal{F}'$. This implies that $\pi'\psi = 0$, where $\pi' : \mathcal{F} \to \mathcal{F}'$ is the orthogonal projection. But, by interchanging P and P', $\langle \pi'\psi, \pi'\psi \rangle = \Delta^{-2}\langle \psi, \psi \rangle$. Hence $\psi = 0$. It follows that $\Delta\pi$ is unitary. This completes the proof of the proposition. ∎

Suppose now that $J_1, J_2, \ldots$ are positive complex structures on (V, ω). We shall denote the corresponding subspaces of $\mathcal{H}$ by $\mathcal{F}_\alpha$ $(\alpha = 1, 2, \ldots)$. Let

$$\Delta_{\alpha\beta} = \sqrt[4]{\det \tfrac{1}{2}(J_\alpha + J_\beta)}$$

and let $\pi_{\alpha\beta}$ denote the projection $\mathcal{F}_\beta \to \mathcal{F}_\alpha$. There is then a complex number τ_{123} of unit modulus such that

$$\Delta_{12}\Delta_{23}\pi_{32}\pi_{21} = \tau_{123}\Delta_{31}\pi_{31}. \tag{9.9.6}$$

This is a consequence of the fact that both sides are unitary, together with the following lemma.

Lemma (9.9.7). If $U : \mathcal{F}' \to \mathcal{F}$ is unitary and $\hat{X}U = U\hat{X}$ for every $X \in V$, then U is a multiple of the projection $\pi : \mathcal{F}' \to \mathcal{F}$.

Proof. By considering $\Delta U^{-1}\pi$, we can reduce the problem to that of showing that if $U' : \mathcal{F} \to \mathcal{F}$ is a unitary operator such that $\hat{X}U' = U'\hat{X}$ for every X, then U' is a multiple of the identity.

Let ψ_0 be the ground state in $\mathcal{F}$. As a function on V, it is determined up to a constant factor by

$$(\hat{W}\psi_0)(Z) = \psi_0(Z)\exp\left(-\frac{1}{4\hbar}(2\overline{w}_a z^a + \overline{w}_a w^a)\right)$$

for every $W \in V$, where the zs are the complex coordinates of Z and the ws are the complex coordinates of W. The constant multiples of ψ_0 are distinguished from the other elements of $\mathcal{F}$ by the property that $\mathrm{e}^{\overline{w}_a w^a/4\hbar}\hat{W}\psi_0$ depends antiholomorphically on the ws. Since $\hat{W}U' = U'\hat{W}$, it follows that $U'\psi_0$ must have the same property, and hence that $U'\psi_0 = \tau\psi_0$ for some $\tau \in \mathbb{C}$ such that $|\tau| = 1$. But then $U'\psi_W = \tau\psi_W$ for every coherent state in $\mathcal{F}$; and since the coherent states span $\mathcal{F}$, it follows that U' is τ times the identity operator. ∎

If we take $J_1 = J_3$, then (9.9.6) reduces to

$$\Delta_{12}^2\pi_{12}\pi_{21} = \tau_{121}.$$

However, if ψ_1 is the ground state in $\mathcal{F}_1$, then

$$\langle \psi_1, \pi_{12}\pi_{21}\psi_1 \rangle = \langle \psi_1, \pi_{21}\psi_1 \rangle = \Delta_{12}^{-2}$$

by using the first part of lemma (9.9.5). Therefore $\tau_{121} = 1$. It follows $\Delta_{12}\pi_{12}$ is the inverse of $\Delta_{21}\pi_{21}$, and hence that

$$\tau_{123} = \tau_{231} \quad \text{and} \quad \tau_{213} = \tau_{123}^{-1}.$$

We can summarize these results as follows.

Proposition (9.9.8). Let J_1, J_2 and J_3 be positive complex structures on (V, ω) and let $\mathcal{F}_1, \mathcal{F}_2, \mathcal{F}_3 \subset \mathcal{H}$ be the corresponding Fock spaces. Then

$$\Delta_{12}\Delta_{23}\pi_{32}\pi_{21} = \tau_{123}\Delta_{31}\pi_{31}$$

where $\pi_{\alpha\beta}$ is the projection $\mathcal{F}_\beta \to \mathcal{F}_\alpha$,

$$\Delta_{\alpha\beta} = \sqrt[4]{\det \tfrac{1}{2}(J_\alpha + J_\beta)},$$

and $\tau_{\alpha\beta\gamma} \in T \subset \mathbb{C}$. Moreover $\tau_{112} = 1$; and τ is symmetric under even permutation of its subscripts, and goes to its inverse under odd permutations.

We shall obtain an explicit formula for τ in the next chapter.

Example. *Particle creation by external fields.* A direct application of these results is to the problem of particle creation by external fields. Consider, for example, a scalar field governed by the Klein-Gordon equation

$$\Box\phi + \mu^2\phi = 0$$

in curved space-time, where $\Box = \nabla_a\nabla^a$. If the metric is flat in the past and in the future, then the equation describes the interaction of the scalar field with an external classical gravitational field which is 'turned on' only for a finite time (this is not very plausible, but it is a reasonable approximation to a sîtuation in which the gravitational field satisfies asymptotic conditions). The positive-negative frequency decompositions before and after the interaction give two different complex structures, J' and J, on the solution space. The operator L, which depends on the difference $J - J'$, measures the mixing of positive and negative frequency parts as ϕ propagates through the gravitational field. The vacuum state in the past Fock space $\mathcal{F}'$ becomes, under the pairing map, the state $\Delta^{-2}\Phi e^{-K/2\hbar}$, which is a superposition of states of even particle number in the future Fock space

$\mathcal{F}$. Thus the pairing map describes the creation of particle pairs by the interaction.

This is purely formal, but the theory can be made to work in the infinite-dimensional context, provided that the two decompositions are close enough for $J - J'$ to be a Hilbert-Schmidt operator (Shale 1962; see also Ashtekar and Magnon 1975, Magnon-Ashtekar 1976, Vergne 1977, Woodhouse 1981, Robinson 1991). ∎

There is an analogous treatment of fermion Fock spaces. Here one starts with a real vector space V of dimension $2n$ and a positive-definite quadratic form g, which encodes the anticommutation relations. The prequantum Hilbert space is replaced by $\mathcal{H} = \bigwedge^{\cdot} V_{\mathbb{C}}^*$, the elements of which are formal sums of complex skew forms on V. The inner product on $\mathcal{H}$ is

$$\langle \psi, \psi \rangle = \sum_k \hbar^k k! \psi^{(k)}_{\alpha\beta\ldots\gamma} \overline{\psi}^{(k)\alpha\beta\ldots\gamma},$$

where $\alpha, \beta, \ldots = 1, 2, \ldots, 2n$ and

$$\psi = \sum_k \psi^{(k)} \quad \text{with} \quad \psi^{(k)} \in \bigwedge^k V_{\mathbb{C}}^*.$$

Indices are lowered and raised by $g_{\alpha\beta}$ and its inverse $g^{\alpha\beta}$.

A real linear function $f \in V^*$ is necessarily of the form $f(Y) = g(X_f, Y)$ for some $X_f \in V$. Such an f determines a self-adjoint 'prequantum operator' on $\mathcal{H}$ by

$$\hat{f}(\psi) = \hbar X_f \lrcorner \psi + f \wedge \psi,$$

where the contraction $X_f \lrcorner \psi$ is defined term by term, with $X_f \lrcorner \psi^{(0)} = 0$.

A polarization in this context is a complex structure $J : V \to V$ such that $\omega(\cdot, \cdot) = \frac{1}{2} g(J\cdot, \cdot)$ is a symplectic form on V. Such a J picks out a Fock space $\mathcal{F} \subset \mathcal{H}$ by

$$\psi \in \mathcal{F} \quad \text{whenever} \quad \widehat{Jf}\psi - \mathrm{i}\hat{f}\psi = 0$$

for every $f \in V^*$. Here Jf is defined by $(Jf)(Y) = -f(JY)$. The elements of $\mathcal{F}$ are of the form

$$\psi = \phi \wedge \exp\left(-\frac{\mathrm{i}\omega}{\hbar}\right),$$

where ϕ is a sum of forms of type $(k, 0)$ with respect to J. The exponential of a 2-form is defined by replacing ordinary multiplication by $\wedge$ in the series expansion of exp (note that $\wedge$ commutes for forms of even degree). Thus the standard fermion Fock space, $\bigwedge^{\cdot} V_{(J)}^*$, is embedded in $\mathcal{H}$ in much the same way that the bosonic Fock space is embedded in $L^2(V)$.

The subspace $\mathcal{F}$ is invariant under the operators $\hat{f}$. It therefore carries a representation of the 'canonical anticommutation relations', which is, in fact, irreducible. As in the bosonic case, the orthogonal projection in $\mathcal{H}$, $\mathcal{F}' \to \mathcal{F}$, from the Fock space picked out by a second polarization intertwines the corresponding representations, and is unitary up to a constant factor. One can develop the theory of spinor representations along the same lines as the treatment of the metaplectic representation in the next chapter.

The general problem of extending geometric quantization to include fermions is considered by Kostant (1977).

10

THE METAPLECTIC CORRECTION

10.1 The corrected quantization

THE pairing construction we have developed to relate the Hilbert spaces of different real polarizations is not entirely satisfactory. In the analysis of WKB wave functions, for example, we could not keep track of the phase when the classical trajectories passed through a caustic singularity (§9.6). An awkward and unwanted phase factor also appeared in the derivation of the Schrödinger equation (§9.7). These problems arose because there is no consistent way to introduce a preferred orientation in the space of leaves of every real polarization. The solution is to modify the construction by adopting a new definition of the 'square root of a volume element', or 'half-form' (a section of the bundle δ_P in §9.3), which does not depend on a choice of orientation in M/P. The new construction extends the algebra of classical observables for which Dirac's quantum conditions hold to include functions which are quadratic in the momenta.

The half-form bundles are also defined for complex polarizations. By including them in the quantization construction, we arrive at a more useful pairing, which can be applied to a much wider class of nonnegative polarizations. It seems to give the correct physical relationship between the wave functions in the Hilbert spaces of different polarizations. In the case of two Fock spaces, the new pairing is unitary. When P is complex, δ_P may be nontrivial, and its inclusion in the construction can modify quantization conditions and alter the form of the quantum observables. In the case of the harmonic oscillator, the modification is precisely that needed to get the correct zero-point energy (see §9.2).

We shall look in this final chapter at this half-form 'correction' to geometric quantization, which is described in the chapter heading as the 'metaplectic correction' because of its close connection with the metaplectic representation.

10.2 The metaplectic representation

One of the strongest motivations for the metaplectic correction comes from the way in which the symplectic group, or rather its double cover, acts on Fock space. To understand this and also to develop some ideas that we shall need for the general theory, we return to the example that we considered in §9.2 and to the results summarized in (9.9.1) and (9.9.8). It follows from these that the projection operators between the Fock spaces picked out by three positive complex structures J_α, J_β, and J_γ satisfy

$$\pi_{\gamma\beta}\pi_{\beta\alpha} = \chi_{\alpha\beta\gamma}\pi_{\gamma\alpha} \quad \text{where} \quad \chi_{\alpha\beta\gamma} = \frac{\tau_{\alpha\beta\gamma}\Delta_{\gamma\alpha}}{\Delta_{\alpha\beta}\Delta_{\beta\gamma}}, \tag{10.2.1}$$

acting as a multiplication operator. Here $\Delta_{\alpha\beta} = \sqrt[4]{\det \frac{1}{2}(J_\alpha + J_\beta))}$ and $\tau_{\alpha\beta\gamma}$ is a complex number of unit modulus.

The symplectic group $SP(V,\omega)$ acts on $\mathcal{H}$ by the reducible unitary representation $\psi \mapsto \hat{\rho}\psi = \psi \circ \rho$ in the gauge of the invariant potential

$$\theta_0 = \tfrac{1}{2}(p_a \mathrm{d}q^a - q^a \mathrm{d}p_a).$$

It also acts on the positive Lagrangian Grassmannian L^+V by $J \mapsto \rho(J) = J'$, where $J' = \rho^{-1}J\rho$. The two actions are related by $\hat{\rho}\mathcal{F} = \mathcal{F}'$, where $\mathcal{F}$ and $\mathcal{F}'$ are the Hilbert subspaces of $\mathcal{H}$ picked out by J and J'.

Proposition (10.2.2). Let $J_0 \in L^+V$ and let $\mathcal{F}_0$ be the corresponding subspace of $\mathcal{H}$. For $\rho \in SP(V,\omega)$, put $\mathcal{F}' = \hat{\rho}\mathcal{F}_0$ and $\tilde{\rho} = \pi \circ \hat{\rho}$, where $\pi : \mathcal{F}' \to \mathcal{F}_0$ is the orthogonal projection. Then $\rho \mapsto \tilde{\rho}$ is a projective representation[1] of $SP(V,\omega)$ on $\mathcal{F}_0$.

Proof. Let $\rho_1, \rho_2 \in SP(V,\omega)$. Put $\rho_3 = \rho_1\rho_2$ and put $\mathcal{F}_\alpha = \hat{\rho}_\alpha\mathcal{F}$ ($\alpha = 1,2,3$). Then $\hat{\rho}_3 = \hat{\rho}_2\hat{\rho}_1$ and it follows from the diagram

$$\begin{array}{ccccccc}
 & \mathcal{F}_0 & \overset{\hat{\rho}_1}{\rightarrow} & \mathcal{F}_1 & \overset{\hat{\rho}_2}{\rightarrow} & \mathcal{F}_3 \\
\tilde{\rho}_1 & \downarrow & \swarrow & & \swarrow & \\
 & \mathcal{F}_0 & \overset{\hat{\rho}_2}{\rightarrow} & \mathcal{F}_2 & & \\
\tilde{\rho}_2 & \downarrow & \swarrow & & & \\
 & \mathcal{F}_0 & & & &
\end{array}$$

that $\tilde{\rho}_2\tilde{\rho}_1 = \chi_{320}\tilde{\rho}_3$. The diagonal arrows are orthogonal projections. The diagram commutes as a consequence of the definitions of the various maps, and because $\hat{\rho}_2$ is unitary on $\mathcal{H}$. ∎

This is the projective version of the metaplectic representation. The $\tilde{\rho}$s are not unitary operators, although we know from (9.9.8) that they become unitary when rescaled by Δ. Even then, however, the representation is still only projective since we are left with the phase factor τ as a nontrivial cocycle (or 'projective multiplier'). The transformations of $\mathcal{F}_0$ generated by the $\tilde{\rho}$s make up the group $MP^c(V)$ (Robinson and Rawnsley 1989).

Example. *Projective representation of the symplectic Lie algebra.* We shall consider only $n = 1$, since the higher-dimensional cases are very similar. The action of the symplectic Lie algebra on $\mathbb{R}^2$ is generated by the functions

$$f_1 = \tfrac{1}{2}(p^2 + q^2), \quad f_2 = pq \quad \text{and} \quad f_3 = \tfrac{1}{2}(p^2 - q^2). \tag{10.2.3}$$

We want to find the corresponding generators $\tilde{f}_1$, $\tilde{f}_2$, and $\tilde{f}_3$ of the projective representation on $\mathcal{F}_0$, when J_0 is the standard complex structure—the complex structure for which $z = p+iq$ is a holomorphic coordinate. We shall use the θ_0-gauge, so that the elements of $\mathcal{F}_J$ are represented by functions of the form $\psi = \phi(z)\mathrm{e}^{-z\overline{z}/4\hbar}$, where ϕ is holomorphic.

We know already that the canonical flow of f_1 preserves J_0, and that the $\tilde{f}_1$ is the operator $\phi \mapsto \hbar z \partial\phi/\partial z$ (p. 176).

The operator on $\mathcal{H}$ generated by f_2 is

$$\hat{f}_2 = \mathrm{i}\hbar\left(z\frac{\partial}{\partial \overline{z}} + \overline{z}\frac{\partial}{\partial z}\right).$$

To obtain $\tilde{f}_2$, we must make this act on ψ and then project the resulting element of $\mathcal{H}$ back into $\mathcal{F}_0$. If ϕ, ϕ' are holomorphic, then, by integrating by parts,

$$\mathrm{i}\hbar\int \overline{\phi}'\mathrm{e}^{-z\overline{z}/4\hbar}\left(z\frac{\partial}{\partial \overline{z}} + \overline{z}\frac{\partial}{\partial z}\right)\left(\phi\mathrm{e}^{-z\overline{z}/4\hbar}\right)\,\mathrm{d}p\mathrm{d}q = \int \overline{\tilde{f}_2(\phi')}\phi\mathrm{e}^{-z\overline{z}/2\hbar}\,\mathrm{d}p\mathrm{d}q,$$

where

$$\tilde{f}_2(\phi) = \mathrm{i}\hbar^2\frac{\partial^2\phi}{\partial z^2} + \frac{z^2}{4\mathrm{i}}\phi.$$

This is precisely the operator given by the canonical quantization rule $\overline{z} \mapsto 2\hbar\partial/\partial z$. Similarly,

$$\tilde{f}_3(\phi) = \hbar^2\frac{\partial^2\phi}{\partial z^2} + \frac{z^2}{4}\phi.$$

That the representation is only projective is reflected in the fact that $[f_2, f_3] = -2f_1$, while $[\tilde{f}_2, \tilde{f}_3] = 2\mathrm{i}\hbar(\tilde{f}_1 + \frac{1}{2}\hbar)$. ■

Our aim is to absorb χ into the definition of the projection. We shall do this by replacing the Fock space $\mathcal{F}$ associated with $J \in L^+V$ by $\mathcal{F}\otimes\delta$, where δ is the one dimensional space of 'half-forms' on P_J. This has no effect on the physical interpretation of the wave functions in $\mathcal{F}$, but alters the relationship between the phases of the wave functions in different Fock spaces, and consequently changes the way in which $SP(V,\omega)$ acts on $\mathcal{F}_0$.

Parallel transport

There is another way of looking at the projective representation that may make the introduction of half-forms appear more natural.[2] This is to regard the individual Fock spaces as the fibres of a Hermitian vector bundle $F \to L^+V$. The fibres are all subspaces of $\mathcal{H}$, so F is embedded in $L^+V\times\mathcal{H}$, and consequently has a natural connection ∇ (§A.7): if J and J' are nearby, then parallel transport from $F_J = \mathcal{F}$ to $F_{J'} = \mathcal{F}'$ is given to the first order in $J - J'$ by the orthogonal projection $\mathcal{F}\to\mathcal{F}'$ in $\mathcal{H}$.

The holonomy around a closed loop in L^+V is a unitary operator on Fock space. However, it is an immediate consequence of (10.2.1) that its values are all multiples of the identity operator, so the curvature of the connection is, in fact, an ordinary real 2-form. By the general theory outlined in §A.7, it can be calculated directly from the function $\chi(J_1, J_2, J_3) = \chi_{123}$ by restricting

$$\mathrm{i}\mathrm{d}_3 \wedge \mathrm{d}_2 \log \chi \in \Omega^2(L^+V \times L^+V \times L^+V) \tag{10.2.4}$$

to the diagonal $L^+V = \{J_1 = J_2 = J_3\}$. (Here d_2 is the exterior derivative with respect to J_2, with J_1 and J_3 held fixed, and so on.)

Because the holonomy is given by scalar multiplication, ∇ is 'projectively flat', and so we can identify all the projective spaces $\mathbb{P}F_J$ by parallel transport. Let us denote the resulting projective space by $\mathbb{PF}$. The projective version of the metaplectic representation determines a natural action of $SP(V, \omega)$ on $\mathbb{PF}$. To obtain a corresponding linear representation, we must take the tensor product of F with a line bundle $\delta \to L^+V$ with connection which has the opposite curvature to that of F. The tensor product will have a flat connection, and so we shall have a candidate for the representation space in the set $\mathbb{F}$ of covariantly constant sections.

To see that the appropriate line bundle is the bundle of 'half-forms', we must first calculate χ, and hence the curvature of ∇.

The cocycle χ

By choosing $J_0 \in L^+V$ and a symplectic frame $\{X^a, Y_b\}$ in which J_0 takes the standard form,

$$J_0 = \begin{pmatrix} 0 & -1 \\ 1 & 0 \end{pmatrix},$$

each $J \in L^+V$ can be labelled by an $n \times n$ complex symmetric matrix z such that $X^a - z^{ab}Y_b$ spans P_J (p. 93). Let $z_\alpha = z(J_\alpha)$. Then we have the following.

Lemma (10.2.5). Let $\zeta_{\alpha\beta} = \det \frac{1}{2}\mathrm{i}(\bar{z}_\alpha - z_\beta)$. Then

$$(\Delta_{\alpha\beta})^4 = \frac{\zeta_{\alpha\beta}\zeta_{\beta\alpha}}{\zeta_{\alpha\alpha}\zeta_{\beta\beta}} \quad \text{and} \quad (\tau_{\alpha\beta\gamma})^4 = \frac{\zeta_{\beta\alpha}\zeta_{\gamma\beta}\zeta_{\alpha\gamma}}{\zeta_{\alpha\beta}\zeta_{\beta\gamma}\zeta_{\gamma\alpha}}.$$

Proof. Note first that τ satisfies the cocycle relation

$$\tau_{123}\tau_{301} = \tau_{230}\tau_{012} \tag{10.2.6}$$

as a consequence of its definition (§A.7). So if we can establish the formula for τ with $\alpha = 0$, then we can deduce the general result.

Let ψ_α be the ground state in $\mathcal{F}_\alpha$. Then, by using Lemma (9.9.5),

$$\langle \psi_\alpha, \psi_\beta \rangle = \langle \psi_\alpha, \pi_{\alpha\beta}\psi_\beta \rangle = (\Delta_{\alpha\beta})^{-2}.$$

On the other hand,

$$\langle \psi_\alpha, \psi_\beta \rangle = \langle \psi_\alpha, \pi_{\alpha\beta}\psi_\beta \rangle = \frac{\Delta_{0\alpha}\Delta_{0\beta}}{\Delta_{\alpha\beta}\tau_{0\alpha\beta}} \langle \psi_\alpha, \pi_{\alpha 0}\pi_{0\beta}\psi_\beta \rangle.$$

By using Lemma (9.9.5) again, and the unitarity of $\Delta_{\alpha 0}\pi_{\alpha 0}$,

$$\langle \psi_\alpha, \pi_{\alpha 0}\pi_{0\beta}\psi_\beta \rangle = \langle \pi_{0\alpha}\psi_\alpha, \pi_{0\beta}\psi_\beta \rangle = \left(\frac{1}{\Delta_{0\alpha}\Delta_{0\beta}}\right)^2 \int_V \mathrm{e}^{-Q_{\alpha\beta}/\hbar}\epsilon.$$

Here $Q_{\alpha\beta}(X) = \omega(X, A_{\alpha\beta}X)$, where

$$A_{\alpha\beta} = J_0 - \tfrac{1}{2}J_0\big(L_\alpha(1 + \mathrm{i}J_0) + L_\beta(1 - \mathrm{i}J_0)\big),$$

with $L_\alpha = (J_0 + J_\alpha)^{-1}(J_0 - J_\alpha)$ and $L_\beta = (J_0 + J_\beta)^{-1}(J_0 - J_\beta)$.

Now $\omega(\cdot, A_{\alpha\beta}\cdot)$ is a symmetric quadratic form with positive real part, so the integral is equal to $(\det A_{\alpha\beta})^{-1/2}$. Therefore

$$(\tau_{0\alpha\beta})^4 = \left(\frac{\Delta_{\alpha\beta}}{\Delta_{0\alpha}\Delta_{0\beta}}\right)^4 \left(\frac{1}{\det A_{\alpha\beta}}\right)^2. \tag{10.2.7}$$

We shall calculate the determinant of $A_{\alpha\beta}$ in the basis $Z^a, \overline{Z}^a$, where $Z^a = X^a - \mathrm{i}Y_a$. In this basis,

$$J(z) = N\begin{pmatrix} \mathrm{i} & 0 \\ 0 & -\mathrm{i} \end{pmatrix} N^{-1} \quad \text{where} \quad N = \tfrac{1}{2}\begin{pmatrix} 1 - \mathrm{i}z & 1 - \mathrm{i}\overline{z} \\ 1 + \mathrm{i}z & 1 + \mathrm{i}\overline{z} \end{pmatrix}.$$

Hence

$$L_\alpha = \begin{pmatrix} 0 & \mathrm{i} + \overline{z}_\alpha \\ \mathrm{i} - z_\alpha & 0 \end{pmatrix} \begin{pmatrix} \mathrm{i} + z_\alpha & 0 \\ 0 & \mathrm{i} - \overline{z}_\alpha \end{pmatrix}^{-1},$$

and therefore

$$A_{\alpha\beta} = \begin{pmatrix} -1 + \mathrm{i}z_\beta & -\mathrm{i} - \overline{z}_\alpha \\ -\mathrm{i} + z_\beta & 1 + \mathrm{i}\overline{z}_\alpha \end{pmatrix} \begin{pmatrix} \mathrm{i} + z_\beta & 0 \\ 0 & \mathrm{i} - \overline{z}_\alpha \end{pmatrix}^{-1}.$$

It follows that $\det A_{\alpha\beta} = \zeta_{\alpha\beta}/\zeta_{0\beta}\zeta_{\alpha 0}$. By taking $\alpha = \beta$ in (10.2.7) and by using $\tau_{0\alpha\alpha} = 1 = \zeta_{00}$, we deduce the formula for $\Delta_{0\alpha}$. The general formula for Δ follows by taking the modulus of (10.2.7). The formula for $\tau_{0\alpha\beta}$, and hence the general formula for τ, then follow from (10.2.7) itself. ∎

By substituting these into (10.2.1), we obtain

$$\chi_{123} = \left(\frac{\zeta_{22}\zeta_{13}}{\zeta_{12}\zeta_{23}}\right)^{1/2} .$$

The square root is fixed by $\chi_{111} = 1$, together with the requirement that it should vary continuously with z_1, z_2, z_3: this determines it unambiguously since L^+V is simply connected. Only ζ_{23} makes a nonzero contribution to $\mathrm{i}\mathrm{d}_3 \wedge \mathrm{d}_2 \log \chi$. Therefore the curvature of ∇ is $\frac{1}{2}\Omega$, where Ω is the symplectic structure on L^+V introduced in §5.4. That is

$$\Omega = -\mathrm{i}\partial\overline{\partial}(\log \det y)$$

where $y = \frac{1}{2}\mathrm{i}(\overline{z} - z)$.

Half-forms

We saw in §9.2 that the determinant bundle $L \to L^+V$ is a Hermitian line bundle, and that it has a connection with curvature Ω. Its dual is the *canonical bundle* $K \to L^+V$, of which the fibre at J is $K_J = K_{P_J}$, where K_P is defined for any Lagrangian subspace of $V_{\mathbb{C}}$ by

$$K_P = \left\{\mu \in \bigwedge^n V_{\mathbb{C}}^* \,|\, X \lrcorner \mu = 0 \; \forall \; X \in \overline{P}\right\}. \tag{10.2.8}$$

We shall see that the canonical bundle is Hermitian and has a connection with curvature $-\Omega$, which arises from its embedding in the trivial bundle $L^+V \times \bigwedge^n V_{\mathbb{C}}^*$. The half-form bundle is $\delta = \sqrt{K}$. The square root exists, and is unique up to equivalence, because L^+V is contractible. Before considering δ, we shall first look at K in more detail since most of properties of half-forms follow directly from the geometry of the canonical bundle.

The middle exterior power $\bigwedge^n V_{\mathbb{C}}^*$ has dimension $(2n)!/(n!)^2$, where $n = \frac{1}{2}\dim V$. It has an indefinite inner product given by

$$\mathrm{i}^n(\mu, \mu')\epsilon = \overline{\mu} \wedge \mu' \qquad (\mu, \mu' \in \Lambda^n V_{\mathbb{C}}^*)$$

which is positive on K_P for any positive Lagrangian subspace P. As usual, ϵ is the volume element on V given by $\epsilon = (2\pi\hbar)^{-n}\mathrm{d}^n p \mathrm{d}^n q$ in any linear canonical coordinate system.

For each $J \in L^+V$, define $\mu(J) \in K_J$ by

$$\mu(J) = \left(\frac{1}{4\pi\hbar}\right)^{n/2} (\overline{Z}^1 \lrcorner \omega) \wedge \cdots \wedge (\overline{Z}^n \lrcorner \omega),$$

where $Z^a = X^a - z^{ab}Y_b$. Let K_1, K_2, K_3 be the fibres of K at J_1, J_2, J_3, and let $\mu_\alpha = \mu(J_\alpha)$. Then $(\mu_1, \mu_2) = \det \frac{1}{2}\mathrm{i}(\overline{z}_2 - z_1) = \zeta_{21}$. Therefore[3]

$$\frac{(\mu_2, \mu_1)(\mu_3, \mu_2)}{(\mu_3, \mu_1)(\mu_2, \mu_2)} = \frac{\zeta_{12}\zeta_{23}}{\zeta_{13}\zeta_{22}} = (\chi_{123})^{-2}.$$

It follows that the cocycle of the embedding of K in $L^+V \times \bigwedge^n V^*_{\mathbb{C}}$ is $(\chi_{123})^{-2}$ and that the curvature of the connection on K is $-\Omega$.

The connection and Hermitian structure pass to the square root $\delta = \sqrt{K}$; so does the pairing[4]

$$(\nu, \nu') = \sqrt{(\nu^2, \nu'^2)},$$

with the sign fixed by continuity together with $(\nu, \nu) > 0$. The cocycle of the pairing is $(\chi_{123})^{-1}$.

Definition (10.2.9). Let $P = P_J$, where $J \in L^+V$. A half-form on P_J is an element of δ_J.

For each $J \in L^+V$, put $\tilde{F}_J = F_J \otimes \delta_J$. We shall interpret $\tilde{F}_J$ as the 'corrected' quantum Hilbert space of the polarization P_J. So long as we look only at one Fock space, the correction does not change the quantum phase space because the rays in $\tilde{F}_J$ are the same as the rays in F_J. What is different is the pairing. Given $\tilde{s} \in \tilde{F}_J$ and $\tilde{s}' \in \tilde{F}_{J'}$, we put

$$\langle \tilde{s}, \tilde{s}' \rangle = \langle s, s' \rangle (\nu, \nu'),$$

and we define $\tilde{\pi} : \tilde{F}_{J'} \to \tilde{F}_J$ by $\langle \tilde{s}, \tilde{\pi}\tilde{s}' \rangle = \langle \tilde{s}, \tilde{s}' \rangle$. Then not only is the corresponding connection on $\tilde{F} = F \otimes \delta$ flat, but parallel transport from $\tilde{F}_{J'}$ to $\tilde{F}_J$ is given by $\tilde{\pi}$ for any J and J', whether or not they are nearby.

Group actions

The translations in V act trivially on K, and hence on δ. Thus each $\tilde{F}_J$ carries a representation $X \mapsto \tilde{X}$ of the Heisenberg group, where $\tilde{X}$ is defined by $\tilde{X}(s\nu) = (\hat{X}s)\nu$.

The symplectic group acts on K by the pullback action $\mu \mapsto \rho^*\mu$. This preserves the pairing and extends to K the action of $SP(V, \omega)$ on L^+V since, for each J, $\rho^* K_J = K_{\rho(J)}$. It does not, however, pass unambiguously to δ. This is because the subgroup $U_J \subset SP(V, \omega)$ that fixes a particular $J \in L^+V$ is isomorphic to the unitary group $U(n)$. If $\rho \in U_J$ and if u_ρ is the corresponding element of $U(n)$, then ρ acts on K_J by multiplication by $\det u_\rho$. We cannot take the square root of the determinant continuously throughout $U(n)$, so we cannot make U_J act on δ_J.

To make the square roots single-valued, we must replace $SP(V, \omega)$ by its double cover, the metaplectic group $MP(V, \omega)$, which does act on δ. We shall also denote this action by $\nu \mapsto \rho^*\nu$, where $\rho \in MP(V, \omega)$, and we shall not be meticulous in maintaining a distinction between elements of the metaplectic group and the corresponding elements of the symplectic group.[5]

Let $\mathbb{F}$ denote the space of parallel sections of $\tilde{F}$. Then $\mathbb{F}$ is a Hilbert space which is isomorphic to $\tilde{F}_J$ for any J (by evaluating sections at J). It carries the *metaplectic representation* of $MP(V,\omega)$, which is defined by $\tilde{s} \mapsto \tilde{\rho}(\tilde{s}) = \hat{\rho}(s)\rho^*\nu$, where $\tilde{s} = s\nu$. Alternatively, we can define the metaplectic representation on $\tilde{F}_J$ for any fixed J by

$$\tilde{s} \mapsto \tilde{\rho}(\tilde{s}) = \tilde{\pi}\big(\hat{\rho}(s)\rho^*(\nu)\big),$$

where $\tilde{\pi} : \tilde{F}_{\rho(J)} \to \tilde{F}_J$ is the projection determined by the half-form pairing.

Example. *The metaplectic representation of the symplectic Lie algebra.* The Lie algebras of the metaplectic and symplectic groups are the same. The effect of the inclusion of half-forms is to correct the quantization of the generators so that quantized operators make up a representation of the Lie algebra, rather than merely a projective representation.

Consider again the case $n = 1$ (p. 219). The elements of $\tilde{F}_J$ are of the form $\phi\nu e^{-z\bar{z}/4\hbar}$ where $\nu = \sqrt{\mu}$, with $\mu = (4\pi\hbar)^{-1/2}dz$. Let X_1, X_2 and X_3 be the Hamiltonian vector fields generated by f_1, f_2, f_3 in (10.2.3). If ρ_t is the flow of one of the Xs, then $\rho_t^* dz = dz + t\mathcal{L}_X dz$ to the first order in t. Now

$$\mathcal{L}_{X_1} dz = i dz, \quad \mathcal{L}_{X_2} dz = -d\bar{z}, \quad \mathcal{L}_{X_3} dz = i d\bar{z}.$$

Therefore,

$$(\mu, \mathcal{L}_{X_1}\mu) = i, \quad (\mu, \mathcal{L}_{X_2}\mu) = 0, \quad (\mu, \mathcal{L}_{X_3}\mu) = 0,$$

since $(\mu,\mu) = 1$ and $(dz, d\bar{z}) = 0$. Hence

$$(\nu, \mathcal{L}_{X_1}\nu) = \frac{i}{2}, \quad (\nu, \mathcal{L}_{X_2}\nu) = 0, \quad (\nu, \mathcal{L}_{X_3}\nu) = 0.$$

It follows that the generators $\tilde{f}_2$ and $\tilde{f}_3$ are the same as before, while $\tilde{f}_1$ is replaced by the operator $\phi\nu e^{-z\bar{z}/\hbar} \mapsto \phi'\nu e^{-z\bar{z}/\hbar}$, where

$$\phi' = \hbar\left(z\frac{\partial}{\partial z} + \frac{1}{2}\right)\phi.$$

This is the operator obtained by writing $\frac{1}{2}z\bar{z} = \frac{1}{4}(z\bar{z}+\bar{z}z)$ before making the substitution $\bar{z} \mapsto 2\hbar\partial/\partial z$. As a classical observable, f_3 is the Hamiltonian of the harmonic oscillator (p. 175). The new quantum operator has the correct eigenvalues $(n + \frac{1}{2})\hbar$.

The operators $\tilde{f}_1$, $\tilde{f}_2$, $\tilde{f}_3$ satisfy the quantum condition, and generate the metaplectic representation on $\tilde{F}_J$.

Example. *Berry's phase.* (Berry 1984, 1985, Simon 1983.) The half-form construction may seem far removed from quantum physics. It is worth bearing in mind, therefore, that the basic problem it addresses—that of comparing the phases of wave functions in different Hilbert spaces—is one that can have physical significance even in a very simple system.

Associated with each $J \in L^+V$, we have two one-dimensional complex spaces—like the lines in Kostant's (1974a) construction of symplectic spinors. First we have $G_J \subset \mathcal{H}$, which consists of all multiples of the ground state in the Fock space $F_J \subset \mathcal{H}$; second, we have $\tilde{G}_J = G_J \otimes \delta_J \subset \tilde{F}_J = \mathbb{F}$, which consists of the multiples of the ground state in $\tilde{F}_J$. We thus have two line bundles $G \to L^+V$ and $\tilde{G} \to L^+V$, which have connections from their respective embeddings in $L^+V \times \mathcal{H}$ and $L^+V \times \mathbb{F}$. Note that $\tilde{G} = G \otimes \delta$.

In the θ_0 gauge, G_J consists of multiples of $\psi_0 = \mathrm{e}^{-K/2\hbar}$ (p. 173). If $\psi_0' = \mathrm{e}^{-K'/2\hbar}$ is the ground state in $\mathcal{F}_{J'} \subset \mathcal{H}$, then (ψ_0, ψ_0') is real (Lemma 9.9.5). Consequently the connection on G is flat, and therefore the connection on $\tilde{G}$ has curvature $-\frac{1}{2}\Omega$.

The difference is significant in Berry's example of the 'generalized harmonic oscillator' with slowly varying parameters. The Hamiltonian is

$$H = \tfrac{1}{2}(aq^2 + 2bpq + cp^2)$$

where a, b, c are the parameters.

We can quantize for each fixed a, b, c by using the complex structure

$$J = \frac{1}{\sqrt{ac - b^2}} \begin{pmatrix} -b & -a \\ c & b \end{pmatrix}, \tag{10.2.10}$$

which is preserved by the flow of H. The ground state is then an element of G_J in the uncorrected quantization scheme, or of $\tilde{G}_J$ in the corrected scheme.

Berry used the adiabatic theorem to show that if the system is originally in the ground state, and if the parameters are varied slowly over a loop in $\mathbb{R}^3$, then the system returns to its original state, but with a phase shift. Simon interpreted the phase shift as the holonomy of the connection constructed by regarding the multiples of the oscillator ground states as the fibres of a line bundle over the parameter space. The connection comes from embedding the lines in the Hilbert space of all states.

In Berry's analysis, the wave functions were in the configuration space representation, and the 'Hilbert space of all states' was $L^2(\mathbb{R})$. When we look at the problem in Fock space, we have a choice: we can use either $\mathcal{H}$, in which case there is no holonomy because the connection on G is flat; or $\mathbb{F}$, in which case the holonomy is that of the curvature on $\tilde{G}$, which is $-\frac{1}{2}\Omega$.

Equation (10.2.10) determines an embedding of the parameter space $\mathbb{R}^3$ in L^+V. The restriction of $-\frac{1}{2}\Omega$ to $\mathbb{R}^3$ is

$$-\tfrac{1}{8}(ac - b^2)^{-3/2}(a\, \mathrm{d}b \wedge \mathrm{d}c + b\, \mathrm{d}c \wedge \mathrm{d}a + c\, \mathrm{d}a \wedge \mathrm{d}b),$$

by recalling that $\Omega(T, T') = \frac{1}{8}\mathrm{tr}(TJT')$ (p. 94). The $\tilde{G}$ phase factor

$$\exp\left(-\tfrac{1}{2}\mathrm{i}\int\Omega\right)$$

is then the same as Berry's (the integral is over any surface spanning the loop). This supports the view that the half-form pairing gives the correct relationship between the phases of the wave functions in the different Fock spaces. ∎

10.3 Nonnegative Lagrangian subspaces

The pairing of half-forms extends to general nonnegative Lagrangian subspaces of $V_{\mathbb{C}}$. As in the holomorphic case, it introduces additional phase factors into the transformations between the Hilbert spaces of polarizations with real directions.

The set of all nonnegative Lagrangian subspaces of $V_{\mathbb{C}}$ is a manifold with boundary (and corners), which we shall denote by LV: its interior is L^+V and its boundary is made up of the subspaces with real directions, including the real Lagrangian subspaces.[6]

The canonical bundle $K \to L^+V$ extends to LV, with fibres defined by (10.2.8). The embedding of $K \to L^+V$ in the trivial bundle $L^+V \times \bigwedge^n V_{\mathbb{C}}^*$ and the square root $\delta = \sqrt{K}$ also extend; but $\overline{z} - z$ becomes singular on the boundary, so the connection, curvature, and Hermitian structure are defined only on the interior.

However, the inner product in $\bigwedge^n V_{\mathbb{C}}^*$ still gives a pairing between K_P and $K_{P'}$ even when P and P' are not positive, provided that $\overline{P}$ and P' are transverse. By taking the square root continuously, we can pair δ_P and $\delta_{P'}$ for any nonnegative P and P' such that $\overline{P}$ and P' are transverse; when P and P' are also real, this is the same as the pairing that we used in §9.5, apart from an additional complex factor.

We shall need two properties of half-forms on nonnegative spaces. The first concerns triples of real Lagrangian subspaces. The second is an extension of the pairing to nontransverse spaces.

Real Lagrangian subspaces

The function χ of three positive Lagrangian subspaces is not defined when its arguments become real. But the phase factor $\tau = \chi/|\chi|$ has a well-defined limit provided that the subspaces remain mutually transverse. It will be used in the derivation of the corrected form of the Bohr-Sommerfeld condition. Its value is determined by the signature invariant, which is defined as follows (Leray 1974, Souriau 1976).

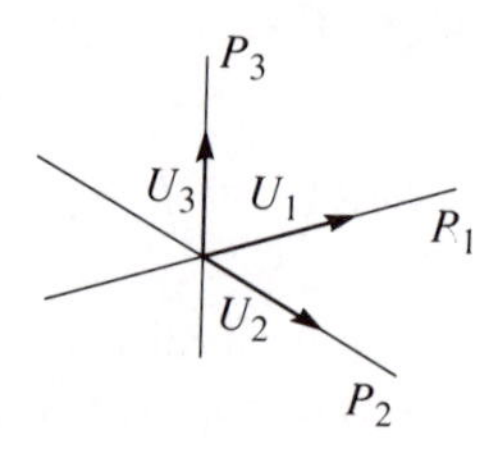

Fig. 10.1. The signature invariant.

Let P_1, P_2, P_3 be mutually transverse real Lagrangian subspaces, and let $U_1, V_1 \in P_1$. Since P_2 and P_3 are transverse, U_1 and V_1 are uniquely of the form

$$U_1 = U_2 + U_3 \quad \text{and} \quad V_1 = V_2 + V_3,$$

where $U_2, V_2 \in P_2$ and $U_3, V_3 \in P_3$ (Fig. 10.1). Because all three subspaces are Lagrangian,

$$0 = \omega(U_1, V_1) = \omega(U_2 + U_3, V_2 + V_3) = \omega(U_2, V_3) - \omega(V_2, U_3).$$

It follows that we can define a symmetric quadratic form g_{123} on P_1 by $g_{123}(U_1, V_1) = 2\omega(U_2, V_3)$.

Choose a symplectic frame $\{X^a, Y_b\}$ such that the span of the Ys is transverse to all three spaces. Then there are three real symmetric $n \times n$ matrices z_1, z_2, z_3 such that

$$X^a - z_\alpha^{ab} Y_b$$

is a basis P_α ($\alpha = 1, 2, 3$). The differences between the zs are nonsingular since the subspaces are mutually transverse. The matrix of g_{123} in the basis for P_1 is

$$(z_1 - z_3)(z_2 - z_3)^{-1}(z_2 - z_1).$$

Consequently g_{123} is also nonsingular.

Definition (10.3.1). The signature invariant σ_{123} of P_1, P_2, P_3 is sign (g_{123}), where 'sign' denotes the signature—the number of positive eigenvalues less the number of negative eigenvalues.

Proposition (10.3.2). Let P_1, P_2, P_3 be real Lagrangian subspaces and let $\nu_1 \in \delta_1, \nu_2 \in \delta_2, \nu_3 \in \delta_3$, where δ_α is the fibre of δ above P_α. Then

$$\frac{(\nu_2, \nu_1)(\nu_3, \nu_2)}{(\nu_3, \nu_1)} = r e^{-\pi i \sigma_{123}/4}$$

where $r > 0$.

Proof. The real symplectic group $SP(V,\omega)$ is transitive on pairs of transverse real Lagrangian subspaces. So nothing is lost by assuming that P_1 and P_2 are given by $z_1 = 1$ and $z_2 = 0$. Then g_{123} has matrix

$$(z_1 - z_3)(z_2 - z_3)^{-1}(z_2 - z_1) = z_3^{-1} - 1$$

and so $\operatorname{sign}(g_{123}) = \operatorname{sign}(z_3^{-1} - 1) = \operatorname{sign}(z_3) + \operatorname{sign}(1 - z_3) - n$.

In the notation of (10.2.5), choose J_0 to be the complex structure that takes the standard form in the frame $\{X^a, Y_b\}$. We have to show that the limit of

$$\frac{1}{\tau_{123}} = r\sqrt{\zeta_{12}}\sqrt{\zeta_{23}}\sqrt{\zeta_{31}},$$

as $z_1 \to 1$, $z_2 \to 0$, and z_3 becomes real, is $\exp(-\frac{1}{4}\pi \mathrm{i}\sigma_{123})$. Here $\zeta_{\alpha\beta} = \det \frac{1}{2}\mathrm{i}(\overline{z}_\alpha - z_\beta)$ and $r = |\zeta_{12}\zeta_{23}\zeta_{31}|^{-1/2} > 0$. The square roots are continuous as z_1, z_2, z_3 vary with nonnegative imaginary part, and are positive when $z_1 = z_2 = z_3 = \mathrm{i}$.

Now if we define $\sqrt{\lambda}$ continuously for complex λ with $\mathrm{Re}\lambda \geq 0$ by taking the positive square root when $\lambda = 1$, then $\sqrt{\lambda}$ is a positive multiple of $\mathrm{e}^{\mathrm{i}\pi/4}$ when $\lambda = \mathrm{i}y$ with $y > 0$, and a positive multiple of $\mathrm{e}^{-\mathrm{i}\pi/4}$ when $\lambda = \mathrm{i}y$ with $y < 0$. By considering the behaviour of the square root in the special case $\mathrm{i}(\overline{z}_\alpha - z_\beta) = \operatorname{diag}(\lambda_1, \ldots, \lambda_n)$, and exploiting continuity, we can deduce that in general

$$\sqrt{\zeta_{\alpha\beta}} \to r_{\alpha\beta}\mathrm{e}^{\mathrm{i}\pi \operatorname{sign}(z_\alpha - z_\beta)/4}$$

as the zs become real, where $r_{\alpha\beta} > 0$, provided that $z_\alpha - z_\beta$ remains nonsingular.

Hence the limit of $(\tau_{123})^{-1}$ is a positive real multiple of

$$\exp\left(\frac{\mathrm{i}\pi}{4}\Big(n + \operatorname{sign}(-z_3) + \operatorname{sign}(z_3 - 1)\Big)\right) = \exp\left(-\frac{\mathrm{i}\pi}{4}\sigma_{123}\right). \qquad \blacksquare$$

Aside. The cocycle property of τ (see eqn 10.2.6) is reflected in the relation $\sigma_{123} + \sigma_{301} = \sigma_{230} + \sigma_{012}$. ∎

We shall need another expression for g_{123}. Let Q be an n-dimensional manifold and let Λ_1 and Λ_3 be the Lagrangian submanifolds of T^*Q generated by $S_1, S_3 \in C^\infty(Q)$. Suppose that these meet at $m \in T^*Q$. Let Λ_2 be the fibre of T^*Q through m and let P_α ($\alpha = 1, 2, 3$) be the tangent space to Λ_α at m. Then P_1, P_2, and P_3 are spannned, respectively, by

$$\frac{\partial}{\partial q^a} + \frac{\partial^2 S_1}{\partial q^a \partial q^b}\frac{\partial}{\partial p_b}, \quad \frac{\partial}{\partial p_a}, \quad \text{and} \quad \frac{\partial}{\partial q^a} + \frac{\partial^2 S_3}{\partial q^a \partial q^b}\frac{\partial}{\partial p_b},$$

where the qs are coordinates on Q and the ps are the conjugate momenta. In the basis for P_1, g_{123} has components

$$\frac{\partial^2(S_1 - S_3)}{\partial q^a \partial q^b}. \tag{10.3.3}$$

Nontransverse subspaces

If P and P' are nonnegative, but $\overline{P} \cap P' \neq 0$, then $\overline{\mu} \wedge \mu' = 0$ and the pairing vanishes. We can take the limit in a different way, however, by interpreting the result of the half-form pairing not as a scalar, but as a density on V/D, where $\overline{P} \cap P' = D_{\mathbb{C}}$ (Blattner 1973, 1977, Rawnsley 1978a). This makes no difference in the case we have already considered because V has a natural volume element, so when $D = 0$, densities on $V/D = V$ are the same as scalars.

First we need to show that the intersection of $\overline{P}$ and P' is the complexification of a real subspace of V.

> **Lemma (10.3.4).** Let P and P' be nonnegative Lagrangian subspaces of $V_{\mathbb{C}}$. Then there is a real isotropic subspace $D \subset V$ such that $\overline{P} \cap P' = D_{\mathbb{C}}$.

Proof. Let $D = \overline{P} \cap P' \cap P \cap \overline{P}' \cap V$. Then D is real and isotropic, so it is sufficient to show that $\overline{P} \cap P' = P \cap \overline{P}'$.

To do this, note first that since P is nonnegative, $\mathrm{i}\omega(\overline{X}, X) \geq 0$ for every $X \in P$, with equality if and only if $X \in \overline{P} \cap P$.

Suppose that $X \in \overline{P} \cap P'$. Then $\mathrm{i}\omega(\overline{X}, X) \geq 0$ since $X \in P'$ and P' is nonnegative, and $\mathrm{i}\omega(X, \overline{X}) \geq 0$ since $\overline{X} \in P$ and P is nonnegative. Therefore $\mathrm{i}\omega(\overline{X}, X) = 0$, and so $X \in P$ and $X \in \overline{P}'$. It follows that $\overline{P} \cap P' \subset P \cap \overline{P}'$. Similarly $P \cap \overline{P}' \subset \overline{P} \cap P'$. Hence $\overline{P} \cap P' = P \cap \overline{P}'$. ∎

We want to construct a sesquilinear pairing $\overline{\delta}_P \otimes \delta_{P'} \to \Delta(V/D)$, where $\Delta(V/D)$ is the one-dimensional space of complex densities on V/D. Following Rawnsley (1978a), we shall do this in two stages: first by constructing $\overline{K}_P \otimes K_{P'} \to \Delta(V/D)^2$, and then by taking a square root.

Let $k = n - \dim D$ and let K_D be the one-dimensional subspace of $\bigwedge^{n+k} V_{\mathbb{C}}^*$ of forms α such that $X \lrcorner \alpha = 0$ for every $X \in D$. Each such α is the pull-back of a complex volume form on V/D under the projection $V \to V/D$. Thus

$$K_D = \bigwedge^{n+k} (V_{\mathbb{C}}/D_{\mathbb{C}})^*$$

and so $\overline{K}_D \otimes K_D = \Delta(V/D)^2$ (note that V/D has dimension $n + k$).

Let $\mu \in K_P$. Then $\mu = \Gamma \lrcorner \alpha$ for some $\Gamma \in \bigwedge^k \overline{P}$ and $\alpha \in K_D$ (see §A.1 for the definition of the contraction $\lrcorner$). These are determined by μ up to

$$\Gamma \mapsto \lambda\Gamma + X \wedge \Psi, \quad \alpha \mapsto \lambda^{-1}\alpha, \tag{10.3.5}$$

where $\lambda \in \mathbb{C}$, $X \in D$, and $\Psi \in \bigwedge^{k-1} \overline{P}$ (if $k = 0$, then $\Gamma \in \mathbb{C}$ and $\Psi = 0$).

We similarly write $\mu' \in K_{P'}$ as $\Gamma' \lrcorner \alpha'$, where $\Gamma' \in \bigwedge^k \overline{P}'$ and $\alpha' \in K_D$. The tensor product $\overline{\alpha} \otimes \alpha'$ is an element of $\Delta(V/D)^2$. The pairing of K_P and $K_{P'}$ into $\Delta(V/D)^2$ is defined by

$$(\mu, \mu') = (-1)^{k(k-1)/2} \frac{(\overline{\Gamma} \wedge \Gamma') \lrcorner \omega^k}{k!(2\pi \mathrm{i} \hbar)^k} \overline{\alpha} \otimes \alpha'.$$

This is nonsingular and invariant under the transformation (10.3.5) since

$$(X \wedge \Psi \wedge \Gamma') \lrcorner \omega^k = 0$$

because $\omega(X, Y) = 0$ for every $Y \in P + \overline{P}'$. It is also unchanged by similar transformations of Γ' and α'. The factors of i and -1 are chosen so that the pairing is sent to its complex conjugate under interchange of P and P', and is positive when $P = P'$.

By taking the square root, we obtain $\overline{\delta}_P \otimes \delta_{P'} = \Delta(V/D)$. Thus we have a sesquilinear pairing that assigns a complex density (ν, ν') on V/D to each $\nu \in \delta_P$ and $\nu' \in \delta_{P'}$. The only issue is the sign of the square root. This we can fix, as before, by continuity, together with the condition $(\nu, \nu) \geq 0$ when, keeping D fixed, we move P and P' into coincidence with a nonnegative polarization with $n - k$ real directions.

Example. Let $V = \mathbb{R}^2$ with its standard symplectic structure $\mathrm{d}p \wedge \mathrm{d}q$. If $P = P'$ is the real subspace $q = 0$, then $D = P$ and the elements of the quotient V/P are labelled by q. We can represent elements of K_P by complex multiples of $\mathrm{d}q$ and elements of $\Delta(V/D)^2$ by real multiples of $|\mathrm{d}q|^2$. The isomorphism $\overline{K}_P \otimes K_P \to \Delta(V/P)^2$ is given $\mathrm{d}q \otimes \mathrm{d}q \mapsto |\mathrm{d}q|^2$.

If $P = P'$ is the complex subspace given by $\overline{z} = p - \mathrm{i}q = 0$, then $D = 0$ and $V/D = V$. Since V is oriented, there is no difference between volume elements on V and densities. Thus elements of $\Delta(V/D)^2$ in this case are simply real multiples of ϵ^2. Elements K_P are complex multiples of $\mu = (4\pi\hbar)^{-1/2}\mathrm{d}z$, and the isomorphism is $\overline{\mu} \otimes \mu \mapsto \epsilon^2 = \epsilon \otimes \epsilon$. ∎

In general, if $\overline{P}$ and P' are transverse, then the isomorphism $\overline{\delta}_P \otimes \delta_{P'} \to \Delta(V)$ sends $\overline{\nu} \otimes \nu'$ to $(\nu, \nu')\epsilon$: in the transverse case, the pairing into densities simply includes the volume element on V in the pairing into scalars.

10.4 Half-form quantization

In linear systems, the inclusion of half-forms seems to give the correct relationship between the phases of the wave functions in the Hilbert spaces of different polarizations. This suggests that we should always include them, if only to get the correct linear approximation near a position of equilibrium. We shall in fact see stronger evidence for this 'metaplectic correction'.

Let (M, ω) be a $2n$-dimensional symplectic manifold and let LM be the set of pairs (m, P), where $m \in M$ and P is a nonnegative Lagrangian subspace of $(T_m M)_{\mathbb{C}}$; LM is a manifold with boundary (and corners), and it has the structure of a bundle over M with projection $\mathrm{pr} : (m, P) \mapsto m$ and fibre LV (p. 227).

Over LM, we have the canonical line bundle $K \to LM$, which has fibre K_P at (m, P).

Definition (10.4.1). A metaplectic structure on M is a square root of K. That is, a line bundle $\delta \to LM$ such that $\delta^2 = K$.

Metaplectic structures do not always exist, and when they do, they need not be unique. The problem is the standard one of the existence of global topological obstructions. The restriction of δ to the fibre of LM over some $m \in M$ is simply the half-form bundle on the nonnegative Lagrangian Grassmannian of $T_m M$. So there is no problem in defining δ locally in a small neighbourhood U of m in which we can identify the corresponding portion of LM with $U \times L_m M$ by choosing a symplectic frame at each point of U. But when we try to piece together these local metaplectic structures, we run into the difficulty that it is not the symplectic group that acts on half-forms, but the metaplectic group. On the overlap of two neighbourhoods in M, we have to choose one of the two elements of the metaplectic group corresponding to the symplectic transformation between the two local frames. It may not be possible to do this in a consistent way on all the overlaps.

Because the fibres $L_m M$ of LM are contractible, we can find a system of trivializations for K of the form $(\mathrm{pr}^{-1}U, \tau)$, $U \subset M$, such that the transition functions are constant along each $L_m M$. The introduction of a metaplectic structure is equivalent to choosing square roots of the transition functions in such a way that they still satisfy the cocycle relations (A.3.3) and (A.3.4). It is sufficient to do this at one point of $L_m M$ for each $m \in M$. Therefore the introduction of a metaplectic structure is equivalent to the construction of a square root of $K|_\Sigma$ where $\Sigma = \sigma(M)$ is the graph of any smooth section of LM.

A section $\sigma : M \to LM$ is a complex distribution on M made up of nonnegative Lagrangian subspaces. The pull-back $\sigma^* K$ is simply the canonical bundle of the distribution. So finding a square root of $K|_\Sigma$ is the same as finding a square root of the canonical bundle of any nonnegative Lagrangian distribution.

In particular, we can use this when the distribution is a nonnegative polarization P_0. A square root δ_{P_0} of the canonical bundle of P_0 determines a square root δ of K and hence a square root δ_P of K_P for any other nonnegative polarization P; δ_P is called the *half-form bundle* of P. If P_0 is real and oriented (i.e. its leaves have a preferred orientation), then the transition functions of K_{P_0} are all positive and there is a natural choice for δ_{P_0}.

The partial connection and Lie derivative that we defined on sections of K_P for real polarizations (p. 185) also exist in the case of general non-negative polarizations. If β is section of K_P (an n-form on M such that $\overline{Y} \lrcorner \beta = 0$ for every $Y \in V_P(M)$), then

$$\nabla_{\overline{X}}\beta = \overline{X} \lrcorner \mathrm{d}\beta \quad \text{and} \quad \mathcal{L}_Z\beta = Z \lrcorner \mathrm{d}\beta + \mathrm{d}(Z \lrcorner \beta),$$

for any $X \in V_P(M)$ and for any real vector field Z such that the flow of Z preserves P. If $X \in V_P(M)$ is real, then its flow preserves P and ∇_X and $\mathcal{L}_X$ coincide. As in the real case, ∇ and $\mathcal{L}$ pass to δ_P.

Now suppose that $B \to M$ is a prequantum bundle and that P is a nonnegative polarization. The connection on B combines with the partial connection on δ_P to give a partial connection on $B_P = B \otimes \delta_P$. In the corrected quantization scheme, the P wave functions are defined to be the smooth sections of B_P such that

$$\nabla_{\overline{X}}(\tilde{s}) = 0$$

for every $X \in V_P(M)$. These are locally of the form $\tilde{s} = s\nu$, where s is a section of B and ν is a section of δ_P such that

$$\nabla_{\overline{X}}s = 0 = \nabla_{\overline{X}}\nu \quad \text{for every} \quad X \in V_P(M). \tag{10.4.2}$$

If $f \in C^\infty(M)$ generates a flow preserving P, then the corresponding quantum observable $\tilde{f}$ acts on the wave functions by

$$\tilde{f}(\tilde{s}) = \hat{f}(s)\nu - \mathrm{i}\hbar s\mathcal{L}_{X_f}\nu.$$

We have remarked already that if $M = V$ and if P is determined by a positive complex structure on V, then the $\tilde{f}$s corresponding to quadratic polynomials on V satisfy the quantum conditions. The same is true if P is constructed from any nonnegative Lagrangian subspace of $V_{\mathbb{C}}$.

The inclusion of the 'half-forms' introduces a new constraint into the quantization construction: it is necessary that M should admit a metaplectic structure. In fact, one can sometimes get away with less than this: it may be that B_P can be constructed as the square root of $B^2 \otimes K_P$ even when B^2 and K_P do not individually have square roots (Czyz 1978).

Suppose that P is strongly integrable and that the isotropic distribution D associated with P is reducible, so that M/D is a Hausdorff manifold. Let s and ν be as in (10.4.2). Then at each $m \in M$, the half-form pairing (ν, ν) determines a density at the corresponding point of M/D, which remains unchanged as m varies over the leaf of D through m. This is because strong integrability implies the existence of Hamiltonian vector fields spanning the leaves of D (the vector fields $\partial/\partial p_a$ in 5.4.7). Their flows preserve P, ω and ν, and consequently leave invariant the density on M/D constructed from (ν, ν). The function (s, s) is also constant on the leaves of D. Therefore we can define an inner product on the wave functions by

$$\langle \tilde{s}, \tilde{s} \rangle = \int_{M/D} (s, s)(\nu, \nu).$$

In the same way, we can use the half-form pairing to pair the wave functions of two nonnegative polarizations P and P' provided that they satisfy the following compatibility conditions. By (10.3.4), at each $m \in M$ there is a real isotropic subspace $D_m \subset T_m M$ such that $\overline{P}_m \cap P'_m = D_{m\mathbb{C}}$. We require that D should be a distribution (it is then necessarily integrable); that it should be reducible; and that $E = D^\perp$ should be integrable. If these hold and if ν and ν' are sections of δ_P and $\delta_{P'}$, covariantly constant along $\overline{P}$ and $\overline{P}'$ respectively, then as before, the pairing (ν, ν') is a density on M/D. We can therefore pair a P wave function $\tilde{s} = s\nu$ with a P' wave function $\tilde{s}' = s'\nu'$ by

$$\langle \tilde{s}, \tilde{s}' \rangle = \int_{M/D} (s, s')(\nu, \nu').$$

When P and P' are positive, $D = 0$, and the integral is over M.

The conditions require, in particular, that the dimension of D should be constant everywhere. This is often too strong: in the case of real polarizations, the interesting phemonena arise when P and P' are transverse almost everywhere. We can still define the pairing in this case by excluding the singular points of M where they are not transverse, and taking the pairing to be an integral over the rest of M.

Blattner's original construction (Blattner 1973) used a weaker condition than the integrabilty of $D^\perp$.

Example. *Kähler manifolds.* The canonical bundle of a strongly integrable polarization P is polarized: it has a preferred class of local trivializations with transition functions which are constant along $\overline{P}$. These are defined by sections which are covariantly constant along $\overline{P}$ with respect to the flat partial connection—for example, if q^a, p_b, z^α is the coordinate system of Proposition (5.4.7), then

$$\mathrm{d}q^1 \wedge \cdots \wedge \mathrm{d}q^n \wedge \mathrm{d}z^1 \wedge \cdots \wedge \mathrm{d}z^{n'}$$

is such a section. Consequently, the half-form bundle is also polarized.

When M is $2n$-dimensional and P is a Kähler polarization, 'polarized bundles' are holomorphic. The canonical bundle is the bundle of holomorphic n-forms and the half-form bundle has holomorphic sections that can be represented as $f(z)\sqrt{\mathrm{d}^n z}$ in local holomorphic coordinates. The Hilbert space of the corrected quantization scheme is the space of square-integrable holomorphic sections of $B \otimes \delta_P$.

The canonical bundle of a Kähler manifold has Chern class $-c$, where c is the first Chern class of M as a complex manifold (Morrow and Kodaira 1971, p. 128). So the condition for $B^2 \otimes K_P$ to have a square root is that

$$\frac{\omega}{2\pi\hbar} - \frac{c}{2}$$

should satisfy the integrality condition.

For example, consider the projective space $\mathbb{CP}_n$, with some multiple of the Fubini-Study symplectic structure. A local section of the canonical bundle is represented by a holomorphic n-form on $\mathbb{CP}_n$. Its pull-back by the projection map $\mathbb{C}^{n+1} \to \mathbb{CP}_n$ is of the form

$$f(z)\epsilon_{\alpha\beta\ldots\gamma} z^\alpha \mathrm{d}z^\beta \wedge \cdots \wedge \mathrm{d}z^\gamma,$$

where f is holomorphic and homogeneous of degree $-n-1$, ϵ is the $(n+1)$-dimensional alternating symbol, and the subscripts run over $0, 1, \ldots, n$. Thus sections of K_P can be identified with homogeneous functions on $\mathbb{C}^{n+1}$ of degree $-n-1$. Sections of $\sqrt{K_P}$, therefore, would be represented by functions homogeneous of degree $-\frac{1}{2}(n+1)$, which exist even locally in $\mathbb{CP}_n$ only if n is odd. Thus there is a metaplectic structure only when n is odd.

When ω is the Fubini-Study symplectic structure itself, we can use Proposition (8.4.9), with

$$\theta' = -\mathrm{i}\hbar(\bar{z}^0 \mathrm{d}z^0 + \bar{z}^1 \mathrm{d}z^1 + \cdots \bar{z}^n \mathrm{d}z^n),$$

to represent the holomorphic sections of B by homogeneous functions on $\mathbb{C}^{n+1}$ of degree 1. So when the symplectic structure is $k\omega$, we can construct a square root of $B^2 \otimes K_P$ whenever $2k - n - 1$ is an even integer.

In the case $n = 1$, there is a metaplectic structure and the local sections of δ_P are homogeneous functions on $\mathbb{C}^2$ of degree -1. If the symplectic structure is that of a particle of spin s, then the sections of $B \otimes \delta_P$ are homogeneous polynomials $\psi_{AB\ldots C} z^A z^B \ldots z^C$ of degree $2\hbar^{-1}s - 1$. Thus the metaplectic correction produces a shift between the classical and quantum values of the spin.

Such shifts occur generally for the phase spaces of compact groups G: they are given by 'half the sum of the positive roots'. For example, suppose that f is a regular element of $\mathcal{G}^*$ and that (M', ω') is the reduction of (G, ω_f). Let $R_1, R_2, \ldots, R_n$ be the right-invariant vector fields on G which are equal at the identity to

$$Z_{\alpha_1}, Z_{\alpha_2}, \ldots, Z_{\alpha_n},$$

where the αs are the positive roots (p. 105). Let μ' be a section of K_P and let μ be its pull-back to G. Then $f = \mu(R_1, R_2, \ldots, R_n)$ is a complex function on G such that, for any $Z \in \mathcal{T} = \mathcal{G}_f$,

$$R_Z \lrcorner \mathrm{d}f = -\mathrm{i}\Delta(Z)f \quad \text{where} \quad \Delta = \alpha_1 + \alpha_1 + \cdots + \alpha_n,$$

since $R_{[A,B]} = -[R_A, R_B]$ for any $A, B \in \mathcal{G}$. Thus, when it exists, the sections of the half-form bundle on M' are represented by functions on G such that $R_Z(f) = -\frac{1}{2}\mathrm{i}\Delta(Z)f$ for every $Z \in \mathcal{T}$. ■

10.5 The WKB approximation

In §9.6, we considered the geometric interpretation of the semiclassical eigenfunctions of a quantum Hamiltonian constructed by the Schrödinger prescription. We interpreted the phase $\mathrm{e}^{\mathrm{i}w/\hbar}$ as a covariantly constant section s of B_Λ, where B is the prequantum bundle and Λ is the Lagrangian submanifold of T^*Q generated by w, and the square of A as a volume form κ on Λ. The triple (Λ, s, κ) determines, up to a sign ambiguity in A, a semiclassical wave function in the Hilbert space of any real polarization transverse to Λ. Provided that we ignore the phase factor $\mathrm{e}^{\mathrm{i}\pi\sigma/4}$ in (9.6.2), the wave functions of a triple in the Hilbert spaces of different polarizations are related by the pairing construction.

The metaplectic construction improves on this. It gives us a natural definition of 'a square root of a volume form' on Λ, and allows us to replace κ by a section α of $\sqrt{\Delta_\Lambda}$ such that $\alpha^2 = \kappa$, where Δ_Λ is the line bundle of complex volume forms on Λ. The triple (Λ, s, α) encodes the amplitude and phase unambiguously. The corrected construction takes account of the possibility that the classical trajectories may have caustic singularities and that Λ may not be everywhere transverse to P.

We shall now look at how this works. The ideas are taken from Maslov (1972), Hörmander (1971), and Guillemin and Sternberg (1977).

Let (M, ω) be a $2n$-dimensional symplectic manifold with a prequantum bundle $B \to M$ and a metaplectic structure $\delta \to LM$. Let Λ be a Lagrangian submanifold. At each $m \in \Lambda$, $T_m\Lambda$ is a Lagrangian subspace of T_mM. We therefore have two line bundles over Λ: the bundle $\Delta_\Lambda = \bigwedge^n T^*_{\mathbb{C}}\Lambda$ of complex volume forms on Λ, and the bundle K_Λ, of which the fibre at m is $K_{T_m\Lambda}$, which is the subspace of $\bigwedge^{2n} T_mM$ of n-forms such that $X \lrcorner \beta = 0$ for every X tangent to Λ. These are connected by a sesquilinear pairing. If β is a section of K_Λ and κ is a section of Δ_Λ, then $\overline{\kappa} \wedge \beta$ is a multiple of ω^n, so we can define (κ, β) by[7]

$$\mathrm{i}^n(\kappa, \beta)\epsilon = \overline{\kappa} \wedge \beta.$$

It follows that $\Delta_\Lambda = \overline{K}^*_\Lambda$.

This suggests that we define $\surd\Delta_\Lambda \to \Lambda$ to be the bundle with fibre $\overline{\delta}^*_{T_m\Lambda}$ at $m \in \Lambda$. We then have $\surd\Delta_\Lambda = \delta_P|_\Lambda$ for any positive polarization transverse to Λ. This is because the fibres of δ_P and $\surd\Delta_\Lambda$ at $m \in \Lambda$ are both isomorphic to $\overline{\delta}^*_{T_m\Lambda}$, in the first case by the metaplectic pairing between δ_{P_m} and $\delta_{T_m\Lambda}$, in the second case by definition. A section α of $\surd\Delta_\Lambda$ is a square root of a complex volume form on Λ. Its Lie derivative along a vector field X tangent to Λ is defined by

$$\alpha \mathcal{L}_X \alpha = \mathcal{L}_X \alpha^2.$$

The right-hand side is well defined since α^2 is an n-form on Λ.

Our previous definition of a WKB wave function is now replaced by the following.

> **Definition (10.5.1).** A WKB wave-function is a triple (Λ, s, α), in which $\Lambda \subset M$ is a Lagrangian submanifold, s is a covariantly constant section of B_Λ such that $(s, s) = 1$, and α is a section of $\surd\Delta_\Lambda$. Such a triple is a 'WKB eigenfunction of $H \in C^\infty(M)$' if H is constant on Λ (the Hamilton-Jacobi condition) and $\mathcal{L}_{X_H}\alpha = 0$ (the transport equation).

From (Λ, s, α), we can recover a wave function for any real polarization P of which Λ is a section: we identify α with a section of $\delta_P|_\Lambda$. We then extend s and α to sections of B and δ_P, respectively, by parallel transport along the leaves of P. The product $s\alpha$ is a P wave function.

Phase and signature

The effect of the modification is to absorb the factor $\mathrm{e}^{-\mathrm{i}\pi\sigma/4}$ in (9.6.2) into the definition of the pairing, as is shown by the following argument based on Proposition (10.3.2). The notation is the same as in §9.6, except in the case of A and A', which differ from their (ambiguous) values in the unmodified construction by certain powers of $\surd\mathrm{i}$. We assume that $Q = M/P$ and $Q' = M/P'$ are Hausdorff manifolds, that P and P' are transverse, and that they have complete simply-connected leaves; but M need no longer be a cotangent bundle, and there is no need to assume that Q and Q' are orientable. The local coordinates on Q and Q' must be chosen so that

$$D = \det\left(\frac{\partial^2 S}{\partial q'^a \partial q^b}\right) > 0,$$

where S is a local generating function of P' relative to P; this involves no loss of generality.

Choose (local) sections ν_2 and ν_3 of δ_P and $\delta_{P'}$ such that $\nu_2^2 = \mathrm{d}^n q$ and $\nu_3^2 = \mathrm{d}^n q'$. Let $m \in \Lambda$. Put $P_1 = T_m\Lambda$, $P_2 = P_m$, $P_3 = P'_m$ and take δ_1, δ_2,

δ_3 to be the corresponding fibres of δ. Let $\nu_1 \in \delta_1$ be such that $(\nu_1, \alpha) = 1$ (this pairing is well defined since the fibre of $\sqrt{\Delta_\Lambda}$ at m is $\overline{\delta}_1^*$).

The values at m of the P wave function and the P' wave function determined by (Λ, s, α) are respectively

$$A e^{iw/\hbar} \nu_2 \quad (\theta \text{ gauge}) \text{ and } \quad A' e^{iw'/\hbar} \nu_3 \quad (\theta' \text{ gauge}),$$

where now $A = (\nu_1, \nu_2)^{-1}$ and $A' = (\nu_1, \nu_3)^{-1}$. These are made into functions on M by taking A and w to be constant along P, and A' and w' to be constant along P'. The amplitudes A and A' differ in argument from their previous definitions in §9.6, but have the same moduli. As before, therefore,

$$|A'| \sqrt{|\det g|} = |A| \sqrt{D}. \tag{10.5.2}$$

The pairing of the P wave function and an arbitrary P' wave function $\psi' \nu_3$ is given by (9.6.2), except that $\sqrt{D}$ is now replaced by $(\nu_2, \nu_3) D$. This change introduces an extra phase factor into the integral. The modulus $|(\nu_2, \nu_3) D|$ is still equal to $\sqrt{D}$; but the argument is determined by Proposition (10.3.2), from which it follows that

$$(\nu_2, \nu_3) = r \frac{(\nu_1, \nu_3)}{(\nu_1, \nu_2)} e^{i\pi\sigma_{123}/4} = r A A'^{-1} e^{i\pi\sigma_{123}/4}$$

where $r > 0$ and σ_{123} is the signature invariant of P_1, P_2, P_3. By taking the modulus, $D^{-1/2} = r|A||A'|^{-1}$, and by taking $S_2 = w$ and $S_3 = S$ in (10.3.3), $\sigma_{123} = \sigma$. So, by using (10.5.2),

$$(\nu_2, \nu_3) \overline{A} D = \sqrt{|\det g|}\, \overline{A}' e^{i\pi\sigma/4}.$$

The factor $e^{i\pi\sigma/4}$ on the right-hand side cancels the unwanted power of i in (9.6.2). With the metaplectic pairing, therefore, the image under the pairing map of $A e^{iw/\hbar} \nu_2$ is $A' e^{iw'/\hbar} \nu_3$ to the first order in $\hbar$. In the semiclassical limit, the pairing map coincides with the geometric procedure for transferring WKB wave functions from Q to Q' via Lagrangian submanifolds of M.

Distributional wave functions

The metaplectic pairing allows us to associate a P wave function with (Λ, s, α) even when Λ is not everywhere transverse to P, provided that we extend the notion of 'P wave function' to include objects that have a distributional dependence on q, and accept that the wave function should be determined by (Λ, s, α) only to the first order in $\hbar$. The definition exploits the fact that the leading term in the asymptotic expansion of the inner product of two WKB wave functions in $\mathcal{H}_P$ can be calculated directly from the corresponding triples.

Consider two WKB wave functions (Λ, s, α) and $(\Lambda'', s'', \alpha'')$, where Λ and Λ'' are both transverse to P. These determine the two P wave functions

$$Ae^{iw/\hbar}\nu_2 \quad \text{and} \quad A''e^{iw''/\hbar}\nu_2$$

in the θ gauge. In the coordinates q^a (pulled back to Λ and Λ''), $\alpha^2 = A^2 d^n q$ and $\alpha''^2 = A''^2 d^n q$.

Their inner product is given by

$$\begin{aligned}
&(2\pi\hbar)^{n/2}\langle Ae^{iw/\hbar}\nu_2, A''e^{iw''/\hbar}\nu_2\rangle \\
&= \left(\frac{1}{2\pi\hbar}\right)^{-n/2}\int \overline{A}A'' f e^{i(w''-w)/\hbar} d^n q \\
&= \sum_q e^{-i\pi\sigma(q)/4}\frac{\overline{A}(q)A''(q)}{\sqrt{|\det J(q)|}} e^{i(w''(q)-w(q))/\hbar} + O(\hbar),
\end{aligned}$$

where $J_{ab} = \partial_a\partial_b(w - w'')$, σ is the signature of J, and the sum is over the critical points of $w - w''$. We assume that the critical points are isolated and nondegenerate. They are the projections into Q of the intersection points of Λ and Λ''. Nondegeneracy is the condition that the intersections should be transverse.

The idea is to express the leading term on the right-hand side in terms of the sum of $(s, s'')(\alpha, \alpha'')$ over the intersection points. Here the pairing of s and s'' is simply the inner product in the fibre of B. The pairing of α and α'' is defined as follows.

Let m be one of the intersection points, and put $P_1 = T_m\Lambda$, $P_2 = P_m$, and $P_3 = T_m\Lambda''$. These are mutually transverse Lagrangian subspaces of T_mM, with $\sigma_{123} = \sigma$. Let $\delta_\alpha = \delta_{P_\alpha}$, $\alpha = 1, 2, 3$. Then $\overline{\delta}_1^*$ and $\overline{\delta}_2^*$ are the fibres of $\sqrt{\Delta_\Lambda}$ and $\sqrt{\Delta_{\Lambda''}}$ at m. Let $\nu_2 \in \delta_2$ be a square root of $d^n q$. Since $\alpha(m) \in \overline{\delta}_1^*$ and $\alpha''(m) \in \overline{\delta}_3^*$, we can define $\nu_1 \in \delta_1$ and $\nu_3 \in \delta_3$ by $(\nu_1, \alpha(m)) = 1$ and $(\nu_3, \alpha''(m)) = 1$, where $(\cdot,\cdot)$ is the natural pairing $\overline{\delta}^* \otimes \delta \to \mathbb{C}$. We define $(\alpha(m), \alpha''(m))$ to be $(\nu_3, \nu_1)^{-1}$, where $(\cdot,\cdot)$ is the metaplectic pairing between δ_3 and δ_1.

By construction, $A = (\nu_1, \nu_2)^{-1}$ and $A'' = (\nu_3, \nu_2)^{-1}$. Therefore,

$$(\alpha(m), \alpha''(m)) = \frac{1}{(\nu_3, \nu_1)} = r\overline{A}A''e^{-\pi i\sigma/4}$$

for some $r > 0$, by (10.3.2). This determines the argument of (α, α''). The modulus is determined by looking at the pairing of $\nu_1^2 \in K_{T_m\Lambda}$ and $\nu_3^2 \in K_{T_m\Lambda''}$. These are complex n-forms in $\bigwedge^n(T_m^*M)_\mathbb{C}$, annihilated, respectively, by vectors tangent to Λ and Λ''. If the ps are the momenta conjugate to the qs, then to within a power of i, ν_1^2 is

$$\frac{1}{\overline{A}^2}(dp_1 - \partial_1\partial_a w\, dq^a) \wedge (dp_2 - \partial_2\partial_b w\, dq^b) \wedge \cdots \wedge (dp_n - \partial_n\partial_c w\, dq^c);$$

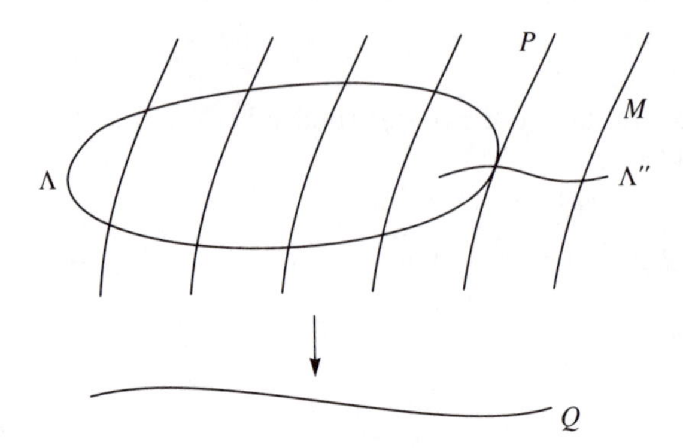

Fig. 10.2. The pairing of a WKB wave function with an oscillatory test function.

ν_3^2 is given similarly, with A and w replaced by A'' and w''. Therefore

$$|(\nu_1, \nu_3)|^2 = \left| \frac{\det J}{A^2 A''^2} \right|,$$

and so as $\hbar \to 0$,

$$\left(\frac{1}{2\pi\hbar} \right)^{n/2} \langle A\mathrm{e}^{\mathrm{i}w/\hbar}, A''\mathrm{e}^{\mathrm{i}w''/\hbar} \rangle \to \sum (s, s'')(\alpha, \alpha''), \tag{10.5.3}$$

where the sum is over the intersection points of Λ and Λ''.

The right-hand side of (10.5.3) is defined even when Λ and Λ'' are not transverse to P. We can therefore use it to extend $A\mathrm{e}^{\mathrm{i}w/\hbar}$ to a distributional wave function by *defining* the right-hand side to be the inner product of the P wave function associated with (Λ, s, α) and an 'oscillatory test function' of the form $(2\pi\hbar)^{-n/2} A''\mathrm{e}^{\mathrm{i}w''/\hbar}$, where A'' has compact support, and w'' is real and generates a Lagrangian submanifold which intersects Λ transversally (Fig. 10.2). We think of the test function as picking out the component that oscillates like $\mathrm{e}^{\mathrm{i}w''/\hbar}$ in the support of A''. There is some looseness of terminology in describing the resulting object as a 'distribution' since the dependence on the test function is only linear in the limit $\hbar \to 0$: what we should do, but have not done, is to establish the existence of a distribution on Q, which depends on $\hbar$ and whose value on an oscillatory test function is given by the right-hand side of (10.5.3) in the limit $\hbar \to 0$. This is easy to do only in special cases.[8] For example, if Λ is a leaf of P and $\alpha^2 = \mathrm{d}^n p$, where the ps are the momenta conjugate to the qs, then the distribution is a δ-function on Q.

Nevertheless, the construction gives a way of carrying the amplitude of a WKB eigenfunction through caustic singularities in the classical trajectories since the Lie dragging of α along X_H is well-defined even at points where Λ is not transverse to P. Since σ changes discontinuously at such points, the phase of the wave function also changes discontinuously, by a multiple of $\pi/2$ from one side of a caustic to the other. The phase jump is, in fact, precisely that discovered by Gouy (1890) in the context of geometric optics (Guillemin and Sternberg 1977).

The fact that the leading term in the inner product of two WKB wave functions can be calculated directly from the geometric data contained in the corresponding triples also implies that the pairing map $\mathcal{H}_P \to \mathcal{H}_{P'}$ between the Hilbert spaces of two transverse real polarizations must be unitary to the leading order in $\hbar$.

The corrected Bohr-Sommerfeld condition

Consider the problem of finding a simultaneous WKB eigenstate (Λ, s, α) of a set of classical observables $h_1, \ldots, h_n$ in involution. For simplicity, we shall assume that the corresponding Hamiltonian vector fields are everywhere independent. Then Λ must be of the form $h_a = E_a$, where the Es are constant, and α^2 must be a constant multiple of the volume form $\kappa \in \Omega^n(\Lambda)$ defined by

$$\kappa(X_{h_1}, X_{h_2}, \ldots, X_{h_n}) = \frac{1}{n!}$$

since the Lie derivative of α along all the Hamiltonian vector fields must vanish.

The bundle Δ_Λ has a flat connection, given by Lie differentiation along the X_hs. It is trivial as a locally constant bundle since κ is a global constant section. The connection passes to the square-root $\sqrt{\Delta_\Lambda}$, which is still trivial as a line bundle, but may be nontrivial as a locally constant bundle.

If Λ is not simply connected, then it may not be possible to make s and α single-valued. This is not in itself an obstruction to the existence of a simultaneous eigenfunction since we have the freedom to replace (Λ, s, α) by $(\Lambda, \lambda s, \lambda^{-1}\alpha)$, where $\lambda \in \mathbb{C}$, without changing the wave functions associated with the triple. What is required for the wave functions to be well defined is that $s\alpha$ should be a single valued global section of $\tilde{B}_\Lambda = B_\Lambda \otimes \sqrt{\Delta_\Lambda}$. This bundle is also locally constant, and we should properly regard a WKB eigenfunction with eigenvalues E_a as a constant section of $\tilde{B}_\Lambda$, where Λ is the Lagrangian torus $h_a = E_a$.

The existence of such a section for given values of the Es is the *corrected Bohr-Sommerfeld condition*. It is the condition that $\tilde{B}_\Lambda$ should be trivial as a locally constant bundle.

Suppose that $B = M \times \mathbb{C}$, with connection $\mathrm{d} - \mathrm{i}\hbar^{-1}\theta$, where θ is a symplectic potential. Then the corrected condition is that

$$\exp\left(\frac{2\mathrm{i}}{\hbar}\int\theta\right)\kappa,$$

which is a section of Δ_Λ, should have a single-valued square root with values in $\sqrt{\Delta_\Lambda}$. Here the integral is taken from any fixed point on Λ.

One can put this in a more explicit form when M is a symplectic vector space with linear canonical coordinates p_a, q^b and $\theta = p_a\mathrm{d}q^a$. Let J be the standard complex structure, with respect to which $z^a = p_a + \mathrm{i}q^a$ is a holomorphic coordinate system. Put

$$\mu_0 = \mathrm{d}^n z = \mathrm{d}z^1 \wedge \cdots \wedge \mathrm{d}z^n.$$

Then $\mu_0 \in K_J$ and we can pick out a nonzero $\mu(P) \in K_P$ for each $P \in LV$ by imposing $(\mu(P), \mu_0) = 1$. This defines a nonvanishing global section of $K \to LV$, and enables us to identify each K_P with $\mathbb{C}$.

The metaplectic structure in this case is unique. If we take $\nu_0 \in \delta_J$ to be one of the square roots of μ_0, and define $\nu(P) \in \delta_P$ for each $P \in LV$ by $(\nu(P), \nu_0) = 1$, then we obtain a similar trivialization of $\delta \to LV$.

Let α_0 be the section of $\sqrt{\Delta_\Lambda}$ such that $\big(\nu(T_m\Lambda), \alpha_0\big) = 1$ for each $m \in \Lambda$; α_0 is a global, nonvanishing section, but in general it is not constant with respect to the flat connection. Define $\chi \in C^\infty_{\mathbb{C}}(\Lambda)$ by $\kappa = \chi\,\mathrm{d}^n z|_\Lambda$. Then $\alpha = \sqrt{\chi}\,\alpha_0$ is a square root of κ, and so is locally constant, but it may not be single-valued. The corrected Bohr-Sommerfeld condition is that

$$\chi\exp\left(\frac{2\mathrm{i}}{\hbar}\int\theta\right) \in C^\infty_{\mathbb{C}}(\Lambda)$$

should have a continuous square root over the whole of Λ. Thus the standard Bohr-Sommerfeld condition is corrected by terms involving the winding numbers[9] of $\chi : \Lambda \to \mathbb{C} - \{0\}$ around the closed loops in Λ.

Example. (Simms 1973). Consider the problem of constructing a WKB eigenfunction of the action variables

$$j_a = \tfrac{1}{2}\left((p_a)^2 + (q^a)^2\right)$$

on one of the Lagrangian tori $\Lambda = \{j_a = E_a\}$, where the Es are constants. Let ϕ^a be the angle variables (§4.7). Then $\omega = \mathrm{d}j_a \wedge \mathrm{d}\phi^a$ and

$$z^a = \sqrt{2j_a}\,\mathrm{e}^{\mathrm{i}\phi^a}$$

(without summation). The angle variables are coordinates on Λ and $\kappa = \mathrm{d}^n\phi$. In this case $\mathrm{d}^n z|_\Lambda = \mathrm{i}^n \prod_1^n z^a \mathrm{d}^n\phi$ and

$$\theta|_\Lambda = \sum_{a=1}^{n} 2E_a \cos^2\phi^a \mathrm{d}\phi^a.$$

Hence $\chi^{-1} = \mathrm{i}^n \prod_1^n z^a$ and the corrected Bohr-Sommerfeld condition is that, for each a,

$$-2\pi\mathrm{i} + \frac{4\mathrm{i}E_a}{\hbar}\int_0^{2\pi} \cos^2\phi\, \mathrm{d}\phi = 4N_a\pi\mathrm{i}$$

for some $N_a \in \mathbb{Z}$. That is $E_a = (N_a + \frac{1}{2})\hbar$, which leads to the correct energy levels for the isotropic oscillator in n dimensions.[10]

Hannay's angles and Berry's phase

Berry's phase has already been mentioned briefly in §10.2. It is a very general phenomenon that can arise whenever a quantum system is subjected to a cyclic change by varying external parameters, such as the direction of a magnetic field.[11] If the parameters are brought back to their original values through a closed loop, then the phase of the wave function does not in general return to its original value. Simon (1983) interpreted this phase shift as the holonomy of a connection on a bundle over the parameter space. In the case considered by Berry, the evolution is adiabatic and the system remains in an eigenstate of the Hamiltonian as the parameters are slowly altered. The fibres of the bundle are the eigenspaces and the connection arises from the embedding in the Hilbert space of states (§A.7). Aharonov and Anandan (1987) set Berry's phenomenon in a more general context, and extended Simon's geometric interpretation.

Hannay discovered a classical counterpart in integrable systems in which the trajectories are given by linear flows on the leaves of a real polarization. If we have a family of such systems depending on some external parameters (such as the family of oscillators in §10.2), and if the parameters are changed slowly through a closed loop, then the angle variables on the classical trajectories are again shifted by an amount that can be interpreted as the holonomy of a connection. Berry pointed out how the two phenomena are related through the WKB approximation.

The relationship can be understood in the context of the our geometric treatment of WKB wave functions by using ideas of Weinstein and Anandan (much of the following is from Weinstein 1990). In fact, we shall see that the passage from Hannay's theory to Berry's is an example of prequantization. The discussion will be very informal and will pay even less heed than usual to analytic niceties.

A pair (Λ, κ), in which Λ is a Lagrangian submanifold of M and κ is a volume element on Λ such that $\int \kappa = 1$, can be thought of as an 'ensemble of classical states' in which a set of observables in involution take definite values, with κ interpreted as the density of states. In Hannay's theory Λ is a toroidal leaf of the polarization and κ is the volume element determined by the angle variables. His analysis uses the averaging principle (Arnol'd 1978), according to which one can determine the motion as the parameters are slowly altered by averaging with respect to κ rather than by looking at trajectories individually.

For simplicity, we shall take $M = T^*Q$, $B = T^*Q \times \mathbb{C}$ to be the natural prequantization, P to be the vertical polarization, and θ to be the canonical 1-form. Then the sections of B are complex functions on M, and for any Lagrangian submanifold $\Lambda \subset M$, covariantly constant unit sections of B_Λ are functions on Λ of the form $\mathrm{e}^{\mathrm{i}w}$, where w is a phase function.

Suppose that Λ satisfies the corrected quantization condition. Then the space $\mathcal{B}_\Lambda$ of covariantly constant sections of $\tilde{B}_\Lambda = B_\Lambda \otimes \sqrt{\Delta_\Lambda}$ is one-dimensional (the flat connection on $\sqrt{\Delta_\Lambda}$ is such that a section α is covariantly constant if α^2 is a constant multiple of κ). We thus have a line bundle $\mathcal{B}$ over the manifold $\mathcal{M}$ of pairs (Λ, κ) that satisfy the corrected quantization condition. The elements of $\mathcal{B}_\Lambda$ can be represented as WKB wave functions (Λ, s, α), where, as before, we regard two triples as the same if the sections $s\alpha$ of $\tilde{B}_\Lambda$ are the same. We shall use this freedom to make α^2 a constant real multiple of κ and s a covariantly constant section of B_Λ with $(s, s) = 1$.

We can represent $\mathcal{X} \in T_{(\Lambda,\kappa)}\mathcal{M}$ by a pair (X, ξ), where X is a section of TM_Λ (the restriction of the tangent bundle to Λ) and $\xi \in \Omega^n(\Lambda)$: if $t\mathcal{X}$ is the displacment vector from (Λ, κ) to a nearby pair (Λ', κ'), then, to the first order in t, Λ' is obtained from Λ by displacing $m \in \Lambda$ through $tX(m)$ to $\rho(m) \in \Lambda'$, and $t\xi = \rho^*\kappa' - \kappa$ (Fig. 10.3). The Lagrangian condition on Λ' implies that $X \lrcorner \omega|_\Lambda$ is closed. It must also be exact since both Λ and Λ' satisfy the corrected quantization condition: this implies that the values of $\oint \theta$ are the same on Λ and Λ', and hence that

$$\oint X \lrcorner \omega = 0$$

for any closed curve in Λ vanishes. It follows that

$$X \lrcorner \omega|_\Lambda = \mathrm{d}x \quad \text{for some } x \in C^\infty(\Lambda).$$

Also since both (Λ, κ) and (Λ', κ') have unit volume,

$$\int_\Lambda \xi = 0.$$

The representation of $\mathcal{X}$ by (X, ξ) is not unique: we can replace $(\mathcal{X}, \xi)$ by $(X + Y, \xi + \mathcal{L}_Y\kappa)$, where Y is any tangent vector field to Λ.

The association of P wave functions with triples (Λ, s, α) embeds $\mathcal{B}$ in $\mathcal{H}_P$ (ignoring singularities, the fact that the P wave functions are defined only in the limit $\hbar \to 0$, and so on). The Berry connection on $\mathcal{B}$ is induced by this embedding. To find it explicitly, we choose a nonvanishing local section σ of $\mathcal{B}$ and take the inner product in $\mathcal{H}_P$ of nearby values of σ.

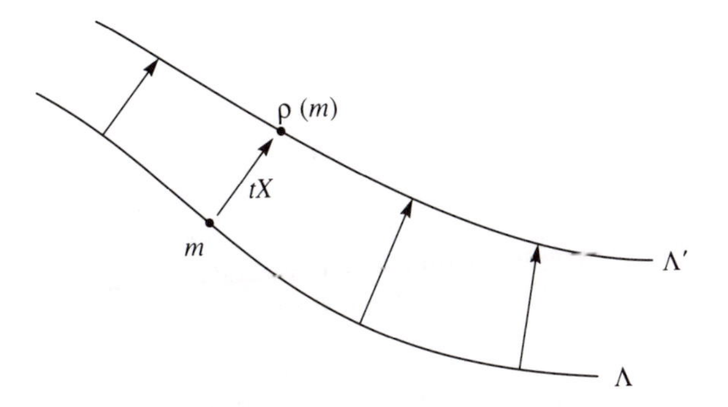

Fig. 10.3. A tangent vector to $\mathcal{M}$.

We define σ as follows. Pick $(\Lambda, s, \alpha) \in \mathcal{B}$, such that $\kappa = \alpha^2$ and $s = 1$ at some point $m_0 \in \Lambda$. Choose an open subset Λ_0 of the leaf of P through m_0 so that Λ_0 intersects each Lagrangian submanifold near Λ in exactly one point. We take the value of σ at (Λ, κ) to be (Λ, s, α), and its value at nearby points $(\Lambda', \kappa') \in \mathcal{M}$ to be (Λ', s', α'), where $\alpha' = \sqrt{\kappa'}$ changes continuously with κ' and s' is fixed by $s' = 1$ on Λ_0, together with the condition that $s'\alpha'$ should be covariantly constant on Λ'.

Away from its singularities, the value at q of the P wave function associated with (Λ, s, α) is a sum of the form

$$\sum A e^{iw/\hbar} \sqrt{d^n q}$$

with the different branches corresponding to the various intersection points of Λ with the leaf of P above q (Fig. 10.4). When, following Berry (1985), we take the inner product with the wave function

$$\sum A' e^{iw'/\hbar} \sqrt{d^n q}$$

determined by $\sigma(\Lambda', \kappa') = (\Lambda', s', \alpha')$, where $(\Lambda', \kappa') \in \mathcal{M}$ is near (Λ, κ), the cross-terms from different branches cancel (by the stationary phase principle), and we are left with

$$\sum \int \overline{A} A' e^{i(w'-w)/\hbar} d^n q,$$

where again the sum is over the different branches. This can be expressed as an integral over Λ, as follows. Suppose that the displacement vector from

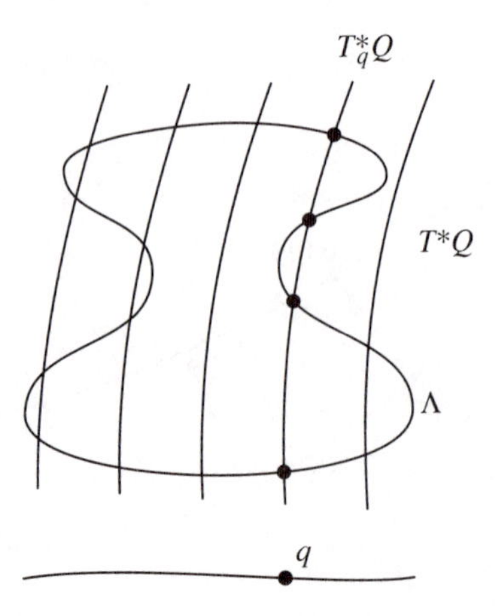

Fig. 10.4. The branches of the P wave function corrrespond to the intersection points of T_q^*Q with Λ.

(Λ, α^2) to (Λ', α'^2) is $t\mathcal{X}$, where t is some small parameter. Let (X, ξ) be a representative of $\mathcal{X}$ such that X is tangent to P (X will be singular at points where Λ is not transverse to P, but this does not affect the argument). Then $X \lrcorner\, \omega|_\Lambda = \mathrm{d}x$ for some $x \in \mathrm{C}^\infty(\Lambda)$, which is fixed uniquely by imposing the condition that $x = 0$ at $\Lambda \cap \Lambda_0$. With this choice, we have $w - w' = tx$ and $(A'^2 - A^2)\mathrm{d}^n q = t\xi$, and the inner product becomes

$$1 + t \int_\Lambda (\tfrac{1}{2}\xi - \hbar^{-1}\mathrm{i}x\kappa) = 1 - \frac{\mathrm{i}t}{\hbar}\int_\Lambda x\kappa$$

to the first order in t. Therefore the connection $\mathcal{M}$ is $\mathrm{d} - \mathrm{i}\hbar^{-1}\Theta$, where

$$\mathcal{X} \lrcorner\, \Theta = \int_\Lambda x\kappa.$$

This depends on our choice for the local section of $\mathcal{B}$, but the curvature $\hbar^{-1}\Omega = \hbar^{-1}\mathrm{d}\Theta$ does not. It is given by

$$2\Omega(\mathcal{X}, \tilde{\mathcal{X}}) = \int_\Lambda (x\tilde{\xi} - \tilde{x}\xi),$$

where the representatives (X, ξ) and $(\tilde{X}, \tilde{\xi})$ are chosen so that X and $\tilde{X}$ are tangent to P. It is not hard to see that Ω is nondegenerate. It is also clear—at least at the rather informal level at which we are working—that Ω does not depend on P since the pairing maps between the Hilbert spaces of different real polarizations are unitary (to the leading order in $\hbar$) and intertwine the embeddings of $\mathcal{B}$: Ω is therefore a natural 2-form on $\mathcal{M}$.

Weinstein (1990) shows this explicitly by giving a formula for Ω that does not involve P.

Thus, at the semiclassical level, Berry's phase can be understood as the holonomy of a prequantum bundle $\mathcal{B}$ over the symplectic manifold $(\mathcal{M}, \Omega)$.

Hannay's angles can also be derived very simply from Ω. Suppose that the system is integrable, and has action-angle coordinates j_a, ϕ^b. The original state of the system is represented by (Λ, κ), where Λ is one of the tori $j =$ constant and $\kappa = \mathrm{d}^n\phi$: there is one such state for each set of values of the js. Under variation of the external parameters, (Λ, κ) passes around a loop in $\mathcal{M}$, and Hannay's angles are given by

$$\Delta\phi^a = -\frac{\partial}{\partial j_a}\int \Omega,$$

where the integral is over a surface spanning the loop. Weinstein's interpretation suggests generalizations that do not involve assumptions about integrability.

There is another way of defining Ω without using P. It avoids the awkwardness of the singularities in X and it is closer to Anandan's approach. Let T be a fixed n-torus with a volume element μ such that $\int \mu = 1$ (we could use some other manifold instead of T, but the torus is what actually arises in Hannay's example). Let $\mathcal{C}$ be the space of embeddings $\rho : T \to M$ such that

(1) $\rho(T)$ is Lagrangian; and

(2) the n action variables

$$j_a = \frac{1}{2\pi}\oint \theta$$

take given values on $\rho(T)$.

A tangent vector $\mathcal{X} \in T_\rho\mathcal{C}$ is represented uniquely by a section X of TM_Λ, where $\Lambda = \rho(T)$, such that $X \lrcorner w|_\Lambda$ is exact.

We define a 1-form Θ' on $\mathcal{C}$ by

$$\mathcal{X} \lrcorner \Theta' = \int_{\rho(T)} (X \lrcorner \theta)\kappa$$

where $\rho^*\kappa = \mu$. Then $\mathrm{d}\Theta' = \Omega'$, where

$$\Omega'(\mathcal{X}, \mathcal{X}') = \int_{\rho(T)} \omega(X, X')\kappa$$

This form is closed, by construction, but degenerate: it is annihilated at ρ by any $\mathcal{X}$ such that X is tangent to $\rho(T)$ and $\mathcal{L}_X\kappa = 0$. Reduction

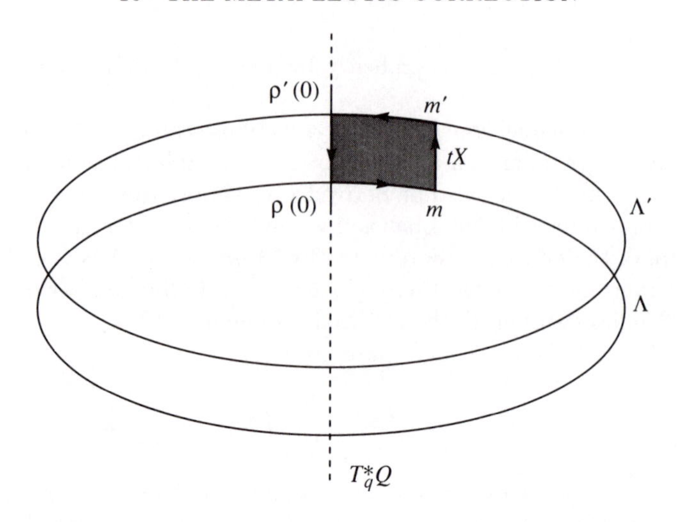

Fig. 10.5. The calculation of the gradient of $\mathcal{F}$.

identifies embedding maps that have the same image Λ and that induce the same volume form κ on Λ. Therefore the reduced phase space of $\mathcal{C}$ is a component of $\mathcal{M}$. The reduced symplectic structure is Ω, as can be seen either from Weinstein's formula for Ω, or as follows.

Introduce cyclic coordinates $\phi^1, \ldots, \phi^n$ on T so that

$$T = [-\pi, \pi] \times \cdots \times [-\pi, \pi],$$

with identification modulo 2π, and $\mu = \mathrm{d}^n\phi/(2\pi)^n$. Let $\mathcal{C}_0$ be the submanifold of $\mathcal{C}$ on which $\rho(O) \in \Lambda_0$, where O is the origin in T. This intersects every leaf of the characteristic foliation in some neighbourhood of Λ. For each $\rho \in \mathcal{C}_0$, let w be the phase function on $\Lambda = \rho(T)$ such that $w = 0$ at $\rho(O)$. We can make w single valued and well defined almost everywhere on Λ by excluding the subsets $\phi^a = \pi$.

Define $\mathcal{F} : \mathcal{C}_0 \to \mathbb{R}$ by

$$\mathcal{F}(\rho) = \left(\frac{1}{2\pi}\right)^n \int_{-\pi}^{\pi} \cdots \int_{-\pi}^{\pi} w \circ \rho \, \mathrm{d}^n\phi.$$

Suppose that $\mathcal{X}$ is tangent to $\mathcal{C}_0$ at ρ. Then X is tangent to Λ_0 at $m_0 = \rho(O)$. We shall calculate $\mathcal{X} \lrcorner \mathrm{d}\mathcal{F}$, as follows.

Let ρ' be the map obtained from ρ by displacement through a small parameter along $\mathcal{X}$. Let $\phi \in T$ and put $m = \rho(\phi) \in \Lambda = \rho(T)$ and $m' = \rho'(\phi) \in \Lambda' = \rho'(T)$. Then, to the first order in t, the displacement vector from m to m' is tX and

$$w(m) - w(m') = tX \lrcorner \theta(m_0) - tX \lrcorner \theta(m) + \oint \theta,$$

where $m_0 = \rho(O)$ and the integral is around the path shown in Fig. 10.5. But $X \lrcorner\, \theta(m_0) = 0$ because X is tangent to Λ_0. Hence

$$w(m) - w(m') = t \int_{m_0}^{m} X \lrcorner\, \omega - tX \lrcorner\, (m) = tx(m) - tX \lrcorner\, \theta(m),$$

where $x \in C^\infty(\Lambda)$ is defined by $X \lrcorner\, \omega = \mathrm{d}x$ and $x(m_0) = 0$. By integrating over Λ,

$$\mathcal{X} \lrcorner\, \mathrm{d}\mathcal{F} = \mathcal{X} \lrcorner\, (\pi^*\Theta - \Theta'),$$

where π is the reduction map from $\mathcal{C}$ to $\mathcal{M}$. It follows that

$$\pi^*\Theta|_{\mathcal{C}_0} = \Theta'|_{\mathcal{C}_0} + \mathrm{d}\mathcal{F} \quad \text{and hence that} \quad \pi^*\Omega|_{\mathcal{C}_0} = \Omega'|_{\mathcal{C}_0}.$$

Therefore the reduced symplectic structure is Ω.

This construction emphasises the analogy between the classical and quantum phenomena: the 2-form Ω arises from the embedding of Lagrangian tori in a symplectic manifold in very much the same way that the Berry connection arises from the embedding of eigenspaces in the Hilbert space of all states.

10.6 Cohomological wave functions

It often turns out that the Hilbert space associated with a polarization is trivial. We have seen an example in §9.1: when M is compact and the polarization is Kähler, but nonpositive, there are no polarized sections of the prequantum bundle. Another example is a real polarization with compact leaves. The wave functions must be covariantly constant along the leaves, but only on the leaves that satisfy the corrected Bohr-Sommerfeld condition is it posssible to have non-vanishing constant sections. These leaves are given by isolated values of the action variables, so the wave functions must vanish almost everywhere and cannot be smooth.

In both these cases, it is possible to construct 'cohomological wave functions' which, in spite of their somewhat abstract definition, seem to represent the physical quantum states.

Let P be an admissible polarization of (M, ω) and let B be a prequantization of M. When P is Kähler, B is a holomorphic bundle and the cohomological wave function spaces are the various Dolbeault cohomology groups of $\overline{\partial}$-closed forms with values in B. For a general polarization, they are defined as follows. Let $\alpha \in \Omega^k_B(M)$; that is, α is a k-form on M with values in B. We shall represent α as an ordinary complex form, which transforms under gauge transformations of the symplectic potential θ by

$$\alpha \mapsto \alpha' = \mathrm{e}^{\mathrm{i}u/\hbar}\alpha \quad \text{when} \quad \theta \mapsto \theta' = \theta + \mathrm{d}u.$$

Definition (10.6.1). Let $X \in V(M)$ and $\alpha \in \Omega_B^k(M)$. The covariant exterior derivative $\mathrm{D}\alpha \in \Omega_B^{k+1}(M)$ and the covariant Lie derivative $\mathrm{L}_X\alpha \in \Omega_B^k(M)$ are defined by

$$\begin{aligned}\mathrm{D}\alpha &= \mathrm{d}\alpha - \frac{\mathrm{i}}{\hbar}\theta \wedge \alpha \\ \mathrm{L}_X\alpha &= \mathrm{D}(X \lrcorner \alpha) + X \lrcorner \mathrm{D}\alpha = \mathcal{L}_X\alpha - \frac{\mathrm{i}}{\hbar}(X \lrcorner \theta)\alpha.\end{aligned}$$

Note that

$$\mathrm{D}^2\alpha = -\mathrm{i}\hbar^{-1}\omega \wedge \alpha \quad \text{and} \quad \mathrm{L}_X(\mathrm{D}\alpha) - \mathrm{D}(\mathrm{L}_X\alpha) = -\frac{\mathrm{i}}{\hbar}(X \lrcorner \omega) \wedge \alpha. \quad (10.6.2)$$

Although $\mathrm{D}^2 \neq 0$, $\mathrm{D}^2\alpha|_{\overline{P}} = 0$, for any α; and if $\alpha|_{\overline{P}} = 0$, then $\mathrm{D}\alpha|_{\overline{P}} = 0$ since P is integrable. Therefore, so long as we only look at the restrictions of forms to $\overline{P}$, D behaves like an ordinary exterior derivative: it gives the bundles $\bigwedge_B^k \overline{P}^*$ the structure of a differential complex (Rawnsley, Schmid and Wolf 1983, Vaisman 1981, Puta 1986).

Definition (10.6.3). A form $\alpha \in \Omega_B^k(M)$ is said to be $\overline{P}$-closed whenever $\mathrm{D}\alpha|_{\overline{P}} = 0$, and to be $\overline{P}$-exact whenever there exists β such that $(\alpha - \mathrm{D}\beta)|_{\overline{P}} = 0$. The kth $\overline{P}$-cohomology group with coefficients in B is the space $H^k(M, \overline{P}, B)$ of equivalence classes of $\overline{P}$-closed forms in $\Omega_B^k(M)$, where two forms are regarded as equivalent if they differ by a $\overline{P}$-exact form.

We shall think of the elements of $H^k(M, \overline{P}, B)$ as 'cohomological wave functions'. Two $\overline{P}$-closed k-forms with values in B determine the same class in $H^k(M, \overline{P}, B)$ whenever they differ by $\mathrm{D}\beta + \gamma$, where $\beta \in \Omega_B^{k-1}(M)$ and $\gamma \in \Omega_B^k(M)$, with $\gamma|_{\overline{P}} = 0$.

Except in special cases, there is no natural way to define inner products and transition amplitudes in $H^k(M, \overline{P}, B)$. We can, however, extend the definition of the prequantum operator $\hat{f}$ associated with a classical observable $f \in C^\infty(M)$ to $\Omega_B^k(M)$ by putting

$$\hat{f}(\alpha) = -\mathrm{i}\hbar \mathrm{L}_{X_f}\alpha + f\alpha.$$

The calculation in §9.2 shows that $[\hat{f}_1, \hat{f}_2] = -\mathrm{i}\hbar \hat{f}_3$ whenever $[f_1, f_2] = f_3$.

It follows from (10.6.2) that $\mathrm{D}\hat{f}(\alpha) = \hat{f}(\mathrm{D}\alpha)$. Moreover, if the flow of X_f preserves P, then

$$\mathrm{L}_{X_f}\gamma|_{\overline{P}} = 0 \quad \text{whenever} \quad \gamma|_{\overline{P}} = 0.$$

So $\hat{f}$ is well defined as an operator on $H^k(M, \overline{P}, B)$ whenever the flow of X_f preserves P.

Since P is admissible, it is possible to choose the symplectic potentials to be adapted to P. In the corresponding local trivializations, α is $\overline{P}$-closed if $\mathrm{d}\alpha|_{\overline{P}} = 0$, and α is $\overline{P}$-exact if $\alpha|_{\overline{P}} = \mathrm{d}\beta|_{\overline{P}}$ for some β. This observation allows us to extend the definition of $H^k(M, \overline{P}, B)$ to any polarized line bundle. That is, to any line bundle which has a preferred class of local trivializations with transition functions that are constant along $\overline{P}$. The prequantum bundle itself is polarized; so is the half-form bundle δ_P, when it exists, and the tensor product $B_P = B \otimes \delta_P$. In all cases, the preferred local trivializations are given by local sections that are covariantly constant along $\overline{P}$, either with respect to the connection on B, or the partial connection on δ_P or B_P.

The operator corresponding to a real observable $f \in S^1_P(M)$ also extends to an operator $\tilde{f}$ on the 'corrected cohomological wave function spaces' $H^k(M, \overline{P}, B_P)$. Its action on a representative form with values in $B \otimes \delta_P$ is defined by writing the form as $\alpha\nu$ where $\alpha \in \Omega^k_B(M)$ and ν is a section of δ_P, and putting

$$\tilde{f}(\alpha\nu) = \hat{f}(\alpha)\nu - \mathrm{i}\hbar\alpha\mathcal{L}_{X_f}\nu.$$

So far we have defined δ_P only for nonnegative polarizations. We shall not consider the general definition (Blattner and Rawnsley 1986, Robinson 1987), but only remark that in the case of a negative Kähler polarization, that is, the complex conjugate of a positive polarization, $K_P = \overline{K_{\overline{P}}}$ and δ_P is *defined* to be $\overline{\delta}_{\overline{P}}$.

Example. *Real polarizations with compact leaves.* Suppose that M is a bundle over an n-dimensional manifold Q, with fibres which are the leaves of a real polarization P with compact leaves. Then $H^k(M, \overline{P}, B)$ is nontrivial only when $k = n$. The cohomological wave functions in $H^n(M, \overline{P}, B)$ have the character of delta-functions on Q: they are concentrated on the leaves of P that satisfy the Bohr-Sommerfeld condition

$$\frac{1}{2\pi\hbar}\oint \theta \in \mathbb{Z}$$

for every closed curve—in other words, on the leaves on which the holonomy of the prequantum connection is trivial (Śniatycki 1977).

To see how this works, we introduce action-angle coordinates j_a, ϕ^b in a neighbourhood of one of the leaves, and use the symplectic potential $\theta = j_a \mathrm{d}\phi^a$ to trivialize B. The leaves of P are Lagrangian tori, labelled by the js. The ϕs run from 0 to 2π around the generating circles and the Bohr-Sommerfeld leaves are those on which the js are integral multiples of $\hbar$.

Consider first the case $n = 1$, in which (M, ω) is two-dimensional and P is a foliation by circles. There is one action coordinate j and one angle coordinate ϕ, and $\omega = \mathrm{d}j \wedge \mathrm{d}\phi$.

The elements of $H^1(M, \overline{P}, B)$ have representatives $f(j, \phi)\mathrm{d}\phi$, where f is periodic in ϕ with period 2π. Two such forms $f\mathrm{d}\phi$ and $f'\mathrm{d}\phi$ define the same class whenever there exists a smooth function $g(j, \phi)$ with the same periodicity such that

$$(f - f')\mathrm{d}\phi = \mathrm{D}g = \left(\frac{\partial g}{\partial \phi} - \frac{\mathrm{i}j}{\hbar} g\right) \mathrm{d}\phi.$$

The existence of such a g is a local problem in j, so nothing is lost by working in a single coordinate neighbourhood.

The general solution to the differential equation for g is

$$g(j, \phi) = \mathrm{e}^{\mathrm{i}j\phi/\hbar} \left(\int_0^\phi (f(j, \phi') - f'(j, \phi'))\mathrm{e}^{-\mathrm{i}j\phi'/\hbar} \, \mathrm{d}\phi' - C(j) \right)$$

where C is a function of j. To determine whether f and f' are equivalent, therefore, we must decide whether or not it is possible to choose C to make g periodic in ϕ. Away from the Bohr-Sommerfeld leaves—the leaves on which j is an integral multiple of $\hbar$—this can be done by putting

$$C(j) = \frac{1}{1 - \mathrm{e}^{-2\pi \mathrm{i}j/\hbar}} \int_0^{2\pi} (f(j, \phi') - f'(j, \phi'))\mathrm{e}^{-\mathrm{i}j\phi'/\hbar} \, \mathrm{d}\phi'.$$

However, C extends smoothly to the Bohr-Sommerfeld leaves if and only if

$$\int_0^{2\pi} f(j, \phi)\mathrm{e}^{-\mathrm{i}j\phi/\hbar} \mathrm{d}\phi = \int_0^{2\pi} f'(j, \phi)\mathrm{e}^{-\mathrm{i}j\phi/\hbar} \, \mathrm{d}\phi$$

whenever $j = N\hbar$ for some $N \in \mathbb{Z}$.

This condition can be expressed in a gauge-invariant form. If Λ is a Bohr-Sommerfeld circle, then $B|_\Lambda$ is trivial as a line bundle with connection: in the $j\mathrm{d}\phi$ gauge, $\mathrm{e}^{-\mathrm{i}j\phi/\hbar}$ is a global parallel section. There is therefore a natural identification of $B|_\Lambda$ with $\Lambda \times \mathbb{C}$, up to an undetermined constant phase. If we use the identification to define the integral around Λ of a B-valued form, then the condition becomes: two forms $\alpha, \alpha' \in \Omega^1_B(M)$ define the same class in $H^1(M, \overline{P}, B)$ if and only if

$$\oint \alpha = \oint \alpha' \tag{10.6.4}$$

for every Bohr-Sommerfeld circle. It is straightforward to extend this to a general value of n: $\alpha, \alpha' \in \Omega^n_B(M)$ are equivalent whenever (10.6.4) holds for every leaf such that B_Λ is trivial as a bundle with connection. Thus

$H^n(M, \overline{P}, B) = \mathbb{C}^N$, with one copy of $\mathbb{C}$ for each Bohr-Sommerfeld leaf (N may be infinite if Q is not compact).

The same is true if we replace B by $B_P = B \otimes \delta_P$, except that the Bohr-Sommerfeld condition is replaced by the corrected version. The obvious application is to the isotropic oscillator. When $n = 1$, for example, we have a one-dimensional space of cohomological wave functions associated with each circle on which the energy is an odd integral multiple of $\frac{1}{2}\hbar$ (Simms 1977).

With half-forms included, the correspondence between cohomological wave functions and distributions supported on leaves of the polarization can be made precise. Suppose that $\alpha \in \Omega^n_{B_P}(M, \overline{P})$ represents a class in $H^n(M, \overline{P}, B_P)$ and that Λ is a leaf of P such that the corrected Bohr-Sommerfeld condition holds. Then $B_P|_\Lambda$ has a covariantly constant section $s\nu$, which we can use to define a trivialization. By using the freedom in the choice of representatives, we can arrange that in this trivialization, $\alpha|_\Lambda = f\mathrm{d}\phi^1 \wedge \cdots \wedge \mathrm{d}\phi^n$, where f is constant, that is, $\alpha|_\Lambda$ is harmonic. Then $\mu = f\nu\mathrm{d}\phi^1 \wedge \cdots \wedge \mathrm{d}\phi^n$ is a section of $\sqrt{\Delta_\Lambda}$, and (Λ, s, μ) is a WKB distributional wave function which is naturally associated with the class of α in $H^n(M, \overline{P}, B)$. ∎

Example. *Negative Kähler polarizations.* Suppose that M is compact and that P is a negative Kähler polarization. Then B is a holomorphic bundle. Since P and $\overline{P}$ span the tangent space to M at each point, we can arrange that the representative forms of the classes in $H^k(M, \overline{P}, B)$ are $\overline{\partial}$-closed $(0,k)$-forms, where $\overline{\partial}\alpha$ is defined to be the $(0, k+1)$-part of $\mathrm{D}\alpha$. In a local holomorphic trivialization, $\overline{\partial}$ coincides with the standard Dolbeault operator. Therefore $H^k(M, \overline{P}, B)$ is the kth Dolbeault cohomology group of M with coefficients in sections of B.

There is a natural inner product on the $(0, k)$ forms with values in B, given in the gauge of a real symplectic potential by

$$\langle \alpha, \beta \rangle = \mathrm{i}^k(-1)^{\frac{1}{2}k(k-1)+n} \left(\frac{1}{2\pi\hbar}\right)^{n/2} \int \overline{\alpha} \wedge \beta \wedge \omega^{n-k}, \tag{10.6.5}$$

where α, β are $(0, k)$ forms with values in B. It is invariant under gauge transformations between real potentials, which only change the phase of α and β. This enables us to pick a unique *harmonic representative* in each class, which minimizes $\langle \alpha, \alpha \rangle$; equivalently, α is harmonic if $\overline{\partial}^*\alpha = 0$, where $\overline{\partial}^*$ is defined by

$$\langle \beta, \overline{\partial}^*\alpha \rangle = \langle \overline{\partial}\beta, \alpha \rangle$$

for any $(0, k-1)$-form β with values in B. See, for example, Wells 1973, Vaisman 1973, or Griffiths and Harris 1978.

By taking harmonic representatives, we can define an inner product on $H^k(M, \overline{P}, B)$, and hence the normal quantum-mechanical apparatus of inner products and transition amplitudes. The same is true if we replace B by $B_P = B \otimes \delta_P$: the only difference is that in the definition of the inner product (10.6.5), α and β take values in B_P and $\overline{\alpha} \wedge \beta$ is made into an ordinary $2k$-form on M by using the Hermitian structure in the fibre of B_P.

In the case $k = n$, there is a natural isomorphism (Serre duality) between $H^n(M, \overline{P}, B_P)$ and the Hilbert space obtained by applying the ordinary metaplectic quantization to $\overline{P}$, which is a positive polarization. A $(0,n)$-form with values in B_P is a section of

$$B \otimes \delta_P \otimes \bigwedge{}^n \overline{P}^* = B \otimes \delta_P \otimes K_{\overline{P}} = B \otimes (\delta_P \otimes \delta_{\overline{P}}) \otimes \delta_{\overline{P}}.$$

By definition, $\delta_P = \overline{\delta}_{\overline{P}}$. Since δ_P is Hermitian, therefore, $\delta_P \otimes \delta_{\overline{P}} = \mathbb{C}$. Consequently a $(0,n)$-form with values in B_P is the same as a section of $B \otimes \delta_{\overline{P}}$. The harmonic forms precisely correspond to the sections polarized relative to $\overline{P}$. The role played here by half-forms is further evidence for the metaplectic correction.

In the case of the orbits of a compact group, the construction fits into a general pattern of correspondences between the cohomological wave function spaces of the various polarizations of the orbits (Bott 1957). ∎

Example. *The Penrose transform.* We remarked in §6.8 that the phase space M^+_{s0} of a massless particle with helicity $s > 0$ has a conformally invariant nonpositive Kähler polarization. It is the reduction of the polarization of twistor space spanned by the holomorphic vector fields

$$\frac{\partial}{\partial \omega^A} \quad \text{and} \quad \frac{\partial}{\partial \pi_{A'}}.$$

The corresponding cohomological wave functions are central to twistor theory, and are related to the positive frequency solutions of the corresponding massless field equation by the *Penrose transform* (see, for example, Penrose and Rindler 1986, Baston and Eastwood 1989, Ward and Wells 1990). I shall do no more here than sketch the connection with quantization theory.

We begin by relabelling the coordinates on $\mathbb{T}$, by putting

$$(Z^0, Z^1, Z^2, Z^3) = (\omega^0, \omega^1, \pi_0, \pi_1),$$

and introducing a (nonpositive) Hermitian form

$$g_{\alpha\beta} Z^\alpha \overline{Z}^\beta = \omega^A \overline{\pi}_A + \overline{\omega}^{A'} \pi_{A'} = \omega^0 \overline{\pi}_0 + \omega^1 \overline{\pi}_1 + \overline{\omega}^0 \pi_0 + \overline{\omega}^1 \pi_1,$$

where $\alpha, \beta = 0, 1, 2, 3$. The symplectic structure on $\mathbb{T}$ and the constraint (6.8.2) are then given by

$$\Omega = -\mathrm{i} g_{\alpha\beta} \mathrm{d}Z^\alpha \wedge \mathrm{d}\overline{Z}^\beta \quad \text{and} \quad g_{\alpha\beta} Z^\alpha \overline{Z}^\beta = 2s.$$

As a complex manifold, M^+_{s0} is an open subset of $\mathbb{CP}_3$. Apart from the fact that the Hermitian form on $\mathbb{T}$ has signature $++--$, the problem is the same as the quantization of a four-dimensional isotropic oscillator, and much of the treatment of the Fubini-Study symplectic structure (p. 235) also applies here.

We shall denote the polarization of M^+_{s0} by P. The sections of the prequantum bundle $B \to M^+_{s0}$ are holomorphic functions of the Zs of degree $-2s/\hbar$. The sections of the canonical bundle are holomorphic functions of degree -4. Although we have not developed the general theory of half-forms for nonpositive polarizations, it is natural in this context to define δ_P so that its sections are homogeneous functions of degree -2.

When $2s/\hbar > -2$, there are no holomorphic sections of $B_P = B \otimes \delta_P$. To get nontrivial wave functions, therefore, we consider $H^1(M^+_{s0}, \overline{P}, B_P)$. The cohomology classes are represented by $(0,1)$-forms on some neighbourhood of C_{s0} in $\mathbb{T}$ satisfying

$$\overline{\partial}\gamma = 0, \quad \overline{E} \lrcorner \gamma = 0, \quad \text{and} \quad E \lrcorner \partial\gamma = \left(-\frac{2s}{\hbar} - 2\right)\gamma,$$

where $E = Z^\alpha \partial/\partial Z^\alpha$. Such forms project onto $(0,1)$-forms on M^+_{s0} with values in B_P.

The correspondence with positive frequency solutions of the massless field equations is obtained as follows. Let z be a point in the *forward tube* of complex Minkowski space. In other words, z is a complex event with coordinates $z^a = x^a - \mathrm{i}y^a$, where y is timelike and future pointing. Then

$$\omega^A = \mathrm{i}z^{AA'}\pi_{A'}$$

determines a two-dimensional subspace of $\mathbb{T}$, which projects onto a copy L_z of $\mathbb{CP}_1$ in $M^+_{s0} \subset \mathbb{CP}_3$. If γ is a representative of a class in $H^1(M^+_{s0}, \overline{P}, B_P)$, then for each fixed set of values of the $2s/\hbar$ spinor indices $A', B', \ldots, C'$,

$$\pi_{A'}\pi_{B'} \ldots \pi_{C'}(\gamma \wedge \pi^{D'}\mathrm{d}\pi_{D'})$$

is a well-defined 2-form on L_z. The Penrose transform of γ is the spinor field on the forward tube defined by

$$\phi_{A'B'\ldots C'} = \int_{L_z} \pi_{A'}\pi_{B'} \ldots \pi_{C'}(\gamma \wedge \pi^{D'}\mathrm{d}\pi_{D'})$$

(Wells 1979). It is easy to check that ϕ depends holomorphically on z, that it depends only on the class of γ, and that $\nabla^{AA'}\phi_{A'B'\ldots C'} = 0$. Thus ϕ is the continuation to the forward tube of a positive frequency solution of the massless field equations.

Both polarizations—the real one considered in §9.4 and the Kähler one considered here—associate M^+_{s0} with the positive frequency solutions of the massless field equations, but by strikingly different constructions. The second route brings out conformal invariance, but the definition of the inner product on the $H^1(M, \overline{P}, B_P)$ is not straightforward, either by *ad hoc* methods, or as part of a more general theory (Rawnsley, Schmid, and Wolf 1983, Baston and Eastwood 1989). ∎

APPENDIX

THE treatment of the material in this appendix is not intended to be rigorous, exhaustive, or self-contained. Its purpose is to record notation and terminology, which are sometimes used in unusual ways, and to save time spent searching elsewhere by recalling standard definitions.

A.1 Notation and definitions

Tensor algebra. Symmetrization and skew symmetrization are denoted, respectively, by round and square brackets. For example,

$$T^{(ab...c)} = \frac{1}{p!}\sum_{\sigma} T^{\sigma(a)\sigma(b)...\sigma(c)} \tag{A.1.1}$$

$$T^{[ab...c]} = \frac{1}{p!}\sum_{\sigma} \operatorname{sign}(\sigma) T^{\sigma(a)\sigma(b)...\sigma(c)}, \tag{A.1.2}$$

where T is a p-index tensor and the summations are over the permutations σ of $a, b, \dots, c$.

The p-fold exterior power of a vector space V is denoted by $\bigwedge^p V$.

Complex structures. A complex vector space can be made into a real vector space of twice the dimension by 'forgetting' how to multiply by complex scalars, so that when X is a nonzero vector, X and $\mathrm{i}X$ are regarded as independent. Multiplication by i becomes a real linear transformation $J : X \mapsto \mathrm{i}X$ with the property that $J^2 = -1$. A *complex structure* on a real vector space V is a linear transformation $J : V \to V$ such that $J^2 = -1$. Complex structures exist only in spaces of even dimension.

Given a real vector space V and a complex structure $J : V \to V$, we can recover a complex vector space. Multiplication by complex scalars is defined by the rule

$$(x + \mathrm{i}y)X = xX + yJX.$$

When it is necessary to emphasize the role of J, we shall use the notation $V_{(J)}$ to denote the complex space constructed in this way—there are situations in which we want to introduce two different complex structures in the same real vector space. The complex dimension of $V_{(J)}$ is half the real dimension of V.

Any real vector space V—whether or not it has a complex structure—can be 'complexified'. The complexification $V_{\mathbb{C}}$ is the set of formal linear combinations $X+\mathrm{i}Y$, where $X, Y \in V$, with the obvious rule for multiplying by a complex scalar $z = x + \mathrm{i}y$:

$$z(X + \mathrm{i}Y) = xX - yY + \mathrm{i}(xY + yX).$$

It can also be defined from the complex structure $(X, Y) \mapsto (-Y, X)$ on the real vector space $V \oplus V$.

The complex dimension of $V_{\mathbb{C}}$ is the same as the real dimension of V. If $Z = X + \mathrm{i}Y \in V_{\mathbb{C}}$, then $\overline{Z}$ denotes the complex conjugate $X - \mathrm{i}Y$. The map $Z \mapsto \overline{Z}$ is *antilinear*, that is $\overline{zZ} = \overline{z}\overline{Z}$, and the Zs such that $Z = \overline{Z}$ are simply the elements of V, regarded as a subset of $V_{\mathbb{C}}$.

If J is a complex structure on a real vector space V, then the vectors $X \in V_{\mathbb{C}}$ such that $JX = \mathrm{i}X$ ($JX = -\mathrm{i}X$) are called $(1,0)$-vectors ($(0,1)$-vectors). Similarly, a 1-form—an element of $V_{\mathbb{C}}^*$—is called a $(1,0)$-form ($(0,1)$-form) if it annihilates $(0,1)$-vectors ($(1,0)$-vectors). An (r,s)-*form* is a linear combination of exterior products, each containing r $(1,0)$-forms and s $(0,1)$-forms. Any $\alpha \in \bigwedge^p V_{\mathbb{C}}^*$ can be written uniquely as a sum of forms of types $(p,0)$, $(p-1,1)$, ..., $(0,p)$.

The complex conjugate of a complex vector space V is denoted by $\overline{V}$. As a set, $\overline{V}$ is a copy of V. We denote the element of $\overline{V}$ corresponding to $X \in V$ by $\overline{X}$. Addition and scalar multiplication by $z \in \mathbb{C}$ are defined by $\overline{X} + \overline{Y} = \overline{X+Y}$ and $z\overline{X} = \overline{\overline{z}X}$. When $V \subset V'_{\mathbb{C}}$, where V' is a real vector space, $\overline{V}$ is naturally identified with the complex conjugate subspace.

We adopt the physicists' convention that a Hermitian inner product $(\cdot,\cdot)$ on a complex vector space is conjugate linear in the first entry and linear in the second entry.

Densities. A *density* on an n-dimensional real or complex vector space is a map δ that assigns a real or complex number to each basis $\mathcal{X} = \{X_1, X_2, \ldots, X_n\}$ in V with the following property. Suppose that $\mathcal{X}'$ is a second basis, and that $X'_a = L^b{}_a X_b$. Then $\delta(\mathcal{X}') = |\det L|\delta(\mathcal{X})$. The space of densities is one-dimensional, and is denoted by $\Delta(V)$. The k-fold tensor product $\Delta(V)\otimes\cdots\otimes\Delta(V)$ is denoted by $\Delta^k(V)$. There is an obvious natural isomorphism $\Delta^2(V) = \bigwedge^n V \otimes \bigwedge^n V$, in the real case, or

$$\Delta^2(V) = \bigwedge^n V \otimes \overline{\bigwedge^n V},$$

in the complex case. If $\alpha \in \bigwedge^n V$, then we define $|\alpha| \in \Delta(V)$ by $|\alpha|(\mathcal{X}) = |\alpha(X_1, X_2, \ldots, X_n)|$.

Real manifolds. All manifolds in this book are smooth (C^∞), Hausdorff, and paracompact. If M is a real manifold, then

TM and T^*M denote the tangent and cotangent bundles;

$T_{\mathbb{C}}M$ and $T_{\mathbb{C}}^*M$ denote the complexified tangent and cotangent bundles;

$C^\infty(M)$ and $C_{\mathbb{C}}^\infty(M)$ are the spaces of real and complex smooth functions on M;

$V(M)$ and $V_{\mathbb{C}}(M)$ are the spaces of real and complex vector fields on M;

$\Omega^p(M)$ and $\Omega^p_{\mathbb{C}}(M)$ are the spaces of real and complex p-forms.

If $\alpha \in \Omega^p(M)$, $\beta \in \Omega^q(M)$, and $X, Y \in V(M)$, then $\alpha \wedge \beta$ denotes the exterior product, $\mathcal{L}_X\alpha$ denotes the Lie derivative, $X \lrcorner \alpha$ denotes the contraction of X with α, and $[X,Y]$ denotes the Lie bracket of X and Y. In local coordinates x^a,

$$\alpha = \alpha_{ab\ldots c}\mathrm{d}x^a \otimes \mathrm{d}x^b \otimes \cdots \otimes \mathrm{d}x^c = \alpha_{ab\ldots c}\mathrm{d}x^a \wedge \mathrm{d}x^b \wedge \cdots \wedge \mathrm{d}x^c,$$

where the components $\alpha_{ab\ldots c}$ are defined by

$$\alpha_{ab\ldots c} = \alpha(\partial_a, \partial_b, \ldots, \partial_c),$$

with $\partial_a = \partial/\partial x^a$. Note that $\alpha_{ab\ldots c} = \alpha_{[ab\ldots c]}$. If X and Y have components X^a and Y^a, and α and β have components $\alpha_{ab\ldots c}$ and $\beta_{ab\ldots d}$, then

$$\begin{aligned}
(\alpha \wedge \beta)_{ab\ldots f} &= \alpha_{[ab\ldots c}\beta_{de\ldots f]} && \text{(A.1.3)}\\
(\mathrm{d}\alpha)_{abc\ldots d} &= \partial_{[a}\alpha_{bc\ldots d]} && \text{(A.1.4)}\\
(\mathcal{L}_X\alpha)_{bc\ldots d} &= X^a\partial_a\alpha_{bc\ldots d} + p\alpha_{a[c\ldots d}\partial_{b]}X^a && \text{(A.1.5)}\\
(X \lrcorner \alpha)_{b\ldots c} &= pX^a\alpha_{ab\ldots c} && \text{(A.1.6)}\\
[X,Y]^a &= X^b\partial_bY^a - Y^b\partial_bX^a, && \text{(A.1.7)}
\end{aligned}$$

with range and summation conventions. For example,

$$\frac{\partial}{\partial x} \lrcorner (\mathrm{d}x \wedge \mathrm{d}y) = \mathrm{d}y, \qquad \frac{\partial}{\partial x^1} \lrcorner \left(\alpha_{ab}\mathrm{d}x^a \wedge \mathrm{d}x^b\right) = 2\alpha_{1a}\mathrm{d}x^a.$$

The contraction $T \lrcorner \alpha \in \Omega^{p-k}(M)$ of $\alpha \in \Omega^p(M)$ with a k-index skew-symmetric contravariant tensor field T has components

$$(T \lrcorner \alpha)_{c\ldots d} = \frac{p!}{(p-k)!}T^{a\ldots b}\alpha_{a\ldots bc\ldots d},$$

where T has components $T^{a\ldots b}$ ($k \le p$). Note that $(X \wedge Y) \lrcorner \alpha = Y \lrcorner (X \lrcorner \alpha)$.

The following identities are extremely useful. For any $\alpha \in \Omega^p(M)$, $\beta \in \Omega^q(M)$, $X, Y \in V(M)$,

$$\begin{aligned}
\mathrm{d}(\alpha \wedge \beta) &= (\mathrm{d}\alpha) \wedge \beta + (-1)^p\alpha \wedge \mathrm{d}\beta && \text{(A.1.8)}\\
\alpha \wedge \beta &= (-1)^{pq}\beta \wedge \alpha && \text{(A.1.9)}\\
X \lrcorner (\alpha \wedge \beta) &= (X \lrcorner \alpha) \wedge \beta + (-1)^p\alpha \wedge (X \lrcorner \beta) && \text{(A.1.10)}\\
\mathcal{L}_X\alpha &= X \lrcorner \mathrm{d}\alpha + \mathrm{d}(X \lrcorner \alpha) && \text{(A.1.11)}\\
\mathcal{L}_X(Y \lrcorner \alpha) &= [X,Y] \lrcorner \alpha + Y \lrcorner \mathcal{L}_X\alpha && \text{(A.1.12)}\\
\mathcal{L}_X(\mathrm{d}\alpha) &= \mathrm{d}(\mathcal{L}_X\alpha) && \text{(A.1.13)}\\
\mathcal{L}_X\mathcal{L}_Y\alpha - \mathcal{L}_Y\mathcal{L}_X\alpha &= \mathcal{L}_{[X,Y]}\alpha. && \text{(A.1.14)}
\end{aligned}$$

For any 2-form ω such that $\mathrm{d}\omega = 0$,

$$X \lrcorner \mathcal{L}_Y\omega - Y \lrcorner \mathcal{L}_X\omega = [X, Y] \lrcorner \omega + 2\mathrm{d}\big(\omega(X, Y)\big). \tag{A.1.15}$$

For any $\alpha \in \Omega^p(M)$ and for any $X_i, X_j, \ldots \in V(M)$,

$$\begin{aligned} &\mathrm{d}\alpha(X_i, X_j, X_k, \ldots, X_m) = \\ &\qquad X_{[i}(\alpha(X_j, X_k, \ldots, X_{m]}) - \frac{p}{2}\alpha([X_{[i}, X_j], X_k, \ldots, X_{m]}). \end{aligned} \tag{A.1.16}$$

If α is closed ($\mathrm{d}\alpha = 0$), then locally it is exact: in some neigbourhood of each point, there exists a $(p-1)$-form γ such that $\alpha = \mathrm{d}\gamma$. This is the *Poincaré lemma.*

The restriction of α to a submanifold $N \subset M$ is denoted by $\alpha|_N$—that is, $\alpha|_N$ is the restriction to points of N and to vectors tangent to N.

If $\rho : M \to M'$ is a smooth map, then ρ_* and ρ^* denote the push-forward map (derived map) on contravariant tensors on M and the pull-back map on covariant tensors on M'.

Integration. The conventions for integration are as follows. Suppose that $\alpha = f\mathrm{d}x^1 \wedge \mathrm{d}x^2 \wedge \cdots \wedge \mathrm{d}x^p$. Then the integral of α over the p surface $N = \{x^{p+1} = x^{p+2} = \cdots = 0\}$ is given by

$$\int_N \alpha = \int\!\!\int \cdots \int f\, \mathrm{d}x^1 \mathrm{d}x^2 \ldots \mathrm{d}x^p.$$

If S is a submanifold of M of dimension $p+1$ with boundary ∂S and if α is any p-form, then

$$\int_S \mathrm{d}\alpha = \int_{\partial S} \alpha.$$

This is *Stokes' theorem.* There is a useful corollary: let $X \in V(M)$ and let S_t be the one-parameter family of surfaces obtained by dragging S along X. Then

$$\frac{\mathrm{d}}{\mathrm{d}t}\left[\int_{S_t} \alpha\right]_{t=0} = \int_S X \lrcorner \mathrm{d}\alpha + \int_{\partial S} X \lrcorner \alpha. \tag{A.1.17}$$

The integral of a differential form is well defined only on an oriented manifold. On a general manifold, one can integrate densities: a density is represented by a function in a local coordinate system x^a, subject to the transformation rule

$$f \mapsto \tilde{f} = \left|\det\left(\frac{\partial x^{\cdot}}{\partial \tilde{x}^{\cdot}}\right)\right| f$$

under change of coordinates. That is, it is a section of the density bundle, which has fibres $\Delta(T_m M)$. If α is a volume form on M, then $|\alpha|$ is a density. When M is oriented and α is positive, $\int |\alpha| = \int \alpha$.

Groups of transformations. Let G be a Lie group that acts smoothly on M on the right. That is, there is a smooth map $M \times G \to M : (m, g) \mapsto g(m) = mg$ such that the following hold

(R1) for all $g \in G$, $g : M \to M$ is a diffeomorphism;

(R2) for all $g, g' \in G$ and $m \in M$, $(mg)g' = m(gg')$.

Then each $A \in \mathcal{G}$, where $\mathcal{G}$ is the Lie algebra of G, determines a vector field $X_A \in V(M)$. For small t, $tX_A(m)$ is the vector from $m \in M$ to mg where $g = \mathrm{e}^{tA}$. For every $A, B \in \mathcal{G}$,

$$[X_A, X_B] = X_{[A,B]}. \tag{A.1.18}$$

For actions on the left, we write $g(m)$ as gm and replace (R2) by $g(g'm) = (gg')(m)$. Then instead of (A.1.18), we have $[X_A, X_B] = -X_{[A,B]}$.

Let $X \in V(M)$. For $m \in M$, $t \in \mathbb{R}$, let $\rho_t(m) = \gamma(t) \in M$, where γ is the integral curve of X such that $\gamma(0) = m$. We call the family of maps ρ_t the *flow* of X. If $\rho_t(m)$ is defined for every $t \in \mathbb{R}$ and $m \in M$, then for each t, $\rho_t : M \to M$ is a diffeomorphism, and $(m, t) \mapsto \rho_t(m)$ is an action of $\mathbb{R}$ on M. In this case X is said to be complete.

Time-dependent vector fields and differential forms. Details of the following material can be found in Abraham and Marsden (1978).

A *time-dependent* vector field X on M has components that depend on $t \in \mathbb{R}$, as well as on coordinates x^a on M. Thus

$$X = X^a(x, t)\frac{\partial}{\partial x^a}.$$

More formally, X is a smooth map $M \times \mathbb{R} \to TM$ such that $X(m, t) \in T_m M$ for every $m \in M$. We also allow X to be defined only on some subset of $M \times \mathbb{R}$. The obvious example is the velocity field in fluid dynamics.

Associated with X there is an ordinary vector field $\tilde{X} \in V(M \times \mathbb{R})$, given by

$$\tilde{X} = X + T = X^a(x, t)\frac{\partial}{\partial x^a} + \frac{\partial}{\partial t}, \tag{A.1.19}$$

where $T = \partial/\partial t$. Conversely, every $\tilde{X} \in V(M \times \mathbb{R})$ such that $\pi_* \tilde{X} = T$, where π is the projection onto $\mathbb{R}$, determines a time-dependent vector field on M.

The time derivative $\partial_t X$ of X is the time-dependent vector field with components $\partial_t X^a$; and, for fixed t, $X(t)$ denotes the (ordinary) vector field on M with components $X^a(., t)$.

The *Lie bracket* $[X, Y]$ of two time-dependent vector fields is the time-dependent vector field with components

$$X^b \partial_b Y^a - Y^b \partial_b X^a + \partial_t Y^a - \partial_t X^a. \tag{A.1.20}$$

Thus $\widetilde{[X,Y]} = [\tilde{X},\tilde{Y}] + T$. Note also that if $X = 0$, then $[X,Y] = \partial_t Y$.

The *integral curves* $t \mapsto \gamma(t)$ of X are the solutions of the system of ordinary differential equations

$$\frac{\mathrm{d}x^a}{\mathrm{d}t} = X^a(x,t). \tag{A.1.21}$$

Note that if $t \mapsto \gamma(t)$ is an integral curve of X, then $t \mapsto (\gamma(t),t)$ is an integral curve of $\tilde{X}$ in $M \times \mathbb{R}$.

For $m \in M$ and $t,t' \in \mathbb{R}$, we put $\rho_{tt'}(m) = \gamma(t') \in M$, where γ is the integral curve such that $\gamma(t) = m$. This defines the *flow* of X. Again, X is *complete* if the flow is defined for every $m \in M$ and $t,t' \in \mathbb{R}$, in which case it is a family of diffeomorphisms $\rho_{tt'} : M \to M$ labelled by two parameters t and t'. Note that $\rho_{t't''}(\rho_{tt'}(m)) = \rho_{tt''}(m)$.

A time-independent vector field X can be treated as if it were time dependent. Then, with a minor ambiguity of notation, $\rho_{tt'} = \rho_{t'-t}$.

A time-dependent p-form on M is a map $\alpha : M \times \mathbb{R} \to \bigwedge^p(TM)$ such that $\alpha(m,t) \in \bigwedge^p(T_m M)$; its components $\alpha_{ab\cdots c} = \alpha_{[ab\cdots c]}$ are functions of x^a and t. A time-dependent function or 0-form is simply a function on $M \times \mathbb{R}$.

We define $\partial_t\alpha$ and $\alpha(t)$ in the same way as for time-dependent vector fields, so that $\partial_t\alpha$ has components $\partial_t\alpha_{ab\cdots c}$, and $\alpha(t)$ is the (ordinary) differential form on M given by holding t fixed. We define $\alpha \wedge \beta$, $\mathrm{d}\alpha$, and $X \lrcorner \alpha$ (for time-dependent X) by the same formulas as in the time-independent case, (A.1.3), (A.1.4), and (A.1.6).

Associated with α there is the ordinary p-form $\tilde{\alpha} \in \Omega^p(M \times \mathbb{R})$ given by

$$\tilde{\alpha}(m,t) = \alpha_{ab\cdots c}(m,t)\mathrm{d}x^a \wedge \mathrm{d}x^b \wedge \cdots \wedge \mathrm{d}x^c. \tag{A.1.22}$$

Conversely, any p-form $\tilde{\alpha}$ on $M \times \mathbb{R}$ such that $T \lrcorner \tilde{\alpha} = 0$ determines a time-dependent p-form on M. Note that

$$\mathrm{d}\tilde{\alpha} = \widetilde{\mathrm{d}\alpha} + \mathrm{d}t \wedge \widetilde{\partial_t\alpha};$$

and also that $\widetilde{X \lrcorner \alpha} = \tilde{X} \lrcorner \tilde{\alpha}$ for any time-dependent X.

The *Lie derivative* of α along a time-dependent vector field is defined by

$$(\mathcal{L}_X\alpha)_{bc\cdots d} = X^a\partial_a\alpha_{bc\cdots d} + p\alpha_{a[c\cdots d}\partial_{b]}X^a + \partial_t\alpha_{bc\cdots d}. \tag{A.1.23}$$

Eqns (A.1.8)-(A.1.10) also hold for time-dependent vector fields and differential forms; but eqn (A.1.11) must be replaced by

$$\mathcal{L}_X\alpha = X \lrcorner \mathrm{d}\alpha + \mathrm{d}(X \lrcorner \alpha) + \partial_t\alpha. \tag{A.1.24}$$

If $\mathcal{L}_X\alpha = 0$, then $\rho^*_{tt'}(\alpha(t')) = \alpha(t)$.

If $f \in C^\infty(M \times \mathbb{R})$ is a time-dependent function (0-form), then

$$\mathcal{L}_X f = \frac{\mathrm{d}f}{\mathrm{d}t} = X^a \frac{\partial f}{\partial x^a} + \frac{\partial f}{\partial t},$$

where $\mathrm{d}/\mathrm{d}t$ denotes the derivative along the integral curves of X.

Riemannian geometry. The Riemann tensor $R^a{}_{bcd}$ of a Riemannian or pseudo-Riemannian metric g_{ab} is defined by

$$\nabla_c \nabla_d X^a - \nabla_d \nabla_c X^a = R^a{}_{bcd} X^b,$$

where ∇ is the Levi-Civita connection. The Ricci tensor and the scalar curvature are defined by $R_{ab} = R^c{}_{acb}$ and $R = R^a{}_a$. The metric volume element is

$$\frac{1}{n!}\sqrt{|g|}\, \epsilon_{ab\dots c} \mathrm{d}x^a \wedge \mathrm{d}x^b \wedge \cdots \wedge \mathrm{d}x^c,$$

where ϵ is the alternating symbol.

A.2 Vector bundles

Definitions. The theory of fibre bundles is described in several books. Accounts which emphasize physical applications are given by Choquet-Bruhat, DeWitt-Morette, and Dillard-Bleick 1982, Schutz 1980, Nash and Sen 1983, and Crampin and Pirani 1986. Most (but not all) of the bundles in this book are vector bundles.

Let M be a real manifold and let $\mathbb{F}$ denote either the real or the complex numbers. A *vector bundle* over M is a smooth manifold V (the *total space*) together with a smooth map $\pi : V \to M$ (the *projection*) such that the following hold.

(VB1) For each $m \in M$, $V_m = \pi^{-1}(m)$ has the structure of an N-dimensional vector space over $\mathbb{F}$ (V_m is called the *fibre* over m; N is constant throughout M: it is called the *fibre dimension* or *rank* of the bundle).

(VB2) There is a system of local trivializations (U_i, τ_i) such that the Us cover M. A local trivialization is an open set $U \subset M$ together with a diffeomorphism $\tau : U \times \mathbb{F}^N \to \pi^{-1}(U) \subset V$ which, on restriction to $\{m\} \times \mathbb{F}^N$, becomes a linear map $\mathbb{F}^N \to V_m$, for each $m \in U$.

With varying degrees of sloppiness, one denotes the bundle by V, $V \to M$, or $\pi : V \to M$.

The basic operations on vector spaces extend in an obvious way to vector bundles: given V and V', one can form the *direct sum* (Whitney sum) $V \oplus V'$, the *tensor product* $V \otimes V'$, the *dual bundle* V^*, and, in the complex case, the *complex conjugate bundle* $\overline{V}$. We say that a vector bundle $W \to M$ is a sub-bundle of V whenever the fibres of W are subspaces of the fibres of V.

A map $s : U \subset M \to V$ such that $\pi \circ s$ is the identity is called a *section* over U. A local section of V is a section over some open subset of M. The space of smooth sections over U is denoted $C_V^\infty(U)$. If s is a section of V and s' is a section of V', then $s + s'$, ss' (more properly $s \otimes s'$), and $\overline{s}$ are, respectively, sections of $V \oplus V'$, $V \otimes V'$ and $\overline{V}$.

A p-form with values in V is a map α that assigns a smooth section $\alpha(X, Y, Z, \ldots)$ to each collection of p vector fields $X, Y, Z, \ldots$. It must be linear in each entry and skew-symmetric under permutations of $X, Y, Z, \ldots$. For example, if α is a differential form on M (in the ordinary sense) and s is a smooth section of V, then αs is a form with values in V. Forms with values in V can be contracted with vector fields on M. One can also take their exterior products with ordinary differential forms on M.

Connections. A connection ∇ on a vector bundle V is an operator that assigns a 1-form ∇s with values in V to each (smooth) section s of V, such that for any function f and sections $s, s' \in C_V^\infty(M)$,

$$\nabla(s + s') = \nabla s + \nabla s' \quad \text{and} \quad \nabla(fs) = (\mathrm{d}f)s + f\nabla s.$$

If X is a vector field, we denote $X \lrcorner \nabla s$ by $\nabla_X s$.

Connections on V and V' induce connections on $V \oplus V'$, $V \otimes V'$, V^*, and $\overline{V}$ in the obvious way. For example, in the case of $V \otimes V'$, $\nabla(ss') = (\nabla s)s' + s(\nabla s')$.

Hermitian structures. A Hermitian structure on a complex vector bundle V is a Hermitian inner product $(\cdot, \cdot)$ on the fibres which is smooth in the sense that $V \to \mathbb{C} : v \mapsto (v, v)$ is a smooth function. It is *compatible* with a connection ∇ if for all sections s, s' and all real vector fields X,

$$X \lrcorner \mathrm{d}(s, s') = (\nabla_X s, s') + (s, \nabla_X s').$$

Equivalence. Two vector bundles V and V' are equivalent if there exists a diffeomorphism $\iota : V \to V'$ which restricts to a linear isomorphism $V_m \to V'_m$ at each $m \in M$. If they have connections or Hermitian structures, then they are equivalent as bundles with connection or as Hermitian vector bundles if ι can be chosen so that in addition

$$\nabla_X(\iota \circ s) = \iota \circ \nabla_X s \quad \text{or} \quad (\iota \circ s, \iota \circ s) = (s, s)$$

for all sections s and vector fields X. It is important in prequantization (Chapter 8) that it is possible to have inequivalent connections on equivalent vector bundles.

Pull-backs. Let $V \to M$ be a vector bundle and let $\rho : M' \to M$ be a smooth map. Then

$$\rho^* V = \{(m', v') \mid m' \in M',\ v' \in V_{\rho(m')}\}$$

is a vector bundle over M', called the *pull-back* of V. If (U, τ) is a local trivialization of V, then

$$\rho^{-1}(U) \times \mathbb{F}^N \to \rho^* V : (m', \mathbf{z}) \mapsto \big(m', \tau(\rho(m'), \mathbf{z})\big),$$

where $\mathbf{z} \in \mathbb{F}^N$, is a local trivialization of $\rho^* V$.

A section s of V determines a pull-back section $\rho^* s$ of $\rho^* V$ by

$$(\rho^* s)(m') = \big(m', s(\rho(m'))\big).$$

A connection or Hermitian structure on V induces a connection or Hermitian structure on $\rho^* V$ in the obvious way.

When M' is a submanifold of M and ρ is the inclusion map, the pull-back is the *restriction* of V to M', and is denoted by $V|_{M'}$ or $V_{M'}$.

In the next section, we shall look in more detail at the special case of complex vector bundles with one-dimensional fibres, which play a central role in geometric quantization. We shall not develop further the general theory of vector bundles since it is required only in particular examples.

A.3 Line bundles

Curvature. Let $L \to M$ be a line bundle, that is, a complex vector bundle with one-dimensional fibres. Strictly, we should call L a 'complex line bundle', but real line bundles do not appear in this book, so the qualification is unnecessary. Let ∇ be a connection on L.

A local trivialization (U, τ) is determined by its *unit section* $s = \tau(\cdot, 1)$, which is a nonvanishing element of $C_L^\infty(U)$. The *potential 1-form* $\Theta \in \Omega^1_{\mathbb{C}}(U)$ of the connection in this trivialization is characterized by

$$\nabla s = -\mathrm{i}\Theta s. \tag{A.3.1}$$

If s' is any other section, then $s' = \psi s$, where ψ is a complex-valued function, and

$$\nabla_X s' = \big(X(\psi) - \mathrm{i} X \lrcorner \Theta \psi\big) s. \tag{A.3.2}$$

By dropping the distinction between s' and its local representative ψ, we can write this as $\nabla = \mathrm{d} - \mathrm{i}\Theta$.

The 2-form $\Omega = \mathrm{d}\Theta$ is the *curvature*. For any $X, Y \in V(M)$,

$$\Omega(X, Y) s' = \frac{\mathrm{i}}{2} \left(\nabla_X \nabla_Y s' - \nabla_Y \nabla_X s' - \nabla_{[X,Y]} s'\right).$$

Since this holds for any smooth section s', the curvature 2-form is independent of the local trivialization.

Let $\gamma : [0,1] \to M$ be a smooth curve with tangent vector $T = \dot{\gamma}$ and let s' be a section over γ; i.e. $s' : [0,1] \to L$, with $\pi(s'(t)) = \gamma(t)$. Then $\nabla_T s'$ is well defined by

$$\nabla_T s' = \left(\frac{\mathrm{d}\psi}{\mathrm{d}t} - \mathrm{i}T \lrcorner \Theta \right) s.$$

We say that s' is *parallel* along γ if $\nabla_T s' = 0$. There is a unique parallel section for each value of $s'(0)$. If γ is closed (i.e. $\gamma(0) = \gamma(1)$) and s' is parallel, then $s'(1) = \xi s'(0)$ for some $\xi \in \mathbb{C}$; ξ is the *holonomy* of the connection around the curve. The holonomy is independent of s', and also of the parametrization of the curve, as well as of the choice of the base point $\gamma(0)$. If γ can be spanned by a smooth surface S, and if S is given the appropriate orientation, then

$$\xi = \exp\left(\mathrm{i} \int_S \Omega \right).$$

The connection on $\overline{V}$ is given by $\nabla = \mathrm{d} + \mathrm{i}\Theta$ and has curvature $-\Omega$. The tensor product of two line bundles with connection is again a line bundle with connection, and its curvature is the sum of the curvatures of the individual connections.

Transition functions. Let (U_1, τ_1) and (U_2, τ_2) be two local trivializations of L, and let s_1 and s_2 be the corresponding unit sections. The *transition function* from (U_1, τ_1) to (U_2, τ_2) is the function $c_{12} \in C^\infty_{\mathbb{C}}(U_1 \cap U_2)$ such that $s_2 = c_{12} s_1$ on $U_1 \cap U_2$.

It is possible to find a collection $\{(U_j, \tau_j)\}$ of local trivializations such that $\{U_j\}$ is a contractible open cover of M (the subscripts are in some indexing set). That is, each U_j and each nonempty finite intersection $U_j \cap U_k \cap \cdots$ is contractible to a point. Let s_j be the unit section of (U_j, τ_j) and let $c_{jk} \in C^\infty_{\mathbb{C}}(U_j \cap U_k)$ be the transition function from (U_j, τ_j) to (U_k, τ_k). Then $s_k = c_{jk} s_j$ and

$$c_{jk} c_{kj} = 1 \quad \text{whenever} \quad U_j \cap U_k \neq \emptyset \tag{A.3.3}$$

$$c_{jk} c_{kl} c_{lj} = 1 \quad \text{whenever} \quad U_j \cap U_k \cap U_l \neq \emptyset \tag{A.3.4}$$

(there is no summation over the repeated indices). These are the *cocycle relations*.

Conversely, given a contractible cover $\{U_j\}$ and a collection of nonvanishing smooth complex functions $\{c_{jk}\}$ on the nonempty intersections such that (A.3.3) and (A.3.4) hold, it is possible to construct a line bundle L with the cs as transition functions. The total space is the disjoint union of the sets $U_j \times \mathbb{C}$ modulo the equivalence relation $(m_j, z_j) \sim (m_k, z_k)$ whenever $(m_j, z_j) \in U_j \times \mathbb{C}$, $(m_k, z_k) \in U_k \times \mathbb{C}$ and

$$m_j = m_k \quad \text{and} \quad z_j = c_{jk}(m_j) z_k.$$

If L and L' are line bundles with transition functions c_{jk} and c'_{jk} (relative to the same open cover), then

$L \otimes L'$ has transition functions $c_{jk}c'_{jk}$;

L^* has transition functions $(c_{jk})^{-1}$;

$\overline{L}$ has transition functions $\overline{c}_{jk}$.

If there exists a collection of nonvanishing functions $f_j \in C^\infty_{\mathbb{C}}(U_j)$ such that

$$c'_{jk} = \frac{f_j}{f_k} c_{jk} \quad \text{whenever} \quad U_j \cap U_k \neq \emptyset, \tag{A.3.5}$$

then L and L' are equivalent. The equivalence $\iota : L \to L'$ is given by $\iota(\tau_j(m, z)) = f_j \tau'_j(m, z)$ on U_j, where the (U, τ)s and the (U, τ')s are the respective local trivializations of L and L'. The converse is also true.

We denote the tensor product of L with itself by L^2; we also use the notation L^{-1} for L^*. By taking tensor products and duals, we can construct arbitrary integral powers L^n (L^0 is the trivial bundle $M \times \mathbb{C}$). We can also try to construct non-integral powers by taking roots of the transition functions. For example, $\sqrt{L}$ has transition functions $\sqrt{c_{jk}}$. It may not be possible to choose the square roots so that the cocycle relations hold, in which case $\sqrt{L}$ does not exist; when it does exist, there may be inequivalent ways of constructing it. The existence of $\sqrt{L}$ requires the vanishing of an obstruction in $H^2(M, \mathbb{Z}_2)$, and the equivalence classes of square roots are labelled by elements of $H^1(M, \mathbb{Z}_2)$ (see §A.6). If M is contractible, then $\sqrt{L}$ exists and is unique up to equivalence.

Connection potentials. Let ∇ be a connection on L and let Θ_j be the potential 1-form in the local trivialization (U_j, τ_j). On each nonempty $U_j \cap U_k$,

$$\Theta_k - \Theta_j = \mathrm{i}\frac{\mathrm{d}c_{jk}}{c_{jk}}. \tag{A.3.6}$$

Given a collection of 1-forms satisfying these relations, we can recover ∇ by using (A.3.1).

Without upsetting (A.3.6), we can replace the cs by $c'_{jk} = d_{jk}c_{jk}$, where the ds are constants satisfying the cocycle relations, while keeping the Θs unchanged. This gives another line bundle with connection with the same curvature, which is not equivalent to (L, ∇) unless there exist complex *constants* f_j such that (A.3.5) holds.

If L also has a Hermitian structure compatible with ∇, then, by rescaling if necessary, we can choose the local trivializations (U_j, τ_j) so that the unit sections satisfy $(s_j, s_j) = 1$. In this case, the transition functions are of unit modulus and the potential 1-forms are real, since for any $X \in V(M)$,

$$0 = X \lrcorner \mathrm{d}(s_j, s_j) = (\nabla_X s_j, s_j) + (s_j, \nabla_X s_j) = \mathrm{i}(X \lrcorner \overline{\Theta}_j - X \lrcorner \Theta_j).$$

Conversely, if the cs have unit modulus and the Θs are real, then L has a Hermitian structure compatible with ∇.

Connection forms. A useful way to describe a connection on a line bundle is in terms of the connection form on the total space (less the zero section).

Let $\pi : L \to M$ be a line bundle with connection ∇ and let $L^\times$ be the complement of the zero section in L. The connection form $\alpha \in \Omega^1_{\mathbb{C}}(L^\times)$ is defined by

$$\alpha = \pi^*(\Theta) + \mathrm{i}(\tau^{-1})^* \left(\frac{\mathrm{d}z}{z} \right),$$

where $\tau : U \times \mathbb{C} \to \pi^{-1}(U)$ is a local trivialization, Θ is the connection potential, and $z \in \mathbb{C}$. If (U', τ') is another local trivialization, then

$$(\tau^{-1})^* \left(\frac{\mathrm{d}z}{z} \right) - (\tau'^{-1})^* \left(\frac{\mathrm{d}z}{z} \right) = \frac{\mathrm{d}c}{c},$$

where c is the transition function from (U, τ) to (U', τ'). However the corresponding potential 1-forms are related by $\Theta' - \Theta = \mathrm{i}\, \mathrm{d}c/c$, so α does not depend on the choice of local trivialization.

The connection can be recovered from α either by using

$$\nabla s = -\mathrm{i}(s^* \alpha) s,$$

which holds for any nonvanishing smooth section s, or by using $\Theta_j = s_j^* \alpha$, where s_j is the unit section of (U_j, τ_j).

If L also has a Hermitian structure compatible with ∇, then

$$\alpha - \overline{\alpha} = \mathrm{i} \frac{\mathrm{d}H}{H}$$

where $H : L \to \mathbb{R} : l \mapsto (l, l)$. When M is connected, this determines H, and hence $(\cdot, \cdot)$, uniquely up to a constant factor. A necessary and sufficient condition for (L, ∇) to admit a compatible Hermitian structure is that the imaginary part of α should be exact (Kostant 1970a, Proposition 1.9.1).

Locally constant bundles. A locally constant line bundle is a line bundle with a flat connection; that is, with $\Omega = 0$. It has a preferred class of local trivializations with covariantly constant unit sections. The corresponding transition functions are locally constant (constant in a neighbourhood of each point). If L is locally constant and α is a form with values in L, then $\mathrm{d}\alpha$ is well defined, by using one of the preferred local trivializations to represent α as a ordinary complex-valued form, and then applying the usual d operator.

A.4 Distributions and foliations

Real distributions. A real distribution P on a manifold M is a sub-bundle of the tangent bundle. The fibre P_m at $m \in M$ is a subspace of $T_m M$ which varies smoothly with m.

A vector field X is *tangent* to P if $X(m) \in P_m$ for every m. The space of all vector fields tangent to P is denoted by $V_P(M)$. A smooth function is *constant* along P if $\mathrm{d}f$ vanishes on restriction to P. The space of smooth functions constant along P is denoted by $C^\infty_P(M)$.

A submanifold $\Sigma \subset M$ is *transverse* to P if $T_m\Sigma + P_m = T_m M$ at every point of Σ. Two distributions P and P' are transverse if $TM = P + P'$.

An immersed submanifold $\Lambda \subset M$ is an *integral manifold* of P if $P_m = T_m\Lambda$ for every $m \in \Lambda$. A distribution is *integrable* if in some neigbourhood of each point of M, there is a coordinate system x^a such that the surfaces of constant $x^{k+1}, x^{k+2}, \ldots, x^N$ are integral manifolds of P. Here $N = \dim M$ and $N - k$ is the (fibre) dimension of P. An equivalent condition is that P should be *involutive*. That is, $[X, Y]$ is tangent to P whenever X and Y are tangent to P.

An integrable distribution is called a *foliation*. The maximal connected integral manifolds of a foliation—the ones that are not proper subsets of other integral manifolds—are called the leaves. A foliation P is *reducible* (a nonstandard term) if the space M/P of leaves is a Hausdorff manifold. A *section* is a submanifold $\Sigma \subset M$ that intersects each leaf transversally in exactly one point.

Let P be a reducible foliation and let $\pi : M \to M/P$ be the projection map. Given $\alpha \in \Omega^p(M)$, there exist $\alpha' \in \Omega^p(M/P)$ such that $\alpha = \pi^*(\alpha')$ if and only if

$$X \lrcorner\, \alpha = 0 = X \lrcorner\, \mathrm{d}\alpha.$$

Complex distributions. A complex distribution is similarly defined to be a sub-bundle of the complexified tangent bundle $T_{\mathbb{C}}M$. Again, a (complex) vector field with values in P is said to be tangent to P. The space of tangent vector fields is denoted by $V_P(M)$. If $U \subset M$ is open, then the space of smooth complex functions $f \in C^\infty_{\mathbb{C}}(U)$ such that

$$\overline{X} \lrcorner\, \mathrm{d}f = 0 \quad \text{for every } X \in V_P(M)$$

is denoted by $C^\infty_P(M)$. (The reason for the complex conjugation is that we want to associate holomorphic, rather than antiholomorphic, functions with the holomorphic tangent bundle of a complex manifold.)

A complex distribution P is integrable if in some neighbourhood of each point, there are smooth complex functions $f_{k+1}, \ldots, f_N$ with gradients that are independent and annihilate all complex vector fields tangent to P, where $N-k$ is the fibre dimension of P. An involutive complex distribution is integrable if either it is analytic or $P \cap \overline{P}$ has constant dimension and $P+\overline{P}$ is involutive (Newlander and Nirenberg 1957, Nirenberg 1957).

Mappings of distributions. Let P and P' be distributions on M and M', and let $\rho : M \to M'$ be a smooth map. In the real case $P \subset TM$, $P' \subset TM'$, and $\rho_* : TM \to TM'$; in the complex case, $P \subset T_{\mathbb{C}}M$, $P' \subset T_{\mathbb{C}}M'$, and $\rho_* : T_{\mathbb{C}}M \to T_{\mathbb{C}}M'$. We define $\rho_* P$ to be the image under ρ_* of P in TM' or $T_{\mathbb{C}}M'$, and we define $\rho^* P'$ to be $(\rho_*)^{-1}(P')$, which is a subset of TM or of $T_{\mathbb{C}}M$. In general, neither $\rho^* P'$ nor $\rho_* P$ are distributions.

The canonical bundle. Let P be any $(N-k)$-dimensional real or complex distribution on an N-dimensional real manifold M. Associated with P, we have a line bundle $K_P \to M$, called the *canonical bundle* of the distribution. Its fibre at m is the one-dimensional subspace of $\bigwedge^k T^*_{m\mathbb{C}}M$ of forms α such that $\overline{X} \lrcorner \alpha = 0$ for every $X \in P_m$.

A.5 Complex manifolds

Definition. A complex manifold is a smooth real manifold M of even dimension on which there is given a smooth real tensor field J such that

(CM1) J_m is a complex structure on $T_m M$ at each $m \in M$;

(CM2) the complex distribution P spanned by the vector fields $X - \mathrm{i}JX$, $X \in V(M)$, is integrable.

A complex manifold has local holomorphic coordinates $z^a = x^a + \mathrm{i}y^a$ such that P is spanned by the vector fields $\partial/\partial z^a$. Two such coordinate systems are related by a holomorphic coordinate transformation.

A complex form α on M is of *type* (r,s) if it is of this type at every p. In local holomorphic coordinates, an (r,s) form can be written

$$\alpha = \alpha_{ab\ldots cde\ldots f}\,\mathrm{d}z^a \wedge \mathrm{d}z^b \wedge \cdots \wedge \mathrm{d}z^c \wedge \mathrm{d}\overline{z}^d \wedge \mathrm{d}\overline{z}^e \wedge \cdots \wedge \mathrm{d}\overline{z}^f, \qquad \text{(A.5.1)}$$

with r dzs and s d$\overline{z}$s, and

$$\alpha_{ab\ldots cde\ldots f} = \alpha_{[ab\ldots c][de\ldots f]}.$$

An arbitrary complex p-form has a unique decomposition as a sum of forms of types $(p,0)$, $(p-1,1), \ldots, (0,p)$.

The exterior derivative operator similarly decomposes as $\mathrm{d} = \partial + \overline{\partial}$, so that if α is of type (r,s), then $\partial\alpha$ is of type $(r+1,s)$ and $\overline{\partial}\alpha$ is of type $(r,s+1)$. In holomorphic coordinates,

$$\partial = \mathrm{d}z^a \wedge \frac{\partial}{\partial z^a} \quad \text{and} \quad \overline{\partial} = \mathrm{d}\overline{z}^a \wedge \frac{\partial}{\partial \overline{z}^a},$$

where the partial derivatives act on the components.

Holomorphic line bundles. A line bundle $L \to M$ over a complex manifold is holomorphic if it has a preferred class of local trivializations for which the transition functions are holomorphic. A holomorphic line bundle with a Hermitian structure has a natural compatible connection: if s is the unit section of one of the preferred local trivializations, then the corresponding potential 1-form is $\Theta = \mathrm{i}\partial(\log(s,s))$. If s' is the unit section of a second preferred local trivialization, then $s' = cs$, where c is a holomorphic function. If $\Theta' = \mathrm{i}\partial(\log(s',s'))$, then

$$\Theta' - \Theta = \mathrm{i}\partial(\log \bar{c}c) = \frac{\mathrm{i}\mathrm{d}c}{c}$$

since $\partial\bar{c} = 0$ and $\partial c = \mathrm{d}c$. Therefore the potentials transform in the right way to define a connection. It is straightforward to establish compatibility by showing that $\mathrm{d}(\log(s,s)) = \mathrm{i}(\overline{\Theta} - \Theta)$

A.6 Cohomology

Čech cohomology. The following brief and over-simplified outline should be sufficient to unravel the occasional uses of terminology from cohomology theory. For complete accounts, see, for example, Godement (1964), Vaisman (1973), or Bott and Tu (1982).

Let M be a manifold and let $\mathcal{U} = \{U_i\}$ be an open cover of M indexed by some set I. A *p-cochain* relative to $\mathcal{U}$ is a collection $f = \{f_{ij\ldots k}\}$ of functions of some specified type (smooth, locally constant, holomorphic, ...) such that

(C1) $f_{ij\ldots k}$ has $p+1$ indices and is defined on $U_i \cap U_j \cap \cdots \cap U_k$;

(C2) f contains one function $f_{ij\ldots k}$ for each ordered set of $p+1$ indices for which $U_i \cap U_j \cap \cdots \cap U_k$ is nonempty;

(C3) $f_{ij\ldots k}$ is skew-symmetric under permutation of its indices.

The *coboundary operator* δ acts on cochains by

$$\delta f = \{(p+2)\rho_{[i} f_{jk\ldots l]}\}, \qquad \text{(A.6.1)}$$

where $\rho_i f_{jk\ldots l}$ denotes the restriction of $f_{jk\ldots l}$ to $U_i \cap U_j \cap U_k \cap \cdots \cap U_l$, whenever this is nonempty; δ maps p-cochains to $(p+1)$-cochains. A p-cochain f is called a *p-cocycle* if $\delta f = 0$ and a *p-coboundary* if $f = \delta g$ for some $(p-1)$-cochain g. It is clear from (A.6.1) that $\delta^2 = 0$, so every coboundary is also a cocycle.

More generally, the fs might take values in any abelian group. In this case, (C3) reads '$f_{ij\ldots k}$ is unchanged under even permutations of its subscripts, and is sent to its inverse under odd permutations', and the addition in the skew-symmetrization on the right-hand side of (A.6.1) is replaced by the group operation. For example, when $p = 1$ and the group operation is multiplication, $(\delta f)_{jkl} = f_{jk} f_{kl} f_{lj}$ (without summation).

The set of p-cochains is denoted by $C^p(\mathcal{U})$ and the set of p-cocycles by $Z^p(\mathcal{U})$. Both are abelian groups under pointwise operations. The quotient $Z^p(\mathcal{U})/\delta C^{p-1}(\mathcal{U})$ is called the pth cohomology group. It is denoted by $H^p(\mathcal{U})$ and the equivalence class of f in $H^p(\mathcal{U})$ is denoted by $[f]$. Two cocycles that define the same class in $H^p(\mathcal{U})$—that is, that differ by a coboundary—are said to be *cohomologous.* When $p = 0$, we take $\delta C^{p-1}(\mathcal{U}) = 0$, so that $H^0(\mathcal{U})$ is the set of global functions of the specified type.

These objects are rather uninteresting when we consider arbitrary continuous or smooth functions. But they are more useful when the functions are restricted to some more special class. Note, however, that f may be a coboundary in one class of functions, but not in another: the restrictions must be imposed before constructing $Z^p(\mathcal{U})$ and $H^p(\mathcal{U})$. The following are the examples that appear most frequently in this book.

Constants. The cochains are allowed to contain only locally constant functions with values in $\mathbb{R}$, $\mathbb{C}$, $\mathbb{Z}$, or some more general abelian group G. A function is locally constant if it is constant in a neighbourhood of each point; it need not be globally constant if its domain is not connected. The resulting cohomology groups are said to have coefficients in $\mathbb{R}$, $\mathbb{C}$, $\mathbb{Z}$, or G, and are denoted by $H^{\cdot}(\mathcal{U}, \mathbb{R})$, $H^{\cdot}(\mathcal{U}, \mathbb{C})$, $H^{\cdot}(\mathcal{U}, \mathbb{Z})$, or $H^{\cdot}(\mathcal{U}, G)$.

An example of the use of such cohomology groups arises in the construction of the square root of a line bundle. Suppose that $\{(U_j, \tau_j)\}$ is a system of local trivializations for a line bundle $L \to M$, with transition functions c_{jk}. Let $\mathcal{U} = \{U_j\}$, and suppose that $\mathcal{U}$ is a contractible open cover of M. For each j, k, choose $b_{jk} \in C^{\infty}_{\mathbb{C}}(U_j \cap U_k)$ such that $b_{jk}^2 = c_{jk}$. Put

$$f_{jkl} = b_{jk} b_{kl} b_{lj}.$$

Then $f = \{f_{jkl}\}$ is a 2-cocycle with values in $\mathbb{Z}_2 = \{1, -1\}$. Its cohomology class in $H^1(\mathcal{U}, \mathbb{Z}_2)$ depends only on the cs, and not on the choice of square roots. If $[f] = 0$, then we can find a 1-cochain g with values in $\mathbb{Z}_2$ such that $f_{jkl} = g_{jk} g_{kl} g_{lj}$. The functions $g_{jk} b_{jk}$ then satisfy the cocycle relations for the transition functions of a line bundle $\sqrt{L}$. Thus $\sqrt{L}$ exists if $[f] = 0$.

Transition functions. Suppose that the fs are smooth functions with values in $\mathbb{C}^{\times}$, the multiplicative group of complex numbers. Then a 1-cocycle is a set of transition functions for a line bundle.

Holomorphic functions. Here M is a complex manifold and the fs are holomorphic functions or holomorphic sections of a holomorphic line bundle. The $H^{\cdot}(\mathcal{U})$s are then *holomorphic cohomology groups.* When M is compact, they are finite-dimensional (over $\mathbb{C}$).

Refinements. Suppose that $\mathcal{V}$ is a second open cover indexed by Λ. We say that $\mathcal{V}$ is a *refinement* of $\mathcal{U}$ if for each $\alpha \in \Lambda$, there is a $j(\alpha) \in I$ such that $V_\alpha \subset U_{j(\alpha)}$. There is then a natural map $C^p(\mathcal{U}) \to C^p(\mathcal{V})$ for each p, defined by restricting $f_{j(\alpha)j(\beta)\ldots j(\gamma)}$ to $V_\alpha \cap V_\beta \cap \cdots \cap V_\gamma$. This induces a homomorphism $H^p(\mathcal{U}) \to H^p(\mathcal{V})$. There is a way of taking the limit over successive refinements which leads to well-defined cohomology groups which depend only on M and the type of coefficients (that is, the class of functions) and not on the particular cover.

It is not usually necessary to carry out this limiting process in practice since, certainly in all the cases we consider, there exist what are called *Leray covers*, for which the cohomology groups of the cover coincide with the cohomology groups of M. In the first two cases, a contractible cover is a Leray cover. If $\mathcal{U}$ is contractible, then it is legitimate to use the notations $H^p(M, \mathbb{R})$, $H^p(M, \mathbb{C})$, and so on, in place of $H^p(\mathcal{U}, \cdot)$, since the precise choice of cover makes no difference. In the third case, we can take a cover in which the Us are *Stein manifolds* (see, for example, Gunning and Rossi 1965).

The de Rham isomorphism. Let $\mathcal{U}$ be a locally finite contractible open cover of M and let $\{h_i\}$ be a partition of unity subordinate to $\mathcal{U}$. That is, each point has a neighbourhood that intersects only a finite number of the Us, and $\{h_i\}$ is a collection of functions such that

(PU1) $0 \leq h_i \leq 1$;

(PU2) $\operatorname{supp}(h_i) \subset U_i$;

(PU3) $\sum_i h_i = 1$ (this sum is everywhere finite).

It is always possible to find $\mathcal{U}$ and $\{h_i\}$; see, for example, Kobayashi and Nomizu (1963).

Let f be a locally constant p-cochain taking values in $\mathbb{R}$ and put

$$\alpha_f = f_{ij\ldots k} h_i \mathrm{d}h_j \wedge \cdots \wedge \mathrm{d}h_k.$$

Here, and below, there is a summation over the repeated indices $i, j, \ldots$. Sums of this sort make sense since wherever $f_{ij\ldots k}$ is not defined, at least one of the corresponding hs vanishes; so it does not matter what values we assign to $f_{ij\ldots k}$ outside of $U_i \cap U_j \cap \cdots \cap U_k$.

Since $\rho_i h_i = 1$ (with the same summation convention),

$$\mathrm{d}\alpha_f = \alpha_{\delta f}.$$

It follows that if $\delta f = 0$, then α_f is closed, and that if f and f' are cohomologous, then α_f and $\alpha_{f'}$ differ by an exact form. The map $f \mapsto \alpha_f$ induces an isomorphism from $H^p(\mathcal{U}, \mathbb{R})$ to the pth de Rham cohomology group of M, which is the space of equivalence classes of closed real p-forms on M, with two forms regarded as equivalent if they differ by an exact form.

It is not immediately obvious that the map is surjective. We shall not prove this in general, but only look at the special case when $p = 2$. Suppose that α is a closed 2-form. On each U_i, $\alpha = \mathrm{d}\beta_i$ for some 1-form β_i. On each nonempty $U_i \cap U_j$, therefore,

$$\beta_i - \beta_j = \mathrm{d}g_{ij}$$

for some function g_{ij}. On each nonempty triple intersection $U_i \cap U_j \cap U_k$, $f_{ijk} = g_{ij} + g_{jk} + g_{ki}$ is constant. The fs satisfy the cocycle condition $\delta f = 0$. It is claimed that $[f] \in H^2(\mathcal{U}, \mathbb{R})$ is mapped to the class of α. To prove this, we have to show that $\alpha - \alpha_f$ is exact.

We remark first that the cocycle condition on f can be written

$$f_{ijk} - f_{jkl} + f_{kli} - f_{lij} = 0 \quad \text{on} \quad U_i \cap U_j \cap U_k \cap U_l.$$

By multiplying by $h_i \mathrm{d}h_j \wedge \mathrm{d}h_k$ and summing over all i, j, k, we deduce that

$$\alpha_f = f_{ijk} h_i \mathrm{d}h_j \wedge \mathrm{d}h_k = f_{ljk} \mathrm{d}h_j \wedge \mathrm{d}h_k \quad \text{on } U_l,$$

since $\sum \mathrm{d}h_j = 0$. Therefore $\alpha_f = \mathrm{d}(f_{ijk} h_j \mathrm{d}h_k)$ on U_i. Similarly, by multiplying by $h_j \mathrm{d}h_k$ and summing,

$$f_{ijk} h_j \mathrm{d}h_k - f_{ljk} h_j \mathrm{d}h_k = f_{ilk} \mathrm{d}h_k = \mathrm{d}(f_{ilk} h_k)$$

on $U_i \cap U_l$. But $f_{ilk} h_k = g_{il} + g_{lk} h_k - g_{ik} h_k$. Therefore if we put

$$\gamma_i = \beta_i - f_{ijk} h_j \mathrm{d}h_k - \mathrm{d}(g_{ik} h_k) \quad \text{on } U_i,$$

then $\gamma_i = \gamma_l$ on $U_i \cap U_l$. Hence γ_i is the restriction to U_i of a globally defined 1-form γ. On U_i, however, $\mathrm{d}\gamma_i = \alpha - f_{ijk} \mathrm{d}h_j \wedge \mathrm{d}h_k = \alpha - \alpha_f$. It follows that $\alpha - \alpha_f$ is exact.

Exactly the same argument works for $H^{\cdot}(\mathcal{U}, \mathbb{C})$ and complex-valued forms, and also when the cochains are locally constant sections of a locally constant line bundle and the forms take values in the line bundle. With a slight modification, it also gives the *Dolbeault isomorphism*, where the fs are holomorphic, the forms are of type $(0, p)$, and d is replaced by $\bar{\partial}$.

A.7 Pairings and embeddings

Pairings. Let V_1 and V_2 be Hilbert spaces. A (sesquilinear) pairing between V_1 and V_2 is a map that assigns a complex number (v_1, v_2) to each $v_1 \in V_1$, $v_2 \in V_2$; it must be conjugate linear in the first entry, linear in the second entry. In the infinite-dimensional case, $(\cdot, \cdot)$ should also be bounded, although we shall often drop this condition.

A pairing determines a linear map $\pi : V_2 \to V_1$ by

$$(v_1, \pi v_2) = (v_1, v_2),$$

for every $v_1 \in V_1$, $v_2 \in V_2$. The bracket on the left is the inner product in V_1; the one on the right is the pairing. If V_1 and V_2 are both subspaces of a Hilbert space V, and if the pairing is defined by restricting the inner product in V to $V_1 \times V_2$, then π is the orthogonal projection.

Embeddings in a trivial bundle. Suppose that B is a Hermitian vector bundle over M that can be embedded in a trivial bundle $M \times V$, where V is a complex vector space with Hermitian inner product $(\cdot,\cdot)$. In other words, for each $m \in M$, $B_m \subset V$ is a subspace, and the inner product on V restricts to that on B_m. Then B inherits a connection from V, with parallel transport from m to a nearby point m' given to the first order in the displacement from m to m' by orthogonal projection in V from B_m to $B_{m'}$.

Let $m_\alpha \in M$ ($\alpha = 1, 2, \ldots$) be points of M and let B_α be the corresponding fibres of B: these are subspaces of V. Let $\pi_{\alpha\beta} : B_\beta \to B_\alpha$ be the orthogonal projection and put

$$c_{123} = \pi_{13}\pi_{32}\pi_{21}.$$

Then c_{123} is a linear map $B_1 \to B_1$.

For c to be well defined, it is not necessary that the Hermitian form $(\cdot,\cdot)$ on V should be positive definite, only that it should be positive on B.

More generally, instead of the embedding in a trivial bundle, we may be given only the pairing $(\cdot,\cdot) : B_1 \times B_2 \to \mathbb{C}$ for any pair of fibres; this must be smooth as a function on $B \times B$ and reduce to the Hermitian structure on the fibres when $B_1 = B_2$. The connection and the maps c_{123} are then still well defined, with the 'projection' π_{12} determined by $(b_1, \pi_{12} b_2) = (b_1, b_2)$, where $b_\alpha \in B_\alpha$. We call c the *cocycle* of the pairing or the embedding.

Suppose now that B is a line bundle. Then the connection is given by

$$\nabla s = -\mathrm{i}\Theta s \quad \text{where} \quad \Theta = \mathrm{i}\mathrm{d}_1 \log\big(s(m_2), s(m_1)\big)\big|_{m_1 = m_2}. \tag{A.7.1}$$

Here s is a non-vanishing local section of B, and m_1 and m_2 are points of M; the exterior derivative d_1 is taken with respect to m_1 with m_2 held fixed. The curvature is the restriction of

$$\mathrm{i}\mathrm{d}_2 \wedge \mathrm{d}_1 \log\big(s(m_2), s(m_1)\big). \tag{A.7.2}$$

to the diagonal in $M \times M$ (the diagonal is identified with M). This is independent of s. If the line bundle is holomorphic, and the embedding is holomorphic, then the connection coincides with the one determined by the Hermitian structure on B.

The cs are multiplication operators, given by

$$c_{123} = \frac{(s_2, s_1)(s_3, s_2)(s_1, s_3)}{(s_1, s_1)(s_2, s_2)(s_3, s_3)} \tag{A.7.3}$$

where $s_\alpha = s(m_\alpha)$ (note that c does not depend on the choice of s). They satisfy the cocycle relation

$$(c_{143})^{-1} c_{123} = (c_{432})^{-1} c_{412}.$$

The curvature is given by restricting the 2-form

$$\mathrm{id}_3 \wedge \mathrm{d}_2 \log c_{123} = \mathrm{id}_3 \wedge \mathrm{d}_2 \log \left(\frac{c_{123}}{|c_{123}|} \right)$$

on $M \times M \times M$ to the diagonal $m_1 = m_2 = m_3$.

A.8 Stationary phase

Let f be a compactly supported smooth function on $\mathbb{R}^n$, let (g_{ab}) be an $n \times n$ nonsingular real symmetric matrix, and τ be a parameter. Then, as $\tau \to 0$, we have the following asymptotic expansion

$$\begin{aligned} &\left(\frac{1}{2\pi\tau}\right)^{n/2} \int \exp\left(\frac{\mathrm{i}}{2\tau} g_{ab} x^a x^b\right) f(x) \mathrm{d}^n x \\ &\quad \sim \frac{\mathrm{e}^{\mathrm{i}\pi \operatorname{sign} g/4}}{\sqrt{|\det g_{ab}|}} \left[\sum_0^\infty \frac{\tau^k}{k!} \left(\frac{\mathrm{i}}{2} g^{ab} \partial_a \partial_b \right)^k f \right]_{x=0} \end{aligned} \tag{A.8.1}$$

Here $g^{ab} g_{bc} = \delta^a_c$, $\operatorname{sign} g$ is the signature of g_{ab} (the number of positive eigenvalues less the number of negative eigenvalues), and $\partial_a = \partial / \partial x^a$.

If h is a smooth real function on $\mathbb{R}^n$, then, as $\tau \to 0$,

$$\begin{aligned} &\left(\frac{1}{2\pi\tau}\right)^{n/2} \int f \mathrm{e}^{\mathrm{i}h/\tau} \mathrm{d}^n x \\ &\quad \sim \sum_x \frac{\mathrm{e}^{\mathrm{i}\pi \operatorname{sign} g/4}}{\sqrt{|\det g_{ab}|}} f(x) \mathrm{e}^{\mathrm{i}h(x)/\tau} (1 + O(\tau)). \end{aligned} \tag{A.8.2}$$

Here $g_{ab} = (\partial_a \partial_b h)(x)$ and the sum is over the critical points x of h, which are assumed to be nondegenerate, and hence isolated. See, for example, Fedoryuk (1971), Duistermaat (1973), Guillemin and Sternberg (1977).

When the xs are canonical coordinates on a compact symplectic manifold, the integration is over the manifold, $f = 1$, and h is the Hamiltonian of a circle action, the leading term is exact and the $O(\tau)$ terms vanish (Duistermaat and Heckman 1982, Guillemin and Sternberg 1984, p. 262).

NOTES

Chapter 1

(1) The notation and conventions are explained in the first section of the appendix.

(2) The material in this section is developed in more detail by several authors. See, in particular, Souriau (1970), Dixmier (1974), Kirillov (1976), Weinstein (1977), Guillemin and Sternberg (1984), and Libermann and Marle (1987).

(3) The use of the superscript $\perp$ in this context is, perhaps, a little awkward, but is commonplace. The term 'orthogonal complement' is sometimes used instead of 'symplectic complement', but that can suggest misleading analogies with the behaviour of orthogonal complements under inner products. In the first edition, $F^{\perp}$ was called the *annihilator*, and was denoted by F^0, with the implication that ω should be used to identify V with V^*. But that also had its drawbacks.

(4) The group law is defined by $(\rho\rho')(X) = \rho(\rho'(X))$, so $SP(V, \omega)$ acts on V on the left.

Chapter 2

(1) The various spaces associated with a mechanical system are described by Synge (1960). In general, we shall follow his terminology. Formal definitions are given below.

(2) These topics are dealt with by a number of authors. See, in particular, Sternberg (1964), Hermann (1968), Arnol'd (1978), Abraham and Marsden (1978). The account in this chapter owes much to Abraham and Marsden.

(3) The terms 'orbit', 'dynamical trajectory', etc. are used interchangeably to denote an evolution curve of the system in either configuration space or phase space.

(4) Arnol'd (1990) demonstrates the use of this transformation in the derivation of Kepler's law on the ellipticity of orbits. He attributes it to K. Bohlin. The relationship between the central forces r^{-2} and r is an example of a general duality between the central force laws r^a and r^A when a and A are related by $(a + 3)(A + 3) \doteq 4$: the orbits of dual force laws are connected by the transformation $x+iy \mapsto (x+iy)^{\alpha}$, where $\alpha = \frac{1}{2}(a+3)$. Arnol'd remarks that the self-dual cases $a = -1$ and $a = -5$ were singled in Newton's *Principia*.

The transformation is a two-dimensional version of the Kustaanheimo-Stiefel spinor linearization of the Kepler problem. Kummer (1982) describes the relationship between this spinor construction and Moser's stereographic regularization (Moser 1970). See also Pirani (1974) and Souriau (1974b).

The three-dimensional version is very similar to the two-dimensional version, except that α and β are complex, and are connected by a nonholonomic constraint. In three dimensions, the Lagrangian is

$$L = \tfrac{1}{2}(\dot{x}^2 + \dot{y}^2 + \dot{z}^2) + \frac{1}{r},$$

where $r^2 = x^2 + y^2 + z^2$.

The configuration space is $Q = \mathbb{R}^3 - \{0\}$ and the symplectic structure on TQ is $\omega_L = \mathrm{d}\theta_L$ where

$$\theta_L = \dot{x}\mathrm{d}x + \dot{y}\mathrm{d}y + \dot{z}\mathrm{d}z,$$

in the traditional abuse of notation in which $\dot{x}$ is either a coordinate or the time derivative of x, according to context.

The spinor construction lifts the motion from TQ to a hypersurface $C \subset \mathbb{C}^4$ (the 'nonholonomic constraint'). Let $\alpha, \beta, \dot{\alpha}, \dot{\beta}$ denote the complex coordinates on $\mathbb{C}^4$ and put

$$2r = \alpha\overline{\alpha} + \beta\overline{\beta}, \quad \theta = r(\dot{\overline{\alpha}}\mathrm{d}\alpha + \dot{\overline{\beta}}\mathrm{d}\beta + \dot{\alpha}\mathrm{d}\overline{\alpha} + \dot{\beta}\mathrm{d}\overline{\beta}),$$

and $\omega = \mathrm{d}\theta$ (note that θ and ω are real).

The hypersurface C is given by

$$\alpha\dot{\overline{\alpha}} + \beta\dot{\overline{\beta}} - \dot{\alpha}\overline{\alpha} - \dot{\beta}\overline{\beta} = 0. \tag{N.2.1}$$

If we exclude the zeros of r, then C is a circle bundle over TQ with projection $\pi : C - \{r = 0\} \to TQ$ determined by

$$x + \mathrm{i}y = \alpha\overline{\beta}, \quad z = \tfrac{1}{2}(\alpha\overline{\alpha} - \beta\overline{\beta}), \quad \dot{x} + \mathrm{i}\dot{y} = \dot{\alpha}\overline{\beta} + \alpha\dot{\overline{\beta}}, \quad \dot{z} = \tfrac{1}{2}(\dot{\alpha}\overline{\alpha} + \alpha\dot{\overline{\alpha}} - \dot{\beta}\overline{\beta} - \beta\dot{\overline{\beta}})$$

(note that the pull-back of $x^2 + y^2 + z^2$ is r^2, so the notation is consistent). By direct calculation, $\pi^*(\theta_L) = \theta|_C$ and $h \circ \pi = E|_C$, where h is the Hamiltonian $\frac{1}{2}(\dot{x}^2 + \dot{y}^2 + \dot{z}^2) - r^{-1}$ and

$$E = r(\dot{\alpha}\dot{\overline{\alpha}} + \dot{\beta}\dot{\overline{\beta}}) - \frac{1}{r}.$$

Now E is the Hamiltonian of the Lagrangian

$$L' = r(\dot{\alpha}\dot{\overline{\alpha}} + \dot{\beta}\dot{\overline{\beta}}) + \frac{1}{r},$$

and $\theta = \theta_{L'}$. Therefore the canonical flow of E on $(\mathbb{C}^4 - \{r = 0\}, \omega)$ is given by the Lagrange equations of L', which can be written as

$$\frac{\mathrm{d}}{\mathrm{d}t}\left(r\dot{\alpha}\right) - \frac{E\alpha}{2r} = 0, \quad \frac{\mathrm{d}}{\mathrm{d}t}\left(r\dot{\beta}\right) - \frac{E\beta}{2r} = 0,$$

together with their complex conjugates. These are consistent with (N.2.1), and so the Hamiltonian vector field X_E is tangent to C. Since $E|_C$ is the pull-back of h and $\omega|_C = \mathrm{d}\theta|_C$ is the pull-back of $\omega_L = \mathrm{d}\theta_L$, it follows that the integral curves of X_E in $C - \{r = 0\}$ project onto the orbits of the system in TQ: each orbit is the image of a circle's worth of curves in C, related by phase transformations

$$(\alpha, \beta, \dot{\alpha}, \dot{\beta}) \mapsto e^{i\phi}(\alpha, \beta, \dot{\alpha}, \dot{\beta})$$

for constant ϕ.

We now follow exactly the same procedure as in the two-dimensional case: we introduce a nonphysical time parameter s by $\dot{s} = r^{-1}$, and new velocity variables $\mu = r\dot{\alpha}$, $\nu = r\dot{\beta}$. The negative energy orbits are then given by the oscillator equations

$$\frac{d^2\alpha}{ds^2} - \frac{E\alpha}{2} = 0, \quad \frac{d^2\beta}{ds^2} - \frac{E\beta}{2} = 0.$$

The oscillator orbits remain well-defined at $r = 0$, so we can identify the 'regularized' space $\hat{M}$ of negative energy motions with the solutions of the oscillator equations that satisfy (N.2.1), modulo the action of the phase transformations.

The oscillator orbits are labelled by E together with the values of α, β, $\mu = r\dot{\alpha}$, $\nu = r\dot{\beta}$ at $s = 0$. Thus $\hat{M}$ is the quotient of the seven-dimensional submanifold $N \subset \mathbb{C}^4 \times \mathbb{R}$ on which $E < 0$ and

$$\alpha\overline{\mu} + \beta\overline{\nu} - \overline{\alpha}\mu - \overline{\beta}\nu = 0, \quad \mu\overline{\mu} + \nu\overline{\nu} - \tfrac{1}{2}(\alpha\overline{\alpha} + \beta\overline{\beta})E = 1 \qquad \text{(N.2.2)}$$

by the action of the circle group $(\alpha, \beta, \mu, \nu, E) \mapsto (e^{i\phi}\alpha, e^{i\phi}\beta, e^{i\phi}\mu, e^{i\phi}\nu, E)$. As in the two-dimensional case, the regularized radial orbits bounce off the centre of force.

The space of regularized orbits with a fixed negative energy E is the quotient by the circle action of the subset of $\mathbb{C}^4$ on which the two equations (N.2.2) hold. Alternatively, it can be obtained by treating the second equation as a constraint, reducing with respect to the symplectic structure

$$\omega = d\overline{\mu} \wedge d\alpha + d\overline{\nu} \wedge d\beta + d\mu \wedge d\overline{\alpha} + d\nu \wedge d\overline{\beta},$$

and then imposing the first equation, which is homogeneous in the four complex variables. The constraint hypersurface is $S^7 \subset \mathbb{C}^4$, and its reduction is the complex projective space $\mathbb{CP}_3$. Thus the orbits with energy E correspond to points of the real quadric hypersurface $\alpha\overline{\mu} + \beta\overline{\nu} - \overline{\alpha}\mu - \overline{\beta}\nu = 0$ in $\mathbb{CP}_3$.

This is the space of null twistors (§6.8): the identification is made by taking ω^A and $\pi_{A'}$ to be the spinors with components (μ, ν) and $(i\alpha, i\beta)$, respectively. It is the space of null geodesics in compactified Minkowski space, and is diffeomorphic to the unit tangent bundle of S^3.

The simultaneous reduction of both constraints (N.2.2) is a symplectic manifold, the points of which represent unparametrized negative energy orbits of the Kepler system with energy E. It is the classical analogue of a negative energy eigenspace of the hydrogen atom, and one would expect to recover the eigenspace from it by quantization. We can understand its structure by putting

$$z^0 = \kappa\overline{\alpha} - \mathrm{i}\overline{\mu}, \quad z^1 = \kappa\overline{\beta} - \mathrm{i}\overline{\nu}, \quad w^0 = \kappa\alpha - \mathrm{i}\mu, \quad w^1 = \kappa\beta - \mathrm{i}\nu.$$

where $\kappa = \sqrt{-E/2}$. Then

$$\omega = \frac{\mathrm{i}}{2\kappa}(\mathrm{d}z^0 \wedge \mathrm{d}\overline{z}^0 + \mathrm{d}z^1 \wedge \mathrm{d}\overline{z}^1 + \mathrm{d}w^0 \wedge \mathrm{d}\overline{w}^0 + \mathrm{d}w^1 \wedge \mathrm{d}\overline{w}^1)$$

and the two constraint equations are equivalent to $z^0\overline{z}^0 + z^1\overline{z}^1 = 1$ and $w^0\overline{w}^0 + w^1\overline{w}^1 = 1$. Thus the reduction is $\mathbb{CP}_1 \times \mathbb{CP}_1$, the product of the two Riemann spheres with homogeneous coordinates (z^0, z^1) and (w^0, w^1) respectively (by taking $s = 1/4\kappa$ on p. 55). These are quantizable if $1/4\kappa = n\hbar/2$ for some integer n (p. 169); that is, if $E = -1/2n^2\hbar^2$, which is the usual formula for the energy levels of the hydrogen atom. If we use the natural complex polarization, and the metaplectic correction, then the wave function space associated with each $\mathbb{CP}_1$ has dimension n (the dimension space of the space of homogeneous polynomials of degree $n-1$ in z^0, z^1; see p. 235). Therefore the quantum Hilbert space has dimension n^2, which is the degeneracy of the corresponding energy level of the hydrogen atom. Simms (1974) demonstrates how this follows from the Riemann-Roch formula.

The symmetry group $SU(2) \times SU(2)$ in this description is generated by the components of the angular momentum and the Runge-Lenz vector (Pirani 1974).

(5) We can still define θ_L and ω_L by (2.1.6) and (2.1.4). However, they are now time-dependent forms on TQ. The tangent vector field X to the solution curves in TQ of Lagrange's equations is also time-dependent, and no longer satisfies (2.1.3). Instead,

$$X \lrcorner \, \omega_L + \mathrm{d}h + \frac{\mathrm{d}\theta_L}{\mathrm{d}t} = 0.$$

(6)′ If a time-independent Lagrangian $L(q, v)$ is homogeneous of degree one in the vs, then

$$v^a \frac{\partial L}{\partial v^a} = L.$$

By differentiating with respect to v^b,

$$v^a \frac{\partial^2 L}{\partial v^a \partial v^b} = 0,$$

and so L is degenerate.

(7) Let $L = \frac{1}{2} g_{ab} v^a v^b + A_a v^a - V$ and let $t \mapsto \big(q(t), v(t)\big)$ be a path in TQ such that

$$X = \dot{q}^a \frac{\partial}{\partial q^a} + \dot{v}^a \frac{\partial}{\partial v^a}$$

satisfies (2.5.7). Put $w^a(t) = v^a(t) - \dot{q}^a(t)$. Then $\dot{v}^a = \ddot{q}^a + \dot{w}^a$. Equation (2.5.7) gives

$$w^a \frac{\partial^2 L}{\partial v^a \partial v^b} = w^a g_{ab} = 0, \qquad \text{(N.2.3)}$$

and, with the commas denoting partial derivatives with respect to the qs,

$$\begin{aligned} 0 &= \dot{q}^b \frac{\partial^2 L}{\partial v^a \partial q^b} + \dot{v}^b \frac{\partial^2 L}{\partial v^a \partial v^b} - \frac{\partial L}{\partial q^a} + w^b \frac{\partial^2 L}{\partial v^b \partial q^a} \\ &= \dot{q}^b \dot{q}^c (g_{ac,b} - \tfrac{1}{2} g_{bc,a}) + \dot{q}^b w^c g_{ac,b} + \dot{q}^b (A_{a,b} - A_{b,a}) \\ &\quad + (\ddot{q}^b + \dot{w}^b) g_{ab} + \tfrac{1}{2} w^b w^c g_{bc,a} + V_{,a}. \end{aligned}$$

But, by differentiating (N.2.3) with respect to t along the path,

$$\dot{w}^a g_{ab} + w^a g_{ab,c} \dot{q}^c = 0.$$

Hence if $g_{ab,c} = 0$, then

$$\ddot{q}^b g_{ab} + \dot{q}^b (A_{a,b} - A_{b,a}) + V_{,a} = 0,$$

which implies that $q^a(t)$ is a solution of Lagrange's equation. If, on the other hand, $g_{bc,a} \neq 0$, then we still have the term $\frac{1}{2} g_{bc,a} w^b w^c$, so the statement is not true in this case.

Chapter 3

(1) Kirillov (1976), Kostant (1970a), Smale (1970), Souriau (1970), Guillemin and Sternberg (1984).

(2) See Proposition 1.5.3; $[\mathcal{G}, \mathcal{G}]$ consists of linear combinations of elements of $\mathcal{G}$ of the form $[A, B]$.

(3) The terminology is different to that of the first edition. 'Hamiltonian' is 'strongly Hamiltonian' in Libermann and Marle (1987) and 'Poisson' in Arnol'd (1978).

(4) For $\alpha \in \bigwedge^p \mathcal{G}^*$, where $p > 0$, $\delta\alpha \in \bigwedge^{p+1} \mathcal{G}^*$ is defined by

$$\delta\alpha(A_i, A_j, A_k, \ldots, A_m) = \frac{p}{2} \alpha([A_{[i}, A_j], A_k, \ldots, A_{m]}).$$

(5) A vector field Y on Q lifts to T^*Q and TQ by Lie propagation. If $Y = Y^a \partial/\partial q^a$, then the corresponding vector fields on T^*Q and TQ are, respectively,

$$Y^a \frac{\partial}{\partial q^a} - p_b \frac{\partial Y^b}{\partial q^a} \frac{\partial}{\partial p_a} \quad \text{and} \quad Y^a \frac{\partial}{\partial q^a} + v^b \frac{\partial Y^a}{\partial q^b} \frac{\partial}{\partial v^a}.$$

The vector field on T^*Q is necessarily Hamiltonian, and is generated by $Y^a p_a$. If the vector field on TQ preserves L, then it is Hamiltonian with respect to ω_L and is generated by $Y^a \partial L/\partial v^a$.

(6) The Hermitian matrix A generates the one-parameter unitary group e^{itA}.

(7) It is not quite the standard definition, according to which G acts on the left by $(gf)(A) = f(g^{-1}A)$.

(8) Elements of $\bigwedge^{\cdot} \mathcal{G}^*$ can be identified with right-invariant differential forms on G by right translation. The operator δ in note (4) then coincides with the exterior derivative.

(9) This example is really an illustration of an earlier result of Kostant's, (1973) which applies when the torus is a maximal torus in a compact Lie group and M is a coadjoint orbit. In this case, the compact group is $U(3)$.

(10) G/G_f is the set of right cosets $G_f g$. It might be more usual to denote this by $G_f \backslash G$, as in Kirillov (1976), but as we do not consider spaces of left cosets, the notation should not cause confusion.

(11) We think of $SO(3)$ as acting on $\mathbb{R}^3$ on the right by

$$(x\ y\ z) \mapsto (x\ y\ z)H,$$

so that X, Y, Z are identified with the vector fields

$$X = y\frac{\partial}{\partial z} - z\frac{\partial}{\partial y}, \quad Y = z\frac{\partial}{\partial x} - x\frac{\partial}{\partial z}, \quad Z = x\frac{\partial}{\partial y} - y\frac{\partial}{\partial x}.$$

(12) The volume is calculated by using the volume element ϵ in §8.2.

Chapter 4

(1) The material in the next few sections is well known. See, in particular, Weinstein (1977), Guillemin and Sternberg (1984), Abraham and Marsden (1978).

(2) The proposition is due to Nagano (1968). It also appears in Weinstein (1971, 1977) and Guillemin and Sternberg (1977). There are several other methods of proof, for example the following. Let q and Λ_q be as in the proof and let $\mathrm{pr} : M \to Q$ be the projection along P. For each $m \in \Lambda_q$, the kernel of $\mathrm{pr}_* : T_m M \to T_q Q$ is P_m.

We shall use ω to construct a 1-form α on Λ_q with values in $T_q^* Q$, as follows. Let $X \in P_m$ and let $Y \in T_q Q$. Choose $Y' \in T_m M$ such that $\mathrm{pr}_*(Y') = Y$ and put

$$(X \lrcorner \alpha)(Y) = 2\omega(X, Y').$$

Then α is well defined since the addition of an element of P_m to Y' does not alter the value of the right-hand side. It is also closed. To prove this, let X, X' be vector fields on M tangent to P. Then at each $m \in \Lambda_q$, $\mathrm{d}\alpha(X, X')$ is an element of $T_q^* Q$. Therefore $\mathrm{d}\alpha(X, X')(Y)$ is a function on Λ_q. We want to show that it vanishes for every $X, X' \in V_P(M)$ and $Y \in T_q Q$. To do this, extend Y to a vector field on Q and let Y' be a vector field on M which projects onto Y under pr_*. Then $[Y', X]$, $[Y', X']$, and $[X, X']$ are tangent to P and so

$$\begin{aligned}(\mathrm{d}\alpha(X, X'))(Y) &= X(\omega(X', Y')) - X'(\omega(X, Y')) - \omega([X, X'], Y') \\ &= 3\mathrm{d}\omega(X, X', Y') \\ &= 0\end{aligned}$$

by using (A.1.16). Hence $\alpha = \mathrm{d}p$ for some $p : \Lambda_q \to T_q^* Q$, which is determined uniquely if we also impose $p(q) = 0$.

Define a 1-form θ on M by $X \lrcorner \theta = p(\mathrm{pr}_*(X))$. It is straightforward to show that $\mathrm{d}\theta = \omega$ and hence that $m \mapsto (p,q) = (p(m), \mathrm{pr}(m))$ is a map of $M \to T^*Q$ with the required properties.

(3) If γ_i can be spanned by a 2-surface S_i, then we can fix the sign of j_i by taking j_i to be the area of S_i, divided by 2π.

(4) The action variables are the momentum coordinates and the angle variables are the configuration coordinates, which is the reverse of our normal convention for adapted coordinates in which we think of M/P as the 'configuration space'.

(5) The proof is as follows. Since $[f,h] = 0$, and since f is independent of q^1, we have $[f,b] = [f,c] = [f,d] = 0$. Therefore, since f is independent of p_1, $[f,b_0] = [f,c_0] = [f,d_0] = 0$ and hence $[f,k] = 0$. It follows from (4.11.3) that $[f,w] = [k,w] = 0$.

Now, we have

$$w = w(q'^1, \ldots, q'^n) \quad \text{and} \quad p_\alpha = \frac{\partial S_2}{\partial q^\alpha}$$

for $\alpha = 2,3,\ldots,n$. These imply that $q'^a = q'^a(p_2,\ldots,p_n,q^2,\ldots,q^n,w)$. That is, in the p,q coordinate system, the q's depend on p_1, q^1 only through w. Hence if X_a is the Hamiltonian vector field generated by q'^a, then

$$X_a = K_a + \frac{\partial q'^a}{\partial w} X_w$$

where the K_as are combinations of $\partial/\partial p_\alpha$ and $\partial/\partial q^\alpha$ for $\alpha = 2,\ldots,n$. We know that $X_a(h) = X_w(h) = 0$. Therefore

$$K_a(h) = K_a(b)(p_1)^2 + K_a(c)p_1 + K_a(d) = 0,$$

from which we deduce that $K_a(b) = K_a(c) = K_a(d) = 0$, and hence that $K_a(b_0) = K_a(c_0) = K_a(d_0) = 0$. Therefore $X_a(f) = K_a(f) = K_a(c_0/b_0) = 0$. Finally, we then deduce from (4.11.3) that $X_a(k) = 0$ as well.

(6) The construction is in fact more general. It can be applied to 'Hamiltonian structures' in which the Poisson brackets need not arise from symplectic forms.

Chapter 5

(1) This definition allows the metric g to have any signature, although the term 'Kähler manifold' is usually reserved for manifolds on which g is everywhere positive.

(2) Let ∇ be the Levi-Civita connection and let $X, Y \in V_P(M)$, $Z \in V_{\mathbb{C}}(M)$. Then $JX = \mathrm{i}X$, $JY = \mathrm{i}Y$, and

$$\begin{aligned} g(Z, \nabla_X Y) \;&=\; \tfrac{1}{2}\big(X(g(Y,Z)) + Y(g(Z,X)) - Z(g(X,Y)) + g([X,Y],Z) \\ &\quad + g([Z,X],Y) - g([Y,Z],X)\big) \end{aligned}$$

$$\begin{aligned} &= -\mathrm{i}X(\omega(Y,Z)) + \mathrm{i}Y(\omega(Z,X)) - \mathrm{i}\omega([X,Y],Z) \\ &\qquad + \mathrm{i}\omega([Z,X],Y) - \mathrm{i}\omega([Y,Z],X) \\ &= -2\mathrm{i}\big(X(\omega(Y,Z)) - \omega(Y,[X,Z])\big) \\ &= -\mathrm{i}Z \lrcorner \big(X \lrcorner \mathrm{d}(Y \lrcorner \omega)\big) \end{aligned}$$

since $g(\cdot,\cdot) = 2\omega(\cdot, J\cdot)$. But

$$g(Z, \nabla_X Y) = 2\omega(Z, J\nabla_X Y) = -\mathrm{i}Z \lrcorner (\nabla_X Y \lrcorner \omega)$$

since $\nabla J = 0$. Therefore $\nabla_X Y \lrcorner \omega = X \lrcorner \mathrm{d}(Y \lrcorner \omega)$.

(3) This is an abuse of the notation for partial derivatives since, for example, $z^{01} = z^{10}$. The derivatives must be calculated by treating the entries in z as independent variables, and then imposing the symmetry constraint $z^{ab} = z^{ba}$.

(4) The 2-form Ω is the imaginary part of the Levi form. Up to scale, Ω depends only on Q and not on the particular choice of f. It vanishes when f is the real part of a holomorphic function, and is nondegenerate when f is as far as possible from fulfilling this condition.

(5) One can also construct M and P in this way when Q is abstract CR manifold that cannot be embedded as a real hypersurface in a complex manifold. Then P is not integrable, although it is still involutive.

(6) See Chapter 3, note (8).

Chapter 6

(1) This fits in with the particle interpretation that we shall meet later on, in which a null flag represents a photon propagating in the direction of $\mathbf{n}$, with the flag plane representing the polarization. We think of $\mathbf{u}$ as the electric field $\mathbf{E}$, and $\mathbf{v}$ as the magnetic field $\mathbf{B}$. The vector $\mathbf{n}$ is along the Poynting vector $\mathbf{E} \wedge \mathbf{B}$, and is orthogonal to the electric and magnetic fields $\mathbf{E}$ and $\mathbf{B}$. However the action of the full Lorentz group on an electromagnetic field is not quite the same as the action on N derived here.

(2) The fs can be characterized as the solutions of the differential equation

$$\nabla^c f^{ab} + \tfrac{2}{3} g^{c[a} \nabla_d f^{b]d} = 0.$$

(3) Here we regard S as a real vector bundle, so the direct sum is over $\mathbb{R}$.

Chapter 7

(1) For a different, but related, approach based on multisymplectic structures, see, for example, Kijowski and Szczyrba (1976).

(2) We are using the divergence theorem in the form

$$\int_{D_X} \nabla_a Y^a \epsilon = \int_{\partial D_X} Y^a \, \mathrm{d}\sigma_a.$$

(3) We should be more precise here about signs: the part of D to the future of Σ and to the past Σ' counts positively; the part to the past of Σ and to the future of Σ' counts negatively.

(4) There is a subtle point here concerning dimensions. The action integral and $\omega(\phi,\phi')$ must have the same dimensions as $\hbar$, namely $\mathrm{ML^2T^{-1}}$, which is equivalent to ML since we are working in relativistic units in which $c=1$. Thus ϕ here must have dimensions of $\mathrm{M^{\frac{1}{2}}L^{-\frac{1}{2}}}$ and μ must be identified with $m/\hbar$, where m is the rest mass. The Klein-Gordan field in §9.4, on the other hand, has dimensions of M since $\psi\overline{\psi}\mathrm{d}\tau$ is dimensionless and $\mathrm{d}\tau$ has dimensions M^2. To go between the two forms, one must absorb a factor of $\sqrt{\hbar}$ into the definition of ϕ.

(5) One can also use this general framework to show that h does not depend on the choice of Σ. Suppose that Σ and Σ' are two Cauchy surfaces. Denote the corresponding functions on M by f, f', h, and h'. Then

$$X \lrcorner \theta' - X \lrcorner \theta = X \lrcorner \mathrm{d}S_D = f - f'. \tag{N.7.1}$$

The last equality follows from the fact that if D_{0t}, and D'_{0t} denote, respectively, the appropriately oriented regions of space-time between Σ and $\rho_t(\Sigma)$, and between Σ' and $\rho_t(\Sigma')$, then

$$S_D \circ R_t^{-1} = S_D - S_{D_{0t}} + S_{D'_{0t}}$$

The derivative of the left-hand side at $t=0$ is $-X \lrcorner \mathrm{d}S_D$; the derivative of the right-hand side is $f'-f$. It follows that $h=h'$. In using this argument in a more general setting, one requires that S should be additive. That is, $S_{D\cup D'} = S_D + S_{D'}$ whenever D and D' are disjoint.

(6) Wald (1984, appendix E) gives a concise summary of the gravitational variational problem. A comprehensive account of the various issues is given by Ashtekar (1988); some of his calculations are used here.

(7) In the notation of §6.2, a four-component spinor is an element of $\mathcal{S}\oplus\overline{\mathcal{S}}$. A four-vector X acts on $\mathcal{S}\oplus\overline{\mathcal{S}}$ by

$$\gamma(X) : (\alpha^A, \beta^{A'}) \mapsto \mathrm{i}\sqrt{2}\,(X^{AB'}\beta_{B'}, X^{BA'}\alpha_B).$$

This gives a representation of the Clifford algebra since

$$\gamma(X)\gamma(Y) + \gamma(Y)\gamma(X) = 2X^aY_a\,I,$$

where I is the identity operator.

Chapter 8

(1) There are various reasons for believing this. For example, the Groenwald-van Hove theorem implies that it is impossible to extend the Shrödinger representation of the ps and qs to polynomials in the momenta of degree more than two. There is a discussion in Guillemin and Sternberg (1984); see also Chernoff (1981).

(2) The construction in the form presented here is due to Kostant (1970a). It is equivalent to that given (independently) in a slightly different form by Souriau (1966, 1970). See also Kirillov (1976), and Onofri and Pauri (1972). Some of the ideas also appear in the earlier work of Segal (1960), but essential elements came later.

(3) If $y_{jk} + y_{kl} + y_{lj} = 0$, then $y_{jk} = f_j - f_k$ for some collection of functions f_j. In this case B is unchanged, but ∇ may be replaced by an inequivalent connection. A little thought about the freedom involved here establishes that, given B, the various choices of ∇ are parametrized by $H^1(M,\mathbb{R})/H^1(M,\mathbb{Z})$.

This can be better understood in terms of the short exact sequence

$$0 \to \mathbb{Z} \to \mathbb{R} \xrightarrow{\exp(2\pi \mathrm{i})} T \to 0,$$

which induces the exact sequence of cohomology groups

$$0 \to H^1(M,\mathbb{Z}) \to H^1(M,\mathbb{R}) \to H^1(M,T) \xrightarrow{\tau} H^2(M,\mathbb{Z}) \xrightarrow{\epsilon} H^2(M,\mathbb{R}) \to \ldots$$

A line bundle over M is determined up to equivalence by its Chern class, which is an element of $H^2(M,\mathbb{Z})$. If the bundle admits a flat connection, then its Chern class lies in the kernel of ϵ, and therefore in the image of τ. A Hermitian line bundle with a flat connection corresponds to an element of $H^1(M,T)$. Two such bundles that have the same Chern class are topologically equivalent, but may have different connections. The various different flat connections on the same bundle are parametrized by $H^1(M,\mathbb{R})/H^1(M,\mathbb{Z})$.

(4) Locally B is the product of a neighbourhood $U \subset M$ and $\mathbb{C}$. We have identified X_f, which is a vector field on U, and $\partial/\partial\phi$, which is a vector field on $\mathbb{C}$, with vector fields on $U \times \mathbb{C}$.

(5) We are thinking of G as acting on M on the right and on $\mathcal{H}$ on the left.

(6) The subscript associates θ_0 with the origin in V, which is the only point of V at which $\theta_0 = 0$. The pull-back of θ_0 under translation through $-W$ is a symplectic potential θ_W which vanishes at $W \in V$.

(7) It is close in spirit to Souriau's form of prequantization.

Chapter 9

(1) The proof is as follows. We have to show that $\mathcal{H}_P$ is closed in $\mathcal{H}$. Let $m \in M$ and let s_0 be a nonvanishing polarized section on some open set $U \ni m$. Introduce local holomorphic coordinates $z^a = x^a + \mathrm{i}y^a$ in a neighbourhood of m. By rescaling and adding constants, we can ensure that the coordinates are defined throughout $V = \{|z^a| \leq 2\}$ and that $V \subset U$. Let $D = \{|z^a| < 1\}$ and let $H \neq 0$ be the function defined by

$$(s_0, s_0)\epsilon = H \mathrm{d}x^1 \wedge \cdots \wedge \mathrm{d}x^n \wedge \mathrm{d}y^1 \cdots \wedge \mathrm{d}y^n.$$

Any polarized section s can be expressed in V as $s = \psi(z)s_0$, where ψ is holomorphic. If $k = \inf_V |H|$, then

$$\langle s, s\rangle \geq k \int_V \overline{\psi}\psi \,\mathrm{d}^n x \mathrm{d}^n y. \tag{N.9.1}$$

By Cauchy's theorem, if $w \in D$ and if $0 \leq r^a \leq 1$ for $a = 1, \ldots, n$, then

$$\psi(w)^2 = \left(\frac{1}{2\pi}\right)^n \int_0^{2\pi} \cdots \int_0^{2\pi} \psi(w^1 + r^1 \mathrm{e}^{\mathrm{i}\phi_1}, \ldots, w^n + r^n \mathrm{e}^{i\phi_n})^2 \,\mathrm{d}\phi_1 \ldots \mathrm{d}\phi_n.$$

By integrating with respect to $\prod r\mathrm{d}r$ over $0 \leq r^a \leq 1$,

$$\psi(w)^2 = \left(\frac{1}{\pi}\right)^n \int_{D_w} \psi(z)^2 \,\mathrm{d}^n x \mathrm{d}^n y$$

where $D_w = \{|z^a - w^a| < 1\}$. Hence, by taking the modulus, and by using (N.9.1),

$$\sup_D |\psi|^2 \leq \frac{1}{k\pi^n} \langle s, s\rangle.$$

It follows by Morera's theorem and the dominated convergence theorem that any Cauchy sequence in $\mathcal{H}_P$ converges to an element of $\mathcal{H}_P$.

(2) The interpretation of the limit $\hbar \to 0$ needs some care since M is only quantizable for certain values of $\hbar$. One should start with a value of $\hbar$ for which M is quantizable, replace $\hbar$ by $\hbar/N$, where N is an integer, and then let $N \to \infty$.

(3) In the θ_W gauge (Chapter 8, note 6),

$$\psi_W = \mathrm{e}^{-(\overline{z}_a - \overline{w}_a)(z^a - w^a)/4\hbar}.$$

(4) What I have described here is essentially half-density quantization (Blattner 1973), although the mildly restrictive assumptions about orientation have made it unnecessary to distinguish between forms and densities.

(5) There may be singularities where the dimension of the space of polarized sections of B' changes.

(6) The resulting expression for ω' differs in sign from (6.6.3) when p is past-pointing. This is because the antiparticle part of the phase space is $\overline{M_{sm}^-}$, and not M_{sm}^-. For the same reason, to obtain a nonnegative polarization of M_{sm}, we take the complex conjugate of the nonnegative polarization of M_{sm}^- constructed in Chapter 6.

(7) This is a rather trivial way of obtaining a new polarization since the new Kähler manifold is isomorphic as a Kähler manifold to the original. In the case of a sphere, the complex structure is rigid, and this is the only freedom. A more interesting example is the torus constructed from $\mathbb{R}^2$ by identifying the sides of the unit square $\{0 \leq p \leq 1,\ 0 \leq q \leq 1\}$. A polarization of the symplectic structure $\mathrm{d}p \wedge \mathrm{d}q$ is given by taking P to be the span of

$$z^0 \frac{\partial}{\partial p} + z^1 \frac{\partial}{\partial q},$$

where $z^0, z^1 \in \mathbb{C}$. The polarizations corresponding to different choices of the constants z^0, z^1 are essentially different (i.e. they do not give isomorphic Kähler structures) unless the constants are related by a Möbius transformation

$$\frac{z^0}{z^1} \mapsto \frac{az^0 + bz^1}{cz^0 + dz^1},$$

where a, b, c, d are integers with $ad - bc = 1$. See, for example, Gunning (1966).

(8) A problem arises here if there are three polarizations P, P' and P'': if we use ω to transfer the orientation on Q from Q' and then from Q' to Q'', we may not get the same result as by direct transfer from Q to Q''. This will be resolved by the metaplectic construction.

(9) There is an awkwardness about dimensions, in that there is an implicit assumption that p and q have the same units, so it seems illogical not to set $\hbar = 1$. The $\hbar$s have been kept since we shall be concerned with the semiclassical limit $\hbar \to 0$.

(10) For the corrections that appear in the modern form of the WKB approximation, see Keller (1958) and Maslov (1972).

(11) This is a standard formula. It can be proved by writing out the expression for $R^a{}_{bcd}$ in terms of Christoffel symbols. Alternatively one can use the equation of geodesic deviation. Let v be a tangent vector to $T^*_q Q$ at p, and let $x^a(t)$ and $x'^a(t)$ be the two geodesics through q with initial velocities p^a and $p^a + sv^a$, where s is small. Let $sy^a(t) = x'^a(t) - x^a(t)$. Then y is a vector field along $t \mapsto x^a(t)$ such that $y^a(0) = 0$, $\dot{y}^a(0) = v^a$ and

$$\ddot{y}^a = -R^a{}_{bcd}\dot{x}^b y^c \dot{x}^d,$$

where the dot denotes covariant differentiation along the tangent vector $\dot{x}^a$. By differentiating again and putting $t = 0$, one deduces that

$$\dddot{y}^a(0) = -R^a{}_{bcd}\, p^b v^c p^d,$$

and hence that $(\exp_q^* g)(v, v) = y^a(1)y_a(1) = v^a v_a - \frac{1}{3}v^a p^b v^c p^d R_{abcd}$, plus higher order terms in p. The formula follows.

(12) Some analytical difficulties arise here because the integrand is not compactly supported.

(13) Note that the scalar curvature term is the same irrespective of the dimension of Q. In four dimensions, but not in other dimensions, the left-hand side is the conformally invariant Laplacian: only in four dimensions does $(\det g_{ab})\psi$ scale in the right way under conformal transformation.

(14) See also Bakas and La Roche (1986) for a treatment based on Isham's approach to quantization.

(15) Because of the reproducing property of the coherent states, $d\kappa$ has a well-defined existence as a cylinder measure. That is, a measure against which one can integrate functions on Γ that depend only on the values of γ at a finite number of points in the interval $[t, t']$. Path integrals based on reproducing kernels that arise in geometric quantization have been studied by Gawędzki (1974, 1976).

Chapter 10

(1) Note that $SP(V, \omega)$ acts on $\mathcal{F}_0$ on the right and on V on the left. If we think of (V, ω) as a symplectic manifold, then this is the reverse of our normal convention that groups act on manifolds on the right and on vector spaces on the left. This convention is derived, on the one hand, from the requirement that the bracket in the Lie algebra of the group should coincide with the Lie bracket of vector fields on the manifold, and on the other from the standard 'bra and ket' notation of quantum mechanics. It is hard to maintain consistency, however, when (V, ω) can be regarded either as a symplectic vector space or as a linear symplectic manifold, without introducing notational contortions that would probably obscure more than they would illuminate.

(2) The ideas here are taken from Axelrod, Della Pietra, and Witten (1991), Atiyah (1990), and Hitchin (1990). See also Sternberg (1988).

(3) Note that $|\chi^{-2}| > 1$, which may be slightly surprising as the cocycle is constructed by composing projections. They are, however, projections with respect to an indefinite inner product.

(4) That is, given a nonvanishing section ν of K, we define $\big(\nu(J), \nu(J')\big)$ as a function on $L^+V \times L^+V$ by taking the square root of $\big(\nu(J)^2, \nu(J')^2\big)$ which is positive when $J = J'$. Such sections exist because K and δ are topologically trivial and the square root is well defined because L^+V is simply connected. For the same reason, $\big(\nu(J), \nu(J')\big)$ depends only on the values of ν at J and J', and not on the ν itself.

(5) One can give a concrete realization of the metaplectic group as follows. Let μ be a nonvanishing section of K (for example, the one constructed earlier). For each $\rho \in SP(V, \omega)$, $\rho^*\mu = \lambda_\rho \mu$ for some function $\lambda_\rho : L^+V \to \mathbb{C}$. The metaplectic group is the set of pairs $(\rho, \sqrt{\lambda_\rho})$, where $\sqrt{\lambda_\rho}$ is one of the two square roots of λ_ρ, with the obvious multiplication rule.

(6) Our representation of L^+V by complex symmetric matrices maps it to the 'generalized upper half-plane'. One can instead map LV to the closure of the 'generalized unit disk' by first picking a base point $J_0 \in L^+V$ and a symplectic frame X^a, Y_b in V in which J_0 takes the standard form. One puts $Z^a = X^a - \mathrm{i}Y_a$, $\overline{Z}_a = X^a + \mathrm{i}Y_a$. Then any $P \in LV$ has a unique basis of the form

$$Z^a + M^{ab}\overline{Z}_b,$$

where M is a symmetric $n \times n$ matrix such that $1 - \overline{M}M^t$ is nonnegative definite. The interior, L^+V, is mapped to the set of Ms such that $1 - \overline{M}M^t$ is positive definite. This representation makes it clear that LV is contractible.

The Grassmannian $L_{\mathbb{R}}V$ of real Lagrangian subspaces, on the other hand, has a more complicated topology. For example, R. Penrose has remarked that in the case that V is four-dimensional ($n = 2$), $L_{\mathbb{R}}V$ can be represented as a compactified, three-dimensional Minkowski space, with transversality corresponding to non-null separation. This can be seen as follows. Let $\{X_\alpha\} = \{X_1, X_2, X_3, X_4\}$ be an arbitrary basis in V and let $\omega_{\alpha\beta}$ and $e_{\alpha\beta\gamma\delta}$ be the components of ω and the volume form $\omega \wedge \omega$.

The 2-planes in V are in one-to-one correspondence with the simple real bivectors (the skew-symmetric tensors $F^{\alpha\beta} = F^{[\alpha\beta]}$ of the form $F^{\alpha\beta} = X^{[\alpha}Y^{\beta]}$), modulo the equivalence relation

$$F \sim \lambda F,$$

where $\lambda \neq 0 \in \mathbb{R}$. If F and G are bivectors, then

(1) F is simple if and only if $e_{\alpha\beta\gamma\delta}F^{\alpha\beta}F^{\gamma\delta} = 0$;

(2) if F is simple, then the corresponding 2-plane is Lagrangian if and only if $\omega_{\alpha\beta}F^{\alpha\beta} = 0$;

(3) if F and G are simple, then the corresponding 2-planes are transverse if and only if $e_{\alpha\beta\gamma\delta}F^{\alpha\beta}G^{\gamma\delta} \neq 0$.

The space of real bivectors is six-dimensional, and carries the inner product

$$\langle F, G\rangle = e_{\alpha\beta\gamma\delta}F^{\alpha\beta}G^{\gamma\delta},$$

which has signature $+++---$. The subspace $\omega_{\alpha\beta}F^{\alpha\beta} = 0$ is five-dimensional, with a $+++--$ inner product. The real Lagrangian Grassmannian is simply the projective light-cone in this space. This, in turn, is a compactification of three-dimensional Minkowski space—Minkowski space together with a light-cone at infinity.

(7) We define $\overline{\kappa} \wedge \beta$ as follows. We can represent κ as the restriction to Λ of an n-form α on M: α is determined up to the addition of γ such that $\gamma|_\Lambda = 0$. For such a γ, $\overline{\gamma} \wedge \beta = 0$ at points of Λ. It follows that we can define $\overline{\kappa} \wedge \beta$ at points of Λ by $\overline{\kappa} \wedge \beta := \overline{\alpha} \wedge \beta$.

(8) The full theory of the relationship between distributions and objects on Lagrangian submanifolds is described in Hörmander (1971), Duistermaat (1973,1974), Treves (1980).

(9) Note that the winding numbers of χ^2 depend only on Λ (and the metaplectic structure), and not on the hs, since changing the hs, but keeping Λ fixed simply replaces κ by a real multiple. If γ is a closed curve in Λ, then $\chi^2 \circ \gamma$ is a closed curve in $\mathbb{C} - 0$. Its winding number is the *index* of the curve (Arnol'd 1967, 1978, Appendix 11).

(10) If the points at which $j_a = 0$ for one or more value of a are removed from V (as they should be if the djs are to be independent everywhere) then the metaplectic structure is no longer unique. We get a new metaplectic structure by using the vector fields $\partial/\partial\phi$ and $\partial/\partial j$ to identify the tangent space at each point of the remainder of V with $\mathbb{R}^{2n}$. In this case χ is real and the correction is trivial, so this metaplectic structure gives the wrong energy levels for the isotropic oscillator.

(11) Berry 1984, 1985, Simon 1983, Hannay 1985, Aharonov and Anandan 1987, Anandan 1988, Weinstein 1990. Montgomery (1988) describes another geometric framework for understanding Hannay's angles.

REFERENCES

Abraham, R. and Marsden, J. E. (1978). *Foundations of mechanics*, Second edition. Benjamin, Reading, Massachusetts

Adams, J. F. (1969). *Lectures on Lie groups*. Benjamin, New York.

Aharonov, Y. and Bohm, D. (1959). Significance of electromagnetic potentials. *Phys. Rev.*, **115**, 485–91.

—— and Anandan, J. (1987). Phase change during a cyclic quantum evolution. *Phys. Rev. Lett.*, **58**, 1593–6

Anandan, J. (1988). Geometric angles in classical and quantum physics. *Phys. Lett.*, **129A**, 201–7.

Arens, R. (1971). Classical Lorentz invariant particles. *J. Math. Phys.*, **12**, 2415–22.

Arnol'd, V. I. (1967). Characteristic class entering in quantization conditions. *Functional Anal. Appl.*, **1**, 1–13.

—— (1978). *Mathematical methods of classical mechanics*. Springer, New York.

—— (1990). *Huygens and Barrow, Hooke and Newton*. Birkhäuser, Basel.

Ashtekar, A. (1981). Asymptotic quantization of the gravitational field. *Phys. Rev. Lett.*, **46**, 573–6.

—— (1987). *Asymptotic quantization*. Bibliopolis, Naples.

—— (1988). *New perspectives in canonical gravity* (with invited contributions). Bibliopolis, Naples.

—— and Magnon, A. (1975). Quantum fields in curved space-time. *Proc. Roy. Soc. London ser. A*, **346**, 375–94.

—— and Stillerman, M. (1986). Geometric quantization and constrained systems. *J. Math. Phys.*, **27**, 1319–30.

Atiyah, M. F. (1982). Convexity and commuting Hamiltonians. *Bull. London Math. Soc.*, **14**, 1–15.

—— (1990). *The geometry and physics of knots*. Cambridge University Press, Cambridge.

Auslander, L. and Kostant, B. (1971). Polarization and unitary representations of solvable Lie groups. *Invent. Math.*, **14**, 255–354.

Axelrod, S., Della Pietra, S. and Witten, E. (1991). Geometric quantization of Chern Simons-gauge theory. *J. Diff. Geom.*, **33**, 787–902.

Bacry, H. (1967). Space-time and degrees of freedom of the elementary particle. *Comm. Math. Phys.*, **5**, 97–105.

Bakas, I. and La Roche, H. (1986). Path integral quantisation and coherent states. *J. Phys. A: Math. Gen.*, **19**, 2513–23.

Bargmann, V. (1954). On unitary ray representations of continuous groups. *Ann. of Math.* (2), **59**, 1–46.

—— (1961). On a Hilbert space of analytic functions and an associated integral transform. *Comm. Pure Appl. Math.*, **14**, 187–214.

Baston, R. J. and Eastwood, M. G. (1989). *The Penrose transform.* Oxford University Press, Oxford.

Bergmann, P. G. (1956). Quantization of general covariant field theories. *Helv. Phys. Acta, Suppl. IV*, 79–97.

Berry, M. V. (1984). Quantal phase factors accompanying adiabatic changes. *Proc. Roy. Soc. London A,* **392**, 45–57.

—— (1985). Classical adiabatic angles and quantal adiabatic phase. *J. Phys. A: Math. Gen.,* **18**, 15–27.

Bjorken, J. D. and Drell, S. D. (1964). *Relativistic quantum mechanics.* McGraw-Hill, New York.

Blattner, R. J. (1973). Quantization and representation theory. *Proc. Symp. Pure Math.,* **26**, 147–65.

—— (1977). The meta-linear geometry of non-real polarizations. In *Differential geometrical methods in mathematical physics* (ed K. Bleuler and A. Reetz). Lecture notes in mathematics, Vol. 570. Springer, Berlin.

—— and Rawnsley, R. J. (1986). A cohomological construction of half-forms for nonpositive polarizations. *Bull. Soc. Math. Belg. ser. B* **38**, 109–30.

Blau, M. (1988). On the geometric quantisation of constrained systems. *Class. Quantum Grav.,* **5**, 1033–44.

Bogolubov, N. N., Logunov, A. A. and Todorov, I. T. (1975). *Introduction to axiomatic quantum field theory.* Benjamin, Reading, Massachusetts.

Bott, R. (1957). Homogoeneous vector bundles. *Ann. of Math.* (2), **66**, 203–48.

—— (1972). Lectures on characteristic classes and foliations. In *Lectures on algebraic and differential topology* (ed. S. Gitler). Lecture notes in mathematics, Vol. 279. Springer, Berlin.

—— and Tu, L. W. (1982). *Differential forms and algebraic topology.* Springer, New York.

Brouzet, R. (1989). Sur quelques propriétés géométriques des variétés bihamiltoniennes. *C.R. Acad. Sci. Paris,* **308** (I), 287–92.

Carey, A. L. and Hannabuss, K. C. (1978). Twistors and geometric quantisation theory. *Rep. Math. Phys.,* **13**, 199–231.

Chern, S. S. and Moser, J. K. (1975). Real hypersurfaces in complex manifolds. *Acta Math.,* **133**, 219–71.

Chernoff, P. R. (1981). Mathematical obstructions to quantization. *Hadronic J.,* **4**, 879–98.

—— and Marsden, J. E. (1974). *Properties of infinite dimensional Hamiltonian systems.* Lecture notes in mathematics, Vol. 425. Springer, Berlin.

Chevalley, C. (1946). *Theory of Lie groups.* Princeton University Press, Princeton, New Jersey.

—— and Eilenberg, S. (1948). Cohomology theory of Lie groups and Lie algebras. *Trans. Amer. Math. Soc.,* **63**, 85–124.

Choquet-Bruhat, Y., DeWitt-Morette, C. and Dillard-Bleick, M. (1982). *Analysis, manifolds and physics.* Revised edition. North-Holland, Amsterdam.

Courant, R. and Hilbert, D. (1962). *Methods of mathematical physics*, Vol. 2. Wiley, New York.

Crampin, M. and Pirani, F. A. E. (1971). Twistors, symplectic structure and Lagrange's identity. In *Relativity and gravitation* (ed. C. G. Kuper and A. Peres). Gordon and Breach, London.

—— and —— (1986). *Applicable differential geometry.* London Mathematical Society lecture notes, Vol. 59. Cambridge University Press, Cambridge.

Czyz, J. (1978). On some approach to geometric quantization. In *Differential geometrical methods in mathematical physics II* (ed. K. Bleuler, H. R. Petry, and A. Reetz). Lecture notes in mathematics, Vol. 676. Springer, Berlin.

Dirac, P. A. M. (1925). The fundamental equations of quantum mechanics. *Proc. Roy. Soc. London ser. A*, **109**, 642–53.

—— (1931). Quantised singularities in the electromagnetic field. *Proc. Roy. Soc. London ser. A*, **133**, 60–72.

—— (1950). Generalized Hamiltonian dynamics. *Can. J. Math.,* **2**, 129–48

—— (1958). Generalized Hamiltonian dynamics. *Proc. Roy. Soc. London ser. A*, **246**, 326–32.

—— (1964). *Lectures on quantum mechanics.* Yeshiva University, New York.

Dixmier, J. (1974). *Algèbres enveloppantes.* Gauthier-Villars, Paris.

Dixon, W. G. (1974). Dynamics of extended bodies in general relativity. III. Equations of motion. *Philos. Trans. Roy. Soc. London ser. A*, **277**, 59–119.

Donaldson, S. K. and Kronheimer, P. B. (1990). *The geometry of four-manifolds.* Oxford University Press, Oxford.

Duistermaat, J. J. (1973) *Fourier integral operators.* Courant Institute Lecture Notes. New York University, New York.

—— (1974). Oscillatory integrals, Lagrange immersions and unfolding of singularities. *Comm. Pure Appl. Math.,* **27**, 207–81.

——, and Heckman, J. J. (1982). On the variation in the cohomology of the symplectic form of the reduced phase space. *Invent. Math.,* **69**, 259–68.

Eisenhart, L. P. (1949). *Riemannian Geometry.* Second edition. Princeton University Press, Princeton, New Jersey.

Faddeev, L. D. (1976). Introduction to functional methods. In *Methods in field theory.* Les Houches Summer School 1975. Ed. B. Z. Moroz. North Holland, Amsterdam.

Fedoryuk, M. V. (1971). The stationary phase method and pseudodifferential operators. *Russian Math. Surveys,* **26**, 65–115.

Field, M. (1982). *Several complex variables and complex manifolds, I and II.* London Mathematical Society lecture notes, Vols. 65 and 66. Cambridge University Press, Cambridge.

García, P. L. (1980). Tangent structure of the Yang-Mills equation and Hodge theory. In *Differential geometrical methods in mathematical physics* (ed. P. L. García, A. Pérez-Rendón, J.-M. Souriau). Lecture notes in mathematics, Vol. 836. Springer, Berlin.

Gawędzki, K. (1974). Some applications of functional integration to quantum mechanics and quantum field theory. *Rep. Math. Phys.,* **6**, 327–42.

—— (1976). Fourier-like kernels in geometric quantization. *Dissertationes Math.*, **128**.

—— (1977). Geometric quantization and Feynman path integrals for spin. In *Differential geometrical methods in mathematical physics* (ed. K. Bleuler and A. Reetz). Lecture notes in mathematics, Vol. 570. Springer, Berlin.

Godement, R. (1964). *Topologie algébrique et théorie des faisceaux.* Hermann and Cie, Paris.

Gotay, M. J. (1986). Constraints, reduction, and quantization. *J. Math. Phys.*, **27**, 2051–66.

—— (1987) A class of non-polarizable symplectic manifolds. *Monatsh. Math.*, **103**, 27–30.

—— , Nester, J. M. and Hinds, G. (1978). The Dirac-Bergmann theory of constraints. *J. Math. Phys.*, **19**, 2388–99.

Gouy, L. G. (1890). Sur une propriété nouvelle des ondes lumineuses. *C.R. Acad. Sci. Paris*, **110**, 1251–3.

Griffiths, P. and Harris, H. (1978). *Principles of algebraic geometry.* Wiley, New York.

Guillemin, V. and Sternberg, S. (1977). *Geometric asymptotics.* Mathematical surveys, Vol. 14. Amer. Math. Soc., Providence, Rhode Island.

—— and —— (1978). On the equations of motion of a classical particle in a Yang-Mills field and the principle of general covariance. *Hadronic J.*, **1**, 1-32.

—— and —— (1982a). Convexity properties of the moment mapping. *Invent. Math.*, **67**, 491-513.

—— and —— (1982b). Geometric quantization and the multiplicities of group representations. *Invent. Math.*, **67**, 515-38.

—— and —— (1984). *Symplectic techniques in physics.* Cambridge University Press, Cambridge.

Gunning, R. C. (1966). *Lectures on Riemann surfaces.* Princeton mathematical notes. Princeton University Press, Princeton, New Jersey.

—— and Rossi, H. (1965). *Analytic functions of several complex variables.* Prentice-Hall, Englewood Cliffs, New Jersey.

Hannay, J. H. (1985). Angle variable holonomy in adiabatic excursion of an integrable Hàmiltonian. *J. Phys. A: Math. Gen.*, **18**, 221–30.

Hermann, R. (1968). *Differential geometry and the calculus of variations.* Academic Press, New York.

Hitchin, N. J. (1990). Geometric quantization of spaces of connections. In *Geometry of low-dimensional manifolds. Vol.* 2 (ed. S. K. Donaldson and C. B. Thomas). London Mathematical Society lecture notes, Vol. 151. Cambridge University Press, Cambridge.

Hochschild, G. (1942). Semi-simple algebras and generalized derivations. *Amer. J. Math.*, **64**, 677–94.

Hörmander, L. (1971). Fourier integral operators. I. *Acta Math.*, **127**, 79–183.

Hurt, N. E. (1982). *Geometric quantization in action.* Reidel, Dordrecht.

Isham, C. J. (1984). Topological and global aspects of quantum theory. In *Relativity, groups and Topology*, 2. Les Houches Summer School 1983. Ed. B. S. Dewitt and R. Stora. North Holland, Amsterdam.

Keller, J. B. (1958). Corrected Bohr-Sommerfeld quantum conditions for non-separable systems. *Ann. Physics*, **4**, 180–8.

Kijowski, J. and Szczyrba, W. (1976). A canonical structure for classical field theories. *Comm. Math. Phys.*, **46**, 183–206.

Kirillov, A. A. (1976). *Elements of the theory or representations*. Springer, Berlin.

Klauder, (1960). The action option and a Feynman quantization of spinor fields in terms of ordinary c-numbers. *Ann. Phys.*, **11**, 123–68.

—— (1979). Path integrals and stationary phase approximations. *Phys. Rev.*, **D19**, 2349–56.

Kobayashi, S. and Nomizu, K. (1963). *Foundations of differential geometry*, Vol. 1. Wiley, New York.

—— and —— (1969). *Foundations of differential geometry*, Vol. 2. Wiley, New York.

Kostant, B. (1970a). Quantization and unitary representations. In *Lectures in modern analysis III* (ed. C. T. Taam). Lecture notes in mathematics, Vol. 170. Springer, Berlin.

—— (1970b). Orbits and quantization theory. *Actes Congrès Intern. Math.*, **2**, 395–400.

—— (1973). On convexity, the Weyl group and the Iwasawa decomposition. *Ann. Sci. Éc. Norm. Sup.*, **6**, 413–55.

—— (1974a). Symplectic spinors. In *Symposia mathematica*, Vol. 14. Academic Press, London.

—— (1974b). On the definition of quantization. In *Géométrie symplectique et physique mathématique* (ed. J.-M. Souriau). Colloq. Internat. CNRS, No. 237. CNRS, Paris.

—— (1977). On graded manifolds. In *Differential geometrical methods in mathematical physics* (ed. K. Bleuler and A. Reetz). Lecture notes in mathematics, Vol. 570. Springer, Berlin.

Kummer, M. (1982). On the regularization of the Kepler problem. *Commun. Math. Phys.*, **84**, 133–52.

Künzle, H. P. (1972). Canonical dynamics of spinning particles in gravitational and electromagnetic fields. *J. Math. Phys.*, **13**, 739–44.

Leray, J. (1974). Complément à la théorie d'Arnold de l'indice de Maslov. In *Symposia mathematica*, Vol. 14. Academic Press, London.

Libermann, P. and Marle, C.-M. (1987). *Symplectic geometry and analytical mechanics*. D. Reidel, Dordrecht.

Mackey, G. W. (1963). *Mathematical foundations of quantum mechanics*. Benjamin, Reading, Massachusetts.

—— (1968). *Induced representations and quantum mechanics*. Benjamin, New York.

Magnon-Ashtekar, A. (1976). *Champs quantiques en espace-temps courbe*. Thèse d'état, Université de Marseille.

Magri, F. (1978). A simple model of the integrable Hamiltonian equation. *J. Math. Phys.*, **19**, 1156–62.

Marle, G.-M. (1976). In *Differential geometry and relativity* (ed. M. Cahen and M. Flato). Reidel, Dordrecht, Holland.

Marsden, J. E. and Weinstein, A. (1974). Reduction of symplectic manifolds with symmetry. *Rep. Math. Phys.*, **5**, 121–30.

Maslov, V. P. (1972). *Théorie des perturbations et méthodes asymptotiques.* Dunod/Gauthier-Villars, Paris.

Mason, L. J. and Frauendiener, J. (1990). The Sparling 3-form, Ashtekar variables, and quasi-local mass. In *Twistors in mathematics and physics.* Ed. T. N. Bailey and R. J. Baston. London Mathematical Society lecture notes, Vol. 156. Cambridge University Press, Cambridge.

Mathisson, M. (1937). Neue Mechanik materieller Systeme. *Acta. Phys. Polon.*, **6**, 163–200.

Messiah, A. (1961). *Quantum mechanics,* Vol. 1. North-Holland. Amsterdam.

Meyer, K. P. (1973). Symmetries and integrals in mechanics. In *Dynamical systems* (ed. M. Peixoto). Academic Press, New York.

Milnor, J. (1963). *Morse theory.* Princeton University Press, Princeton, New Jersey.

Moncrief, V. (1980). Reduction of the Yang-Mills equation. In *Differential geometrical methods in mathematical physics* (ed. P. L. García, A. Pérez-Rendón, J.-M. Souriau). Lecture notes in mathematics, Vol. 836. Springer, Berlin.

Montgomery, R. (1988). The connection whose holonomy is the classical adiabatic angles of Hannay and Berry and its generalization to the non-integrable case. *Commun. Math. Phys.*, **120**, 269–94.

Morrow, J. A. and Kodaira, K. (1971). *Complex manifolds.* Holt, Rinehart, and Winston, New York.

Moser, J. K. (1965). On the volume elements on a manifold. *Trans. Amer. Math. Soc.*, **120**, 286–94.

—— (1970). Regularization of Kepler's problem and the averaging method on a manifold. *Comm. Pure Appl. Math.*, **23**, 609–36.

Nagano, T. (1968). 1-forms with exterior derivative of maximal rank. *J. Differential Geom.*, **2**, 253–64.

Nash, C. and Sen, S. (1983). *Topology and geometry for physicists.* Academic Press, London.

Newlander, A. and Nirenberg, L. (1957). Complex coordinates in almost complex manifolds. *Ann. of Math.*, **65**, 391–404.

Nijenhuis, A. (1955). Jacobi-type identities for bilinear differential concomitants of certain tensor fields. I *Neder. Akad. Wetensch. Proc. Ser. A,* **58**, 390–7.

Nirenberg, L. (1957). In *Seminars on analytic functions. I.* Princeton University Press, Princeton, New Jersey.

Onofri, E. and Pauri, M. (1972). Dynamical quantization. *J. Math. Phys.*, **13**, 533–43.

Ozeki, H. and Wakimoto, M. (1972). On polarizations of certain homogeneous spaces. *Hiroshima Math. J.*, **2** 445–82.

Papapetrou, A. (1951). Spinning test particles in general relativity. *Proc. Roy. Soc. London ser. A,* **209**, 248–58.

Pauli, W. (1940). The connection between spin and statistics. *Phys. Rev.,* **58**, 716–22.

Penrose, R. (1960). A spinor approach to general relativity. *Ann. Physics,* **10**, 171–201.

—— (1968a). In *Battelle rencontres* (ed. C. M. DeWitt and J. A. Wheeler). Benjamin, New York.

—— (1968b). Twistor quantization in curved space-time. *Internat. J. Theoret. Phys.,* **1**, 61–99.

—— and MacCallum, M. A. H. (1972). *Phys. Rep.,* **6**, 241–316.

—— and Rindler, W. (1984). *Spinors and space-time. Vol. 1: Two-spinor calculus and relativistic fields.* Cambridge University Press, Cambridge.

—— and —— (1986). *Spinors and space-time. Vol. 2: Spinor and twistor methods in space-time geometry.* Cambridge University Press, Cambridge.

Pirani, F. A. E. (1965). In *Lectures on general relativity* (ed. S. Deser and K. W. Ford). Prentice-Hall, Englewood Cliffs, New Jersey.

—— (1974). Once more the Kepler problem. *Nuovo Cimento,* **19B**, 189–207.

Pressley, A. and Segal, G. B. (1986). *Loop groups.* Oxford University Press, Oxford.

Puta, M. (1986). Old and new aspects of geometric quantization. *Monografii Matematice,* Vol. 28. Universitatea din Timisoara.

Rawnsley, J. H. (1972). *Some applications of quantisation.* D. Phil. thesis, University of Oxford.

—— (1977a). On the cohomology groups of a polarisation and diagonal quantisation. *Trans. Amer. Math. Soc.,* **230**, 235–55.

—— (1977b). Coherent states and Kähler manifolds. *Quart. J. Math. Oxford ser. (2),* **28**, 403–15.

—— (1978a). On the pairing of polarizations. *Comm. Math. Phys.,* **58**, 1–8.

—— (1978b). Some properties of half-forms. In *Differential geometrical methods in mathematical physics II* (ed. K. Bleuler, H. R. Petry, and A. Reetz). Lecture notes in mathematics, Vol. 676. Springer, Berlin.

—— (1979). Flat partial connections and holomorphic structures in C^∞ vector bundles. *Proc. Amer. Math. Soc.,* **73**, 391–7.

—— Schmid, W., and Wolf, J. A. (1983). Singular representations and indefinite harmonic theory. *J. Funct. Anal.,* **51**, 1–114.

Reed, M. and Simon, B. (1972). *Methods of modern mathematical physics, I: Functional analysis.* Academic Press, New York.

Robinson, P. L. (1987). A cohomological pairing of half-forms. *Trans. Amer. Math. Soc.,* **301**, 251–61.

—— (1991). An infinite-dimensional metaplectic group. To be published in *Quart. J. Math. Oxford.*

—— and Rawnsley, J. H. (1989). The metaplectic representation, MP^c structures and geometric quantization. *Mem. Amer. Math. Soc.,* **81**, no. 410.

Schutz, B. F. (1980). *Geometrical methods of mathematical physics.* Cambridge University Press, Cambridge.

Segal, I. E. (1960). Quantization of nonlinear systems. *J. Math. Phys.,* **1**, 468–88.

—— (1974). In *Symposia mathematica,* Vol. 14. Academic Press, London.

Shale, D. (1962). Linear symmetries of free boson fields. *Trans. Am. Math. Soc.,* **103**, 149–67.

Simms, D. J. (1968). *Lie groups and quantum mechanics.* Lecture notes in mathematics, Vol. 52. Springer, Berlin.

—— (1973). Geometric quantisation of the harmonic oscillator with diagonalised Hamiltonian. In *Colloquium on group theoretical methods in physics* (ed. A. Janner and T. Janssen). University of Nijmegen, Holland.

—— (1974). Geometric quantization of the energy levels in the Kepler problem. In *Symposia Mathematica,* Vol. 14. Academic Press, London.

—— (1977). Serre dulaity for polarised symplectic manifolds. *Rep. Math. Phys.,* **12**, 213–17.

—— (1978). On the Schrödinger equation given by geometric quantization. In *Differential geometrical methods in mathematical physics II* (ed. K. Bleuler, H. R. Petry, and A. Reetz). Lecture notes in mathematics, Vol. 676. Springer, Berlin.

—— (1980a). Geometric aspects of the Feynman integral. In *Differential geometrical methods in mathematical physics* (ed. P. L. García, A. Pérez-Rendón, J.-M. Souriau). Lecture notes in mathematics, Vol. 836. Springer, Berlin.

—— (1980b). Geometric quantization for singular Lagrangians. In *Differential geometrical methods in mathematical physics* (ed. P. L. García, A. Pérez-Rendón, J.-M. Souriau). Lecture notes in mathematics, Vol. 836. Springer, Berlin.

Simon, B. (1983). Holonomy, the quantum adiabatic theorem, and Berry's phase. *Phys. Rev. Lett.,* **51**, 2167–70.

Simpson, R. (1984). Coherent states and relativistic path integration. D. Phil. thesis, University of Oxford.

Smale, S. (1970). Topology and mechanics, I and II. *Invent. Math.,* **10**, 305–31, **11**, 45–64.

Śniatycki, J. (1974). Prequantization of charge. *J. Math. Phys.,* **15**, 619–20.

—— (1977). On cohomology groups appearing in geometric quantization. In *Differential geometrical methods in mathematical physics* (ed. K. Bleuler and A. Reetz). Lecture notes in mathematics, Vol. 570. Springer, Berlin.

—— (1980). *Geometric quantization and quantum mechanics.* Springer-Verlag, New York.

—— (1983). Constraints and quantization. In *Nonlinear partial differential operators and quantization procedures* (ed. S. I. Anderson and H.-D. Doebner). Lecture notes in mathematics, Vol. 1037. Springer, Berlin.

Sommers, P. D. (1973a). *Killing tensors and type* {2,2} *space-times.* Ph. D. thesis, University of Texas at Austin.

—— (1973b). On Killing tensors and constants of the motion. *J. Math. Phys.,* **14**, 787–90.

Souriau, J.-M (1966). Quantification géométrique. *Comm. Math. Phys.*, **1**, 374–98.

—— (1970). *Structure des systèmes dynamiques.* Dunod, Paris.

—— (1974a). Modèle de particule à spin dans le champ électromagnétique et gravitationnel. *Ann. Inst. H. Poincaré A (N.S.)*, **20**, 315–64.

—— (1974b). Sur la variété de Képler. In *Symposia mathematica*, Vol. 14. Academic Press, London.

—— (1976). In *Group theoretical methods in physics* (ed. A. Janner, T. Janssen, and M. Boon). Lecture notes in physics, Vol. 50. Springer, Berlin.

Steenrod, N. (1951). *Topology of fibre bundles.* Princeton University Press, Princeton, New Jersey.

Sternberg, S. (1964). *Lectures on differential geometry.* Prentice-Hall, Englewood Cliffs, New Jersey.

—— (1977). Minimal coupling and the symplectic mechanics of a classical particle in the presence of a Yang-Mills field. *Proc. Nat. Acad. Sci.*, **74**, 5253–4.

—— (1988). The pairing method and bosonic anomalies. In *Differential geometrical methods in theoretical physics* (ed. K. Bleuler and M. Werner). Kluwer Academic, Dordrecht.

Streater, R. F. and Wightman, A. S. (1964). *PCT, spin and statistics, and all that.* Benjamin, New York.

—— and Wilde, I. F. (1970). Fermion states of a boson field. *Nucl. Phys.*, **B24**, 561–75.

Synge, J. L. (1960). Classical dynamics. In *Handbuch der physik III/I* (ed. S. Flügge). Springer, Berlin.

Tod, K. P. (1977). Some symplectic forms arising in twistor theory. *Rep. Math. Phys.*, **11**, 339–46.

Torrence, R. J. and Tulczyjew, W. M. (1973). Gauge invariant canonical mechanics for charged particles. *J. Math. Phys.*, **14**, 1725–32.

Trautman, A. (1972). In *General relativity* (ed. L. O'Raifeartaigh). Oxford University Press, Oxford.

Treves, F. (1980). *Introduction to pseudodifferential and Fourier integral operators. Vols.* 1 & 2. Plenum Press, New York.

Vaisman, I. (1973). *Cohomology and differential forms.* Dekker, New York.

—— (1981). Geometric quantization on space of differential forms. *Rend. Sem. Mat. Univ. Politec. Torino*, **39**, 139–52.

Vergne, M. (1977). Groupe symplectique et seconde quantification. *C. R. Acad. Sci. Paris Ser. A-B*, **285**, A191–4.

Verlinde, H. and Verlinde, E. (1990). Conformal field theory and geometric quantization. In *Superstrings* '89. Ed. M. Green, R. Iengo, S. Randjbar-Daemi, E. Sezgin, and A. Strominger. World Scientific, Singapore.

Wald, R. M. (1984). *General relativity.* University of Chicago Press, Chicago.

Ward, R. S. and Wells, R. O. (1990). *Twistor geometry and field theory.* Cambridge University Press, Cambridge.

Weil, A. (1958). *Variétś Kählériennes.* Hermann, Paris.

Weinstein, A. (1971). Symplectic manifolds and their Lagrangian submanifolds. *Adv. in Math.*, **6**, 329–46.

—— (1977). *Lectures on symplectic manifolds.* CBMS regional conference series, Vol. 29. Amer. Math. Soc., Providence, Rhode Island.

—— (1978). A universal phase space for particles in Yang-Mills fields. *Lett. Math. Phys.*, **2**, 417–20.

—— (1990). Connections of Berry and Hannay type for moving Lagrangian submanifolds. *Adv. Math.*, **82**, 135–59.

Wells, R. O. (1973). *Differential analysis on complex manifolds.* Prentice-Hall, Englewood Cliffs, New Jersey.

—— (1979). Complex manifolds and mathematical physics. *Bull. Amer. Math. Soc. (N.S.)*, **1**, 269–336.

Whitehead, J. H. C. (1937). Certain equations in the algebra of a semi-simple infinitesimal group. *Quart. J. Math. Oxford*, **8**, 220–37.

Whittaker, E. T. (1904). *A treatise on the analytical dynamics of particles and rigid bodies.* Cambridge University Press, Cambridge.

Wigner, E. P. (1939). On unitary representations of the inhomogeneous Lorentz group. *Ann. Math.* (2), **40**, 149–204.

Woodhouse, N. M. J. (1981). Geometric quantization and the Bogoliubov transformation. *Proc. Roy. Soc. London ser. A*, **378**, 119–39.

—— (1982). Vector bundles and complex polarizations. *Math. Proc. Camb. Phil. Soc.*, **92**, 489–509.

INDEX OF NOTATION

INDEX